IEE COMPUTING SERIES 11

Series Editors: Dr. B. Carré
Professor S. L. Hurst
Dr. D. A. H. Jacobs
M. W. Sage
Professor I. Sommerville

ADVANCES IN COMMAND, CONTROL & COMMUNICATION SYSTEMS

Other volumes in this series

Volume 1 Semi-custom IC design and VLSI
         P. J. Hicks (Editor)
Volume 2 Software engineering for microprocessor systems
         P. G. Depledge (Editor)
Volume 3 Systems on silicon
         P. B. Denyer (Editor)
Volume 4 Distributed computing systems programme
         D. Duce (Editor)
Volume 5 Integrated project support environments
         J. A. McDermid (Editor)
Volume 6 Software engineering '86
         D. J. Barnes and P. J. Brown (Editors)
Volume 7 Software engineering environments
         I. Sommerville (Editor)
Volume 8 Software engineering: the decade of change
         D. Ince (Editor)
Volume 9 Computer aided tools for VLSI system design
         G. Russell (Editor)
Volume 10 Industrial software technology
         R. Mitchell (Editor)

ADVANCES IN COMMAND, CONTROL & COMMUNICATION SYSTEMS

EDITED BY
C. J. HARRIS
I. WHITE

Peter Peregrinus Ltd., on behalf of the Institution of Electrical Engineers

Published by: Peter Peregrinus Ltd., London, United Kingdom
© **1987: Peter Peregrinus Ltd.**

Except chapters 6.2 and 7.4 which are
© Crown Copyright 1987.

All rights reserved. No part of this publication may be reproduced, stored in a retrieval system or transmitted in any form or by any means—electronic, mechanical, photocopying, recording or otherwise—without the prior permission of the publisher.

While the author and the publishers believe that the information and guidance given in this work is correct, all parties must rely upon their own skill and judgment when making use of it. Niether the author nor the publishers assume any liability to anyone for any loss or damage caused by any error or omission in the work, whether such error or omission is the result of negligence or any other cause. Any and all such liability is disclaimed.

British Library Cataloguing in Publication Data
Advances in command, control and communication systems.—(IEE computing series; 11)
1. Telecommunication systems—Data processing
I. Harris, C.J. (Christopher John)
II. White, I. III. Institution of Electrical Engineers IV. Series
621.38'0413   TK5105.5

ISBN 0 86341 094 4

Printed in England by Short Run Press Ltd., Exeter

Contents

Preface xi

Acknowledgments xvii

List of contributors xviii

Chapter 1 The future for command systems
I. White
1.1 Background 1
1.2 The problems of C^3I 1
    1.2.1 Basic considerations 1
    1.2.2 The theory of command and control 1
    1.2.3 Informational chaos 2
    1.2.4 Technological revolution 2
    1.2.5 Organisational chaos 2
1.3 The route towards a solution 2
    1.3.1 Strategy 2
    1.3.2 Analysis and specification methods 3
    1.3.3 Descriptive methods 3
    1.3.4 Application of prototyping 4
    1.3.5 Application of standards to aid interoperability 5
    1.3.6 Implementation strategy 6
1.4 The technology for the future 6
    1.4.1 Introduction 6
    1.4.2 Command system functionality 6
    1.4.3 Hardware 7
    1.4.4 Software 7
    1.4.5 Man–machine interface 9
    1.4.6 Knowledge-based systems 11
1.5 Concluding comments 14
1.6 References 14

Chapter 2 Design and structure of C^3 systems

- 2.1 The C-process: A model of Command — 19
  D. G. Galley
  - 2.1.1 Introduction — 19
  - 2.1.2 The need for a model of command — 19
  - 2.1.3 The overall command system — 20
  - 2.1.4 Development of the C-process model — 23
  - 2.1.5 The basic proposal assessment loop — 24
  - 2.1.6 Hierarchic control — 26
  - 2.1.7 Complex command — 30
  - 2.1.8 World state space interpretation of command — 34
  - 2.1.9 Distributed command and co-ordination — 37
  - 2.1.10 Command message set — 40
  - 2.1.11 The C-process as a basis for design — 45
  - 2.1.12 Conclusions — 47
  - 2.1.13 References — 49
- 2.2 MOSAIC concepts for the future deployment of air power in European NATO — 50
  D. K. Hitchins
  - 2.2.1 Introduction — 50
  - 2.2.2 The changing European military scene — 51
  - 2.2.3 Air power in context — 51
  - 2.2.4 The threat facing air command and control — 57
  - 2.2.5 Air command and control deficiencies — 60
  - 2.2.6 Introduction to MOSAIC — 60
  - 2.2.7 Expanding the MOSAIC concepts — 68
  - 2.2.8 Implementing MOSAIC — 74
  - 2.2.9 MOSAIC strengths and weaknesses — 76
  - 2.2.10 MOSAIC in action — 81
  - 2.2.11 Conclusions — 83
- 2.3 C^3 effectiveness studies — 84
  G. H. Lammers
  - 2.3.1 Introduction — 84
  - 2.3.2 Effectiveness — 86
  - 2.3.3 Information flows — 90
  - 2.3.4 Transition planning — 100
  - 2.3.5 Concluding remarks — 103
  - 2.3.6 References — 103

Chapter 3 Databases for C^3 systems

- 3.1 Spatial database management for command and control — 107
  C. A. McCann, M. M. Taylor and M. I. Tuori
  - 3.1.1 Introduction — 107
  - 3.1.2 Background — 107
  - 3.1.3 ISIS: the goal — 109

	3.1.4	SDBMS-1: a demonstration prototype	110
	3.1.5	Ongoing and future work	121
	3.1.6	Concluding remarks	127
	3.1.7	References	127
3.2	Systems design and data management problems in the utilisation of local area network architectures		129
	A. S. Cheeseman and R. H. L. Catt		
	3.2.1	Introduction	129
	3.2.2	Historical development	130
	3.2.3	Introduction of a local area network	132
	3.2.4	Distributed system architecture	134
	3.2.5	The combat system design process	139
	3.2.6	Future development	145
	3.2.7	Distributed database management	145
	3.2.8	Maintaining the integrity of distributed data	148
	3.2.9	Performance	149
	3.2.10	Standards	150
	3.2.11	Data definition	150
	3.2.12	ADDAM (area distributed data management system)	150
	3.2.13	Conclusions	153
	3.2.14	References	155

Chapter 4 Communications

4.1	Packet radio: a survivable communications system for the forward area		159
	B. H. Davies and T. R. Davies		
	4.1.1	Introduction	159
	4.1.2	Basic concepts	160
	4.1.3	User services	163
	4.1.4	Channel access, routing and network control	165
	4.1.5	Performance	173
	4.1.6	Signal processing requirements and architecture	176
	4.1.7	Experiences with prototype equipment	180
	4.1.8	Conclusions	181
	4.1.9	References	182
4.2	C^2 communications for the tactical area: the Ptarmigan packet switched network design and development proving		183
	C. S. Warren, S. G. Wells, J. R. Bartlett, B. J. Symons		
	4.2.1	Introduction	183
	4.2.2	The design philosophy	184
	4.2.3	The tactical environment	187
	4.2.4	Packet switching overlay on Ptarmigan	188
	4.2.5	Network characteristics	189
	4.2.6	The protocols	190

		4.2.7	Network management	192
		4.2.8	The need to prove	193
		4.2.9	Test Tools	199
		4.2.10	Ptarmigan packet switched network model	201
		4.2.11	Status	207
		4.2.12	References	208

Chapter 5 Standards

	5.1	\multicolumn{2}{l}{International standards in military communications}	211
		\multicolumn{2}{l}{D. F. Bird}	
		5.1.1	Introduction	211
		5.1.2	Open systems interconnection and the ISO reference model	213
		5.1.3	Military requirements	215
		5.1.4	Gateway issues	217
		5.1.5	Initiatives	218
		5.1.6	Summary	220
		5.1.7	Appendix A: Standards bodies	220
		5.1.8	Appendix B: The seven layers of the ISO reference model	221
		5.1.9	References	222
	5.2	\multicolumn{2}{l}{C^3I and the upper layers of the OSI}	223
		\multicolumn{2}{l}{H. J. Pearson}	
		5.2.1	Introduction	223
		5.2.2	The military use of standards	223
		5.2.3	Military requirements	225
		5.2.4	Structure of the upper layers	227
		5.2.5	Session layer	229
		5.2.6	Presentation layer and abstract syntax notations	230
		5.2.7	Application layer standards	231
		5.2.8	Conclusions	236
	5.3	\multicolumn{2}{l}{Security in military OSI networks}	238
		\multicolumn{2}{l}{T. Knowles}	
		5.3.1	Introduction	238
		5.3.2	Types of security protection	239
		5.3.3	Architecture − placement of security services	242
		5.3.4	Non-OSI aspects	243
		5.3.5	System security	243
		5.3.6	Protocol enhancements	244
		5.3.7	Securing military systems	245
		5.3.8	Conclusions	246
		5.3.9	References	246
	5.4	\multicolumn{2}{l}{Standards for naval systems}	247
		\multicolumn{2}{l}{J. S. Hill and F. A. Richards}	
		5.4.1	Introduction	247

	5.4.2	Background	247
	5.4.3	Constructing a system	249
	5.4.4	Framework for standards	251
	5.4.5	Choosing a local area network	252
	5.4.6	Network/transport layer interface	253
	5.4.7	Presentation layer	256
	5.4.8	Application (and other) layers	258
	5.4.9	Other areas of standardisation	259
	5.4.10	Status and conclusions	260
	5.4.11	References	261

Chapter 6 The man—machine interface

6.1 Man—machine aspects of command and control 265
W. T. Singleton
    6.1.1 Introduction 265
    6.1.2 The commander 266
    6.1.3 The picture builders 267
    6.1.4 Decision aiding 268
    6.1.5 The contribution from psychology 272
    6.1.6 Conclusion 276
    6.1.7 References 276

6.2 An engineering standard for a systematic approach to the design of user—computer interfaces 279
A. Gardner
    6.2.1 Introduction 279
    6.2.2 Inputs to systems design 281
    6.2.3 Tasks 283
    6.2.4 User—computer interfaces 286
    6.2.5 Interface cognitive models 289
    6.2.6 Interface languages 293
    6.2.7 Interface devices 300
    6.2.8 Outputs from systems design 302
    6.2.9 Final comments 302
    6.2.10 References 303

Chapter 7 Advanced processing

7.1 Expert systems in C^2 systems 307
C. J. Harris
    7.1.1 Introduction 307
    7.1.2 C^2 complexity and artificial intelligence 308
    7.1.3 Knowledge-based systems in C^2 systems 308
    7.1.4 A prototype structure for an overall knowledge-based C^2 system 312
    7.1.5 Alternative approaches to C^2 architecture 316

		7.1.6	References	317
		7.1.7	Bibliography	318
	7.2	Some aspects of data fusion		321
		G. B. Wilson		
		7.2.1	Introduction	321
		7.2.2	Positional information	322
		7.2.3	Identity information	326
		7.2.4	Behaviour	335
		7.2.5	Conclusion	337
		7.2.6	References	338
	7.3	An AI approach to data fusion and situation assessment		339
		W. L. Lakin and J. A. H. Miles		
		7.3.1	Introduction	339
		7.3.2	Multisensor data fusion	342
		7.3.3	Expert system approach	346
		7.3.4	Multisensor data fusion experiment: stage 1 demonstrator	350
		7.3.5	Stage 1 demonstrator: results	356
		7.3.6	Multiplatform data fusion	362
		7.3.7	Parameter estimation − or combination of evidence	365
		7.3.8	Situation assessment	366
		7.3.9	Artificial intelligence machines for multisensor data fusion	369
		7.3.10	Conclusions and future work	375
		7.3.11	References	377
	7.4	Air defence threat assessment		378
		S. Middleton		
		7.4.1	Introduction	378
		7.4.2	The problem	379
		7.4.3	Analysis of the problem	380
		7.4.4	Approach taken	382
		7.4.5	Current status	396
		7.4.6	Summary and conclusions	396
		7.4.7	References	397

Index   398

Preface

Command, control, communications and intelligence (C^3I) is the management infrastructure for defence and war or for any other large or complex dynamic resource system such as the health service, the police service, public utilities and international communications. It is intrinsically a diverse range of activities. Looking to the future of command and control, there are four essential problems:

1 There is no theory or design methodology for command and control of complex systems.
2 There is an information chaos associated with the diverse nature, sources and wealth of data available in large-scale systems.
3 There is a technology revolution associated with the enabling technologies of C^3I systems (parallel processors, VLSI, MMI, digital communications, portable real-time large-scale software etc.)
4 There is an organisational chaos associated with diverse uncoordinated projects, lack of interoperability, lack of software portability, use of different management techniques etc.

Substantial research effort in Europe and North America is currently being devoted to these aspects of command and control. However, we believe that a fundamentally radical solution to these problems can be achieved through the concept of a goal architecture with an implementation strategy. A *goal architecture* is a clear description of what the system does, how it is constructed and how it can evolve. There are at least four elements in providing a goal architecture:

Use of formal *specification* requirements and analysis methods
Use of system *prototyping*
Use of *international standards* to aid interoperability
The opportunity of an *implementation strategy*.

System specification will essentially involve human factors, command and control, decision making, data fusion and tactics. Prototyping is more than rigorous analysis and descriptive methods; it involves human as well as machine assessments. A fundamental but lower-level aspect of a goal architecture is the application of international standards, featuring LANs, WANs, data types, communications protocols, database interoperability, security, MMI displays, and high-level programming environments together with their definitions and standards.

The technology of future C^3I systems offers exciting prospects in system design, MMI, supporting software and communications. System design, decision support systems and data fusion, which are based upon the use of intelligent knowledge-based systems (IKBS), will play a crucial role in large-scale C^3I systems that require real-time decisions or resource allocations. Hardware developments will need to accommodate the needs of large fast-access databases and high sensor data rates; currently this is likely to be achieved through array processors or transputers or other high-instruction input architectures such as parallel processors. In large-scale software applications to C^3I there are many unanswered questions associated with Ada, such as a real-time design methodology and software reusability as well as computer security and integrity. Many of the above issues are considered in this book, and some insight into the future development of C^3I systems is provided.

The very scale and complexity of C^3I systems prevent any single book (or even series of books) providing a comprehensive review of current and future developments in C^3I. Instead we have attempted to identify some of the important issues in design and structure, standards, databases, communications, the man—machine interface and the advanced processing of command and control. This book is an outcome of the very successful international C^3 conference sponsored by the Institution of Electrical Engineers at Bournemouth in 1985.

Chapter 1 provides an introduction and review of the fundamental problems of command and control. It identifies the lack of a real theory of command and control and explores a solution route through the goal architecture and implementation strategy ideas introduced above. The chapter concludes that future procurement of C^3I systems will be well founded if, in specifying the system tasks, it is ensured that they are achievable and based upon sound and effective design, analysis and prototyping, as well as by utilising a robust inventory of standards and the management infrastructure to optimise them.

It has to be admitted as a prelude to Chapter 2 that there is no well established method for determining the specification, design and architecture of a command, control and communications (C^3) system. A survey of the range of techniques and approaches, and their effectiveness in producing useful systems, is beyond the scope of this book. In this chapter a very brief sample is offered of approaches which in some respects are poles apart. Section 2.1 describes an approach to C^3 design based on the idea that there is a basic paradigm for command and control and that, with sufficiently careful representational means, both structural and semantic, there results a powerful aid for system design with a wide range of applicability. By way of contrast, Section 2.2 discusses a system design based on a particular requirement within the NATO task, and offers a structure to fulfil that requirement. In this case, therefore, the resultant system is more a consequence of the designer's understanding of the requirement and his design creativity. Section 2.3 addresses a neglected area of command and control (C^2), namely that of anticipating performance. The need to audit the design and development of command and control is essential, not only as a design validation tool, but also for management control of command and control projects. It is salutary to reflect that many of the command systems in service today have not been so audited, if only because the tools for the task did not exist. These systems today provide the command and control for a formidable range of military power. In future systems the range and power of performance assessment during design and development will be far better than in the past, but, as in most of command and control, will lack any clear best method, or any method that is unquestionably adequate.

Database management systems provide a central role in the provision of decision aiding in future command and control systems. In Chapter 3 the first section describes a long-term project, the interactive spatial information system (ISIS), whose objective is to demonstrate how commanders use spatial information for planning and decision aiding. This approach utilises the human's advanced ability to convert complex patterns of graphical and textual information into a global coherent picture, and readily to identify anomalies in data as well as minute yet significant details and trends. Spatial data depends upon locations as well as on functional relationships; it can be geometric or topological. The human's intuitive (spatial) mode of thought complements the logical (linguistic) mode, and together they provide the human with a unique problem-solving capability that by analogy is particularly suited to command and control. In Section 3.1 we discuss these human attributes and the particular relevance of spatial/intuitive reasoning for the human decision maker in C^2 systems. From this a prototype robust military land-based highly interactive spatial database management/information system is developed, which is applicable to all C^2 environments. This system provides a library of spatial and non-spatial information/data on the C^2 situation, which is constructed to dynamically respond to questions in graphical or alphanumerical form through a friendly man—machine interface (MMI). The ISIS project is still at an evolutionary stage, yet provides a significant contribution to improving human—computer interface and interaction in command and control systems.

Local area networks (LANs), like wide area networks, have brought significant architectural design opportunities to the C^3 designer through the increased flexibility, reliability and correctivity offered by system modularity. The use of LANs forces the C^3 designer to make a fundamental choice in the data exchange architecture — centralised or distributed. In Section 3.2 the authors clearly demonstrate the role and power of distributed data exchange architectures and distributed database management systems in C^3 systems. They propose generalised military real-time distributed database management system which views the combat/command system as a single integrated system within a cohesive, albeit decentralised, database. This section strongly emphasises the need to avoid the error of assuming that distributed systems can be designed without considering the total system design; the total system must be conceptualised as an entity and then decomposed into the distributed system architecture accompanied by a database management approach advocated by the authors of this section.

There are special difficulties associated with the provision of mobile tactical secure, survivable and interoperable communications in a hostile environment such as the forward area. There is a need to transmit data (telegraph, facsimile etc.) as well as voice over a limited radio frequency spectrum. Packet radio data communications can offer a larger number of users and information than the equivalent digitised speech channel. In particular, combat net radio (CNR) comprises a network of a number of controlled transceivers operating in a simplex mode, all transmitting and receiving in bursts or packets of data on a single narrowband channel. Chapter 4 deals with such communications. Section 4.1 describes the first steps in the system design and development of a narrowband prototype packet radio system, covering the basic concepts and architectures of packet switching techniques to CNR. The authors provide a detailed introduction to channel access, contention-based routing algorithms through self-configurability, network management, and simulated performance to assess system sensitivity/stability to parameter variations. The strength of their approach to CNR is in the utilisation of advanced signal processing techniques based upon the new technologies of VLSI (such as bit slice processors, transputers etc.) on a restricted single-channel communications system. Most significantly, tactical to strategic communications interoperability has been demonstrated between pocket CNR and the DARPA wideband packet radio, although the mobile CNR exhibits the MMI problem of all C^3 systems — entering data through a keyboard.

Section 4.2 provides an in-depth discussion of, and the philosophy behind, the Ptarmigan packet switched network design and development proving. The Ptarmigan packet switched network has been designed to meet the distributed nature of future C^2 communications for the tactical area, and as such was the world's first tactical packet switched system to enter service (1985); simultaneously the Wavell C^2 system was deployed in BAOR on the Ptarmigan system. This contribution is a combination of papers by C. S. Warren, B. G. Wells, B. J. Symons, P. K. Smith and J. R. Bartlett, given at the IEE international conference on C^3 at Bournemouth in April 1985.

Standards, like morals, are both a source of constraint and a source of guidance for the overall good of the community that uses them. They are also similarly difficult to introduce into that user community. Standards for information systems offer the potentiality for wide-ranging interoperability between current systems (interoperability in time). Historically these degrees of freedom have been conspicuously absent from military command and control systems. As described in Chapter 5, there is now a powerful movement to develop information system standards, covering many of the system features which characterise command and control. A layered approach to communication standards is evolving rapidly, and forms the basis, either explicitly or as a conceptual model, for the communication standards described in Section 5.1.

Aspects of information systems other than communications are also now the subject of standardisation endeavours. Apart from the rather famous development of the high order programming language Ada, there are efforts to formulate standards relating to structured architectures for databases and to man—machine interaction, some along the lines of layering. Sections 5.2 and 5.3 are concerned with the central issues of international standards and open system interconnection for C^3 systems; Section 5.4 provides an illustrative example of standards for a naval combat system.

Historically successful standards have usually been the result of retrospective review and compromise between participating users and vendors. The new style of standard development seeks acceptance of standards in new areas of application, and before they are widely used. Standards are now well established in the civil sector, and the military customer is increasingly accepting the benefits of procurement of products complying with these standards. The potential for future systems which perform better, cost less and are easier to support and upgrade through the system life cycle is considerable. For this potential to be achieved, those formulating standards have to appreciate the general needs of a wide class of customers over several generations of technology. This is not an easy task, and deserves greater resourcing.

There is no systematic description of man, and accordingly the systems designer is always faced with grave difficulties in designing systems in which men play major roles. In Chapter 6 two significantly differing viewpoints are presented. Section 6.1 is an examination of command and control from the perspective of the user. It reviews the contributions made by various branches of science to the understanding of man's role in systems. The theme within this section is perhaps that there is more to man than the engineer dreams of. Section 6.2 describes the progress being made in an attempt to specify procedures for designers to follow when producing MMI systems. It will form the basis for a handbook on human factor aspects of the MMI task, to be used by designers of systems for the Royal Navy, and is intended to be supported by some computer-based tools. The two sections show that, as elsewhere in command and control, there is more than one way of trying to skin a cat.

The majority of command and control (C^2) systems are complex and ill defined and are characterised by non-unique solutions. In particular, a C^2 system as an efficient man—machine domain for effective real-time decision support/planning or data fusion problem solving in a dynamically stochastic environment is not feasible (even if a solution exists) with existing von Neumann computer architectures and the lack of a fundamental theory and design methodology for C^2. However, as described in Chapter 7, artificial intelligence techniques and the emerging technology of concurrent processors now offer the possibility of real-time intelligent decision aids for C^2 applications. Artifical intelligence, through intelligent knowledge-based systems (IKBS), provides the tools to utilise the expert's (or commander's) knowledge for effective use in a real-time computer-based control structure. This offers the potential for the efficient use of a knowledge base of production rules for managing resources, the establishment and evaluation of current and future situations, threat assessment, or lower-level activities such as sensor interpretation and data fusion. Section 7.1 provides an introduction to and a review of expert systems in C^2 systems and introduces a prototype architecture for an overall knowledge-based C^2 system.

Application of IKBS to air defence C^2 tasks is considered in Section 7.4; in particular, the high-level tasks of real-time threat assessment are developed. A threat assessment system must deal with data in real time and in the order in which it arrives; therefore a special expert system framework is introduced. A blackboard system is initially considered; however, there are associated implementation problems such as continuous reasoning, timing and consistency of data, and so an object-oriented language is introduced to support continuous reasoning expert systems as well as simulation.

At a lower level in the C^2 systems architecture, multisensor data fusion has proved to be a viable candidate for an IKBS solution. In Section 7.3 an IKBS for multisensor data fusion is developed; it utilises a rule-based system framework, which allows the knowledge engineer to develop an efficient and integrated real-time blackboard system. The IKBS provides effective real-time multisensor data fusion in the naval environment, coping with a large and diverse quantity of real-time sensor data and non-real-time encylopaedic information. It is a relatively straightforward exercise to extend the data fusion IKBS approach of Section 7.3 to provide the next layer in the C^2 systems architecture — that of threat or situation assessment.

Artificial intelligence is not the only means of solving real-time dynamic systems. Indeed, some would argue that artificial intelligence has not yet performed the inference operations that stochastic decisions demand, or transcended the more traditional decision disciplines. Section 7.2 considers the use of classical Bayesian probability methods in the solution of the data fusion problem; however, the author concludes that IKBS methods to represent a very effective means of coping with a wealth of varying types and sources of real-time data in generating optimal evaluations of command and control situations.

C. J. Harris
I. White
July 1986

Acknowledgments

Acknowledgments for help and support in the preparation of this book are made by the following contributors:

To the UK Ministry of Defence for permission to publish those sections which are the subject of Crown Copyright.

I. White: to the Admiralty Research Establishment for permission to coedit this book.

D. G. Galley: to System Designers International for support in producing Section 2.1 and to ARE (MoD, UK) who sponsored the research on which it is based.

A. S. Cheeseman and R. H. L. Catt: to personnel from ARE, CNSWE and Ferranti Computer Systems Ltd. Their work was sponsored in part by ARE, CNSWE and DGSS.

B. H. Davies and T. R. Davies: to J. Jubin of Rockwell Collins and M. Hazell and L. Hodge of RSRE. Parts of Section 4.1 have been previously published in Milcomp '83 and *Proc. IEEE* 1983.

D. F. Bird: to M. Bailey for his assistance with some aspects of Section 5.1.

J. S. Hill and F. A. Richards: to their colleagues in the MoD, Ferranti Computer Systems and Software Sciences, for contributions made during the studies reported in Section 5.4.

H. Pearson: to Dr John Larmouth for his teachings on the fundamentals of open system interconnection.

A. Gardner: This work is supported by MoD (UK) contract NSW 32A/1113.

W. Lakin and J. A. H. Miles: to R. Irving, J. D. Mongomery and J. Haugh who all made important contributions to Section 7.3.

S. Middleton: to M. Bell, M. Bennett, J. Lumley and R. Zanconato of Cambridge Consultants, and to R. Usher and P. Wetherall of RSRE, for Section 7.4.

Disclaimer

The views expressed in this book are those of the authors and editors only, and should not be taken as representing the policy or viewpoint of their sponsoring organisations.

Contributors

J. S. Hill
W. L. Lakin
J. A. H. Miles
F. A. Richards
I. White

Admiralty Research Establishment MOD (PE)
Portsdown
Portsmouth, Hants PO6 4AA

C. A. McCann
M. M. Taylor
M. I. Tuori

Defence and Civil Institute of Environmental Medicine
Toronto
PO Box 2000
Downsview
Ontario, Canada M3M 3B9

D. K. Hitchins

Easams Ltd
Lyon Way
Frimley Road
Camberley, Surrey GU16 5EX

G. H. Lammers
C. S. Warren
S. G. Wells
J. R. Bartlett
B. J. Symons

Plessey Defence Systems
Grange Road
Christchurch, Dorset BH23 4JE

B. H. Davies
T. R. Davies
S. Middleton

Royal Signals and Radar Establishment MOD (PE)
St Andrews Road
Great Malvern, Worcs WR14 3PS

R. H. L. Catt
A. S. Cheeseman

Software Sciences Ltd
Meudon Avenue
Farnborough, Hants GU14 7NB

D. G. Galley

Systems Designers International
Ferneberga House
Alexandra Road
Farnborough, Hants GU14 6DQ

W. T. Singleton	Aston University Dept of Applied Psychology College House Gosta Green, Birmingham B4 7ET
T. Knowles	CAP Scientific Kings Avenue House New Malden, Surrey
G. B. Wilson	Ferranti Computer Systems Ltd Priestwood Wokingham Road Bracknell, Berks RG12 1PA
D. F. Bird	Linwood Scientific Development Systems Division Bowling Green Lane London
A. Gardner	Loughborough University HUSAT Research Centre Elms Grove Loughborough, Leics LE11 1RG
H. J. Pearson	Smiths Associates Consulting Systems Engineers Ltd 7 High Street Cobham, Surrey
C. J. Harris	Southampton University Dept of Aeronautics and Astronautics Highfield Southampton, Hants SO9 5NH

1: The future for command systems

I. White
(Admiralty Research Establishment)

1.1 Background

Command, control and communications (C^3) or, as it is frequently referred to, command, control, communications and intelligence (C^3I) is the management infrastructure for defense and war. It is intrinsically a diverse range of activities, incorporating a complex mixture of personnel and man-made systems. At present we do not properly understand many aspects of either man or his systems, and yet we rely on both to sustain peace and, if necessary, to efficiently conduct war.

In this introductory chapter the fundamental problem of command and control is elaborated, and some of the opportunities that the future offers are discussed. The prospects are exciting, but the pitfalls are old and have been experienced by many earlier optimists.

1.2 The problems of C^3I

1.2.1 Basic considerations
In looking to the future of command systems there are four problems:

There is no theory of command and control.
There is informational chaos.
There is a technology revolution.
There is organisational chaos.

1.2.2 The theory of command and control
A three-day symposium on command and control organised by the US Office of the Secretary for Defense in 1979 concluded that 'there is no adequate foundation for a theory of command and control, and hence no guiding principles for system design and evolution.' (Wohl, 1981).

There are of course several useful theories relating to large complex systems, with the work of Ashby (1956), de Greene (1970), Checkland (1972), Sage (1981) and Wohl (1981) being good examples. There have been some advances since then, but no breakthroughs. If there is a trend it is for the more rigorous (but still very inexact) methods of software engineering to be extended for complex system design and specification. Since most modern complex systems have a significant software content, this trend is both understandable and valuable. It is deficient, however, in that software engineering is a poor basis for better elucidating many of the more human attributes of command and control — such as perception, option formulation, degree of belief, decision making and learning.

1.2.3 Informational chaos
Providing new systems for command and control is more of a problem than ever before, because of the increasing complexity of modern warfare. The volume of data to be assimilated, and the need for rapid response, now mean that much of what has been in the hands (and the minds) of the men in command has got to be done faster than man unaided can manage. We need to understand how machines can help, and to provide these machines.

1.2.4 Technological revolution
The development, primarily in the civil marketplace, of high-capability information technology constitutes nothing less than a revolution. This needs to be appreciated because it radically influences the way forward for new command system design.

1.2.5 Organisational chaos
Organisational chaos has developed because military equipment is being procured by a large number of unrelated projects working to differing time scales and from different budgets. The contracts are with a wide range of companies using different technologies and with varying management styles and competence. The overall interoperability of these systems, and their evolvability, are not properly understood.

Lest these comments be mistaken for criticism of specific projects, note that the institution of the correct structures for establishing large sets of interrelated projects, and ensuring their satisfactory interoperability over all functions, is an unsolved problem of both technology and management. That is why it is mentioned here.

1.3 The route towards a solution

1.3.1 Strategy
In order to provide the framework within which an individual design can proceed, we need:

A goal architecture
An implementation strategy.

Within these brave words are subsumed many issues concerning the use of specifications and standards, the capacity for evolution and the feasibility of building large systems. What do the terms mean?

In following the principle of 'think first, act second', the needs for a military system can be best met by first determining a goal architecture, which is a clear description of

What the system does
How it is put together
How it can grow.

There are four major aspects in providing a goal architecture:

More use of formal specification and design methods
More use of system prototyping
More use of standards to aid interoperability
The application of an implementation strategy.

1.3.2 Analysis and specification methods

It is obviously essential, when building a new system, to understand what it is required to do and how it is to be constructed to do what is required. This itself is prefaced by the belief that such a system can in fact be built!

In this quest for an effective C^3I system, the following observation (paraphrasing Weizenbaum, 1976) is worth bearing in mind: the concept of a C^3I system as an effective man-machine–complex, functioning efficiently, without conflicts, errors of commission or omission, is a hypothesis.

The software for such a system is an attempt to state the theory supporting that hypothesis, and its execution on a machine (with or without men being involved) represents an attempt to prove that theory.

1.3.3 Descriptive methods

The software engineering principle, that the more you get right early the better, applies with a vengeance to command and control.

To specify future systems we require a descriptive language more rigorous and complete than plain English. There are many analysis and specification techniques coming into increasing use, which use features of formal languages, graph theory and common sense. Their principles are in detail quite complicated, but in outline can be summarised by the following steps:

'Accept only that which is clear and distinct as true.
Divide each difficulty into as many parts as possible.
Start with the simplest elements and move by an orderly procedure to the more complex.
Make complete enumerations and reviews to make certain that nothing was omitted.'

These principles were stated in 1637 by Descartes, an early systems optimist.

Of the design methods available today for system design, none is ideal. The most widely used methods, which derive from the field of software engineering, are based on the work of Jackson (1975), Parnas *et al.* (1983) and Yourdon/deMarco (de Marco, 1978). These methods are not ideal for C^3I systems, because they provide no real guidance on the best way to analyse and specify complex real-time systems. Further, the claim that they represent a clear unambiguous route to good design is to a degree not yet established (Parnas and Clements, 1986).

Some advance in specification is being made with new languages for multi-processor applications, and the British OCCAM language, intended for use with cellular arrays of processors, may have applications here. The development, especially in the USA, of dynamic logic (Wolper, 1981) may well be the best hope in the longer term for a rational and rigorous approach to the description and validation of complex real-time C^3I system designs.

The absence of proven methods for sizing, specifying, validating and costing software and other component aspects of large C^3 systems remains a major headache, but will be less of a problem in the future with an improvement in the specification of real-time systems and hence in the anticipation of their operational performance.

All these advances are likely to reduce significantly the problems of creating new systems. However, the idea of precisely defining such complex man-machine systems is probably unachievable. These classes of problems are sometimes referred to as 'wicked problems' in the literature (Partridge, 1981) because they are characterised by

No definite formulation
No stopping rule (to say the design is complete)
No unique solution.

At the highest level of system specification we are concerned with defining imprecise subjects such as human factors, command and control, decision-making, sensor data fusion, and tactics, most of which qualify for the description 'wicked'!

1.3.4 Application of prototyping

The second development required for the provision of a goal architecture is prototyping. Good specification and design will not be achieved just by the application of more rigorous descriptive methods. We also need to do more initial assessments of systems by prototyping, before they are fully engineered and put into service. This is particularly important where the system fundamentals are not well supported by science. It enables assessments to be made not only of machine-based facilities but also of human factors, through observation of and observations from users of the prototype.

It is intriguing that prototyping in command and control, which is a 'soft' science with little established theory, quite rare, whereas in the hard sciences it has a long history.

In both the UK and USA significant research resources are now being committed to various aspects of C^3I system prototyping, especially the task of determining the functions to be performed, and the adequacy of the machine aids within the system to support those functions.

Prototyping systems which seek to reproduce the function of a C^3I system are expensive in both capital and human resources. To exploit their value it is essential to have effective methods of assessing prototype performance. A key aspect of this is to ascertain the role of and the variability imposed by people as part of the total system.

Prototyping and specification are the two pillars of the principle that, to get it right in the end, you have to have it right in the beginning. The problems of obttaining the resources needed at the start of projects, when management visibility of the outputs is poor, is a recurrent impediment to adequate investment. This is none the less the major management decision which is needed to alleviate problems in the design and procurement of new C^3I systems.

1.3.5 Application of standards to aid interoperability
The third aspect of C^3I design is the application of standards. As elements of an architecture, standards have immense potential application for defining the interoperability features of:

Local area networks
Data types
Semantic relationships
Communication protocols
Standard communication channels
Databases
Security protocols
MMI display and MMI dialogue
High-level programming environments.

A dominant role in the development of standards for communications is now being played by the International Standards Organisation (ISO). Its achievements in agreeing standards and promulgating them have been and continue to be impressive for an international body seeking consensus among many nations with competing economic interests. Although the ISO developments are generally targeted at the civil sector, it is now NATO policy to seek convergence towards these standards via a transition strategy.

The inclusion in C^3I of those aspects of performance necessary for military operation is being achieved by the policy of monitoring the ISO developments and lobbying national ISO representatives to seek their inclusion in new standards.

The model for open system interconnection used by ISO is the seven-layer model, which forms the framework around which a whole range of standards has been developed since 1979 (see Chapter 5). It has been predicted that by the end of the decade the civil community will have the standards to build a future society based on electronic information technology (desJardins, 1985).

Within the military community the adoption of standards has not been especially exciting, and their expanding use in the civil sector will further increase the disparity in price and performance between military and civil systems.

The defense research community is uniquely placed to influence many of the higher-layer activities, for example those relating to multiplayer information networks, databases and security. This is an opportunity which must not be missed.

By the mid 1990s it is anticipated that effective ISO standards will be available for a wide range of data exchange functions extending beyond basic communications, for dynamic high-resolution colour displays, for database interoperability, for some security applications, and for most of the routine data handling and conferencing functions which form a large part of the overall C^3I task.

These and other advances in design methods, and the use of standardised high-level languages, will make systems which perform better and more predictably, and are easier to describe, understand, evolve and steal. The last aspect is especially important for military systems.

It should be part of the C^3I community's task to ensure that these standards are correct and are used, and that the full implications of their use are appreciated.

1.3.6 Implementation strategy
The last aspect of C^3I design and application is the need for management. The related major task in establishing a goal architecture is to have for it a feasible implementation strategy.

To produce new systems requires, as well as a range of mandatory specification methods and standards, the provision of management to ensure that they are used by different procuring projects. This demands two changes:

(a) For specification and prototyping, the willingness of project managers to commit far more resources to the earlier, less visible phases of the project
(b) The need for a management group with authority over many projects, able to mandate levels of expenditure on initial phases of projects, and able to make mandatory the use of standards.

1.4 The technology for the future

1.4.1 Introduction
Current trends reveal a wide range of exciting opportunities in systems design, hardware, software, communications and MMI. Many of these developments are in the civil rather than the military sector of industry.

1.4.2 Command system functionality
A particular feature here is the need for aids for the higher levels of C^3I. Studies of command aiding are showing that there is a great deal of assistance that can be provided to the command, which can be categorised at three levels:

Modern management aids In essence this is the electronic office. Given compatible requirements with civil sector products, these forms of support should be met with civil marketplace products, and seek conformity with the appropriate ISO standards. Much of C^3I is surprisingly deficient at this level.

Decision support systems These are based on understanding basic features of C^3I such as logistics, replenishment, orders of battle and fighting instructions. These types of aid are now feasible.

Decision support/supplement systems These are advice and action systems based on the techniques of artificial intelligence. They potentially allow far more powerful intelligence to be applied rapidly and impartially to complex real-time assessment and decision tasks for which the speed of human thought is inadequate.

1.4.3 Hardware

There is a need to introduce families of processors which are interoperable and admit substantial expansion of processing power (speed and memory), without the need for major reconstruction or replacement.

Apart from the requirement for modern powerful processing with a fast and large addressing range, there is a need for specialist machines to accommodate the needs of

Large, fast-access databases
Machine intelligence
Specialist arithmetic (array processors)
Massively parallel processing.

Machines for all these applications are the subject of active research internationally. Many will be available before the end of the century and will probably be inexpensive compared with current processor costs.

The prognosis for the future is that, in hardware terms, memory will be essentially free (Capers-Jones, 1984). Only processing speed will cost money. A consequence of this is that C^3I will be supported by vast database facilities, which are not generally present in today's systems.

1.4.4 Software

1.4.4.1 Introduction: On software there are many unanswered questions. Major issues are the design and specification of software, the use of high-level languages and reusability. Current levels of achievement in producing large software systems are poor. There are still no generally accepted methods of sizing large software projects in terms of time to procure, cost, machine loading and performance. A report (cited in ALVEY, 1985) on US federal software projects classified the degree of achievement of these as follows:

$<$ 2% used as delivered
3% used after minor modification
19% abandoned or reworked

29% paid for but not delivered
47% delivered but not worked

This position is not unique to the USA, but is perhaps more noticeable than in Europe because of their greater investment in large software systems. There are several facets in the fight to improve this situation, including design and analysis methods, better languages, and the development of standards for reusable software.

1.4.4.2 Specification and design methods: The specification and design of software for large real-time systems has proved over the years to be unremittingly a high-risk business. Over the past ten years there has been a strong development of several design and analysis methodologies of which those of Yourdon/deMarco and Jackson are prominent. In most cases where these methods have been applied there has been a noticeable improvement in software performance and productivity.

For real-time applications, RTL has been widely used. A variety of languages and supporting techniques have been employed in Europe, including in the UK the language CORAL and its real-time design method MASCOT. MASCOT has been quite successful, but has suffered from not being hierarchical in structure. It is now being developed in a hierarchically structured form, and is a candidate component of the Ada APSE.

For the longer term, dynamic logic may lead to quite new and formal methods of analysing real-time programs. This development and the use of arrays of processors are going to become dominant features of many real-time applications, as the old reliance on von Neumann architecture for processors relaxes.

1.4.4.3 Languages
1.4.4.3.1 Ada: It is clear that Ada will be widely employed. What also needs to be appreciated, however, is that it will still remain one of several languages.

The rich, variable typing capability in Ada, and other features, make it far more suitable than existing languages for large-scale software implementation. So powerful are these features that Ada has already found application as a specification language!

Massively parallel machines will become a major feature of fourth- and fifth-generation computing. They have an architecture for which Ada is not suited. Other languages will be used and will be widely employed in defense systems, including some aspects of C^3I.

1.4.4.3.2 Artificial intelligence languages: In this domain the use of declarative or non-procedural languages will become extensive. Developments of the PROLOG language are anticipated as being a major feature of machine intelligence development. These languages are conceptually totally different from procedural languages, which have represented the mainstream of high-level languages, and of which Ada represents the apogee of development.

This is again an area in which language needs are not met by Ada.

1.4.4.3.3 Security: The need for security cannot be met by any current high-level language. In order to achieve a system which has provably secure multilevel security (MLS), it is anticipated that there will have to be both new languages and new machine architectures.

The concept of having systems subdivided into security safe and security unsafe partitions offers the possibility of complete systems which exploit much of the available marketplace hardware and software, and relegate specialist security functions to proven software and hardware.

A serious problem in this area may occur if the developments in security systems enable a far deeper understanding of these issues among a broader range of system users. Consequently many current systems, which have been given MLS clearance on the basis of a set of heuristic tests and checks, will become easily breached. This may necessitate the widespread replacement of substantial parts of our current C^3I systems.

1.4.4.4 Software reusability

'Most software in current projects is built from scratch; and yet it is estimated that, of the current spectrum of new software writing, only about 15% is unique, novel, and specific to new applications. The remaining 85% is the target which researchers into reusable code are attempting to standardise.' (Capers-Jones, 1984). Another estimate is that by 1990 the proportion of reusable code in use will be as high as 50%.

This is an area where the development of standards is crucial, and it is already the subject of intensive research activity. If reusability can be developed as predicted, by the year 2000 software could start to see some of the cost reductions that we have so far only been able to associate with hardware.

1.4.5 Man-machine interface

1.4.5.1 Man: It has been observed that over the next twenty years the capability and intelligence of military staff is unlikely to evolve significantly.

In the higher echelons of C^3I the man's role includes among other things, supervision, correlation, pattern matching and decision making. The C^3I system compiles a perception of the tactical environment, and establishes a defensive/aggressive stance in the deployment of its resources. Currently it is the men in the C^3I system who are the repositories for most of the high-level information for the interpretation of the environmental picture and the forces' battle stance.

The men undertaking this duty are required to function for long periods, perhaps under stress and fatigue, and to retain continuity of operation when one man takes over the role of another.

It is well known that the human operator is prone to several perceptual decision/action idiosyncrasies (especially over sustained periods and under stress), of which the following is a short list (Sage, 1981):

Adjustment and anchoring When faced with a large amoung of data, the person selects a datum (such as the mean) as an initial starting point (or anchor) and then adjusts that value improperly in order to incorporate the rest of the data.

Availability The decision maker uses only easily available information. An event is believed to occur frequently (have high probability) if it is easy to recall similar events.

Conservatism The failure to revise estimates as frequently as necessary.

Data saturation Reaching premature conclusions on the basis of too small a sample, and then ignoring further data.

Self-fulfilling prophesy The decision maker values certain outcomes and acquires and analyses only data which support that conclusion.

Attribution error The decision maker associates success with inherent personal ability, and failure with poor luck in chance events.

Gambler's fallacy The decision maker falsely assumes that the occurrence of one set of events enhances the probability of an event that has not occurred.

Habit Familiarity with one rule can result in its excessive use.

Law of small numbers People often express greater confidence in predictions based on small samples with non-discomforting evidence that in predictions based on large samples with minor discomforting evidence.

Order effects The order in which information is presented affects information retention and weighting.

Outcome irrelevant learning Use of an inferior decision rule can lead the decision maker to believe his results are good, because of his inability to evaluate the choices not selected.

Panic The decision maker under stress, and faced with many options that he is unable to evaluate, takes one at random or fails to act all.
Copyright © 1981 IEEE.

Explicit military examples which can be identified in these terms are reviewed by Dixon (1976).

One of the primary functions of command aids is to take account of these potential failings, and to seek to process and present command information so as to minimise their effect. An important consideration in the development of advanced processing and machine decision support is the extent to which the educational, training and basic skills needed to man future command and control systems will change. The impact this can have on career structures, training establishments, recruiting and total manpower needs is potentially profound.

In most areas of industry where there has been a wide-ranging influx of information system technology, there have been related changes in the manpower requirements. Such changes are inevitable in the future C^3I environment, where there will be an increasing need for personnel with computer skills, and a different style of education than is currently represented in C^3I systems staff.

1.4.5.2 Display technology

1.4.5.2.1 Hardware: The dominance of high-resolution raster technology is now almost complete, and the next generation of displays will be high-resolution

The future for command systems 11

color systems. Resolutions of up to 2000 lines are available today, and by the year 2000 it is expected that 5000 line systems will be commonly and inexpensively available.

The development of extremely inexpensive memory (even today) means that a wide range of image manipulations based on the bit mapped display, with a vast page memory, will enable virtually total freedom in the manipulation and superimposition of images.

Plasma systems are predicted which offer both high resolution and color. In the longer term these systems are likely to replace vacuum tube technology, but it will be many years before their currently high costs make them competitive with the older raster display technology.

1.4.5.2.2 Graphic standards: In the function of displays there is a range of very interesting developments relevant to C^3I. The new ISO standards of graphics kernel system (GKS) and the programmer's hierarchical interactive graphics standard (PHIGS) indicate that future command systems so built could exchange pictures despite a wide disparity of architecture and language within the participating systems.

1.4.6 Knowledge-based systems

1.4.6.1 Introduction: A new facet of computer science is that concerned with the way knowledge is represented in the computer and used. Such systems are sometimes called intelligent knowledge-based systems (IKBS) and include expert systems. These developments have a strong potential application to C^3I (see also Chapter 7).

Man's understanding of how to perform a complex task can be communicated in a natural language such as English. It is worth noting that in C^3I even this language has been rarely tried for an expression of the total C^3I requirement.

1.4.6.2 Expert systems: Transforming a plain English description into a form that is machine executable is already being achieved using expert systems, in which the ideas of an expert are translated into rules. This task is achieved by the joint expertise of a domain expert (our command and control specialist) and an intermediary called a knowledge engineer. This process leads to a collection of machine executable rules — several thousand for some C^3I applications. This process is tedious, iterative and often difficult to control when the knowledge engineer has to work with a group of experts rather than a single expert. This is a problem for C^3I because, unlike those who battle with disease, the C^3I specialist seldom has access to real conflict in learning his trade.

The command and control task is a real-time problem, in which data is not a static set of diagnostic parameters but rather a complex combination of data from sensors and intelligence sources, encyclopaedic data, data on tactics and rules of engagement, etc. Much of this data is changing dynamically, and the expert system must accommodate this.

12 The future for command systems

The blackboard expert system meets this need by placing information on a virtual blackboard, which is continually evaluated and re-evaluated as new information is added and as stale information is removed. This type of system has been applied to submarine sonar detection processing in the USA (see Nii *et al.*, 1982) and to surface data sensor fusion in the UK (Lakin and Miles, 1985). It appears to have wide application for certain command and control tasks.

1.4.6.3 Knowledge by structure: To illustrate the development and extension of KBS to C^3I, consider the following simple paradigm for human decision making (Wohl, 1981):

Stimulus	(Picture compilation)
Hypothesis	(Threat assessment)
Options	(Resource allocation)
Response	(Response)

An analogous C^3I sequence is shown in parentheses. Each of these functions involves complex inferential operations on the input data, using a range of knowledge of military systems and operations. Each inference function could be performed by the sort of inference mechanism of a blackboard system.

Conceptually the knowledge can be isolated and the inference mechanism of the expert system deemed to operate on it. So what is the knowledge base? At the lowest level in this knowledge base will be factual knowledge on ships, aircraft, submarines, weapon systems, sensors etc. Above this will be data defining the relationship between these elements (relational data). Scenario data will give positions, speeds, identities of units, and above this will be inexact interpretive data. Whereas the lower-level data is simply elemental, at the higher levels data is formed inferentially from functional relationships between elements and sets of elements. This hierarchy can extend but, as it does, the capability of KBS methods diminishes!

1.4.6.4 Knowledge by induction: At the higher levels of C^3I, defining the problem and the concepts is not clear cut. None the less there are specific C^3I event-response examples which are considered correct by the specialists. A machine intelligence endeavour which can exploit this aspect of difficult problems, and even help to formulate basic theories, is the obverse of rule enacting.

Rather than extract an expert's perception of the rules and evaluate the result by specific examples, we can elect to take a large set of example actions, which are considered correct and use them as the basis for deducing rules. The conceptual power of the idea needs to be emphasised. What we are doing is stating a series of relationships, deriving rules from them, extrapolating as needed using those derived rules. The derivation and application processes are accounted for within the internal logic of the program executing software.

In this sense the program is being executed by a route which is independent of the programmer; he has only declared a requirement. By contrast, in the orthodox program each step is a preselected procedure and has an identifiable statement or

The future for command systems 13

block in the language of the program code. The deductive systems are not so constrained, and for this reason their implementing languages are called non-procedural or declarative languages.

A longer-term hope for developments of declarative languages is to write down a detailed specification of requirements in a formal language, and for a processor to enact this specification directly. If this is achieved the computer becomes a task obeying machine at a much higher level than is seen in orthodox procedural language systems, however well developed the high-level language.

The point to stress is that this is not merely dreaming. The declarative language PROLOG fulfils many of these requirements, although it still needs much development. None the less its potential is such that the Japanese have selected it as the language cornerstone of their programme for their fifth generation of computing systems.

1.4.6.5. Maintenance: Question and answer functions at the lower levels of KBS will naturally allow extension of the system. However, developments are needed which will allow concepts to be revised or deleted. Where examples are used to define or support a rule base at the higher levels of IKBS, maintenance might well be achieved by adding new examples and removing those that are considered irrelevant or inaccurate.

Another benefit of example-based systems is that they may well be better able to explore their own limitations in terms of completeness (ability to perform in all possible circumstances) and consistency (checking that no dual interpretation of the data can lead to contradiction). It is noteworthy that current C^3I systems are very inadequate in these respects.

A more insidious problem with maintenance may occur due to the diminution of human skills as the machine supplants the man's role. This is in itself not a problem, but it may lead to difficulties in extending the application if no human expert is available. Provided such systems are introduced gradually this may be no more than a transitory problem; old skills can decay as the new emerge. In engineering at least there are many examples of such transitions being successful, although perhaps initially painful.

1.4.6.6 The future of KBS: KBS using computing machinery with new architectures, new languages, and new man-machine interfacing will embrace many remarkable developments and concepts in software, hardware and overall function. Intensive research to provide the KBS environment for future systems is now going on in the USA, in Japan (in their fifth-generation programme) and in Europe (within the Esprit programme).

Many aspects of C^3I should be well within the proposed capabilities of KBS now envisaged by researchers, including force picture compilation, threat evaluation, intelligence collation, resource allocation, task briefing, and equipment testing and fault diagnosis.

In the USA, Professor Edward Feigenbaum (1981) of Stanford University has predicted that the greatest application for machine intelligence in the 1980s will be in military command and control. Is this because it is a rich source of applications, or a rich source of research funding? The answer is clearly both, but the idea of artificial intelligence as a major palliative for many of the problems of command and control needs to be regarded with considerable caution. The problems of better understanding perception, option formulation, degree of belief, decision making, and learning have been cited above as important aspects of command and control. The record of artificial intelligence in these areas, and in the demonstration of method rather than example, is poor (Anderson, 1986) and does not add much to our understanding of how to build command and control systems.

1.5 Concluding comments

Current C^3I is frequently characterised by high cost, poor functionality, poor decision support, inflexible architectures, and high through-life costs. Although technology is clearly going to provide faster computers with much larger memories, and better displays and MMI, the reduction of software costs and provision for the planned evolution of systems are features that will only come from direct effort within the C^3I community itself.

The tasks of specifying functions and ensuring that they are achievable and effective requires sound design, analysis and prototyping as the foundation of any major C^3I system procurement. Many of the features I have outlined will provide this foundation.

Similarly, the ability to achieve safe and cost effective evolution of these systems requires, in addition to the features above, a sound inventory of standards, and the management infrastructure to ensure that individual projects use them. Again the prognosis here is optimistic, with a wide range of exciting opportunities.

If I have not written much on the problems of perception, formation of beliefs and decision making, it is quite simply that I see no clearly discernible road forward for these topics. They are arguably the essence of any effective command and control system. The future of these fields of systems understanding is certainly critical, but I see neither paradigms nor progress here which offers any clear way ahead.

1.6 References

ALVEY (1985) US General Accounting Office statistics, quoted in ALVEY software Engineering Project report 2, 'Requirements for and feasibility of advanced support environments'
ASHBY, W. R. (1956) *Introduction to Cybernetics*, Wiley
CAPERS-JONES, T. (1984) (Reusability in programming: a survey of the state of the art', *IEE Trans.* SE−10 (5)
CHECKLAND, P. B. (1972) 'Towards a systems based methodology for real world problem solving', in *Systems Behaviour* (eds. Beishon, J. and Peters, G.), Harper and Row

De Greene, K.B. (ed.)(1970) *Systems Psychology*, McGraw-Hill
DesJARDINS, R. (1985) 'Tutorial on the upper layers of OSI', NATO Symposium on Interoperability of ADP systems, the Hague, 1985
DeMARCO, T. (1978) *Structured Analysis and Systems Specification,* Yourdon
DIXON, N. F. (1976) *The Psychology of Military Incompetence,* Futura
JACKSON, M. A. (1975) *Principles of Program Design,* Academic Press
LAKIN, W. and MILES J. A. H. (1985) IEE international conference on advances in command control and communications, Bournemouth, England, April 1985
NII, H.P., FEIGENBAUM, E. A., ANTON, J. J. and ROCKMORE, A. J. (1982) 'Signal to symbol transformation: HASP/SIAP case study', *AI Magazine,* Spring 1982
PARNAS, D. L. and CLEMENTS, P. C. (1986) 'A rational design process: how and why to fake it', *IEEE Trans.* **SE-12** (2)
PARNAS, D. L. *et al.* (1983) 'Enhancing reusability with information hiding', Proceedings workshop on reusability in programming
PARTRIDGE, D. (1981) 'Computational theorising as a tool for solving wicked problems', *IEEE Trans.* **SMC-11** (4)
SAGE, A. P. (1981) 'Behavioural and organisational considerations in the design of informationl systems and processing for planning and decision support', *IEEE Trans.* **SMC-11** (9)
SUTHERLAND, J. W. (1986) 'Assessing the artificial intelligence contribution to decision technology', *IEEE Trans.* **SMC-16** (1)
WEIZENBAUM, J. (1976) *Computer Power and Human Reason,* Freeman
WOHL, J. G. (1981) 'Force management decision requirements for air force tactical command and control', *IEEE Trans.* **SMC-11** (9)
WOLPER, P. (1981) 'Temporal logic can be more expressive', Proc. 22nd. Ann. Symp. Foundations of Computer Science, IEEE Computer Society Press, October 1981

2: Design and structure of C^3 systems

Chapter 2.1

The C-process: a model of command

D.G. Galley
(Systems Designers International)

2.1.1 Introduction

A simulation of a command system to support the force-level command of anti-submarine warfare (ASW) is being developed at the Admiralty Research Establishment's Command Systems Laboratory as part of a long-term research programme into naval command and control. During an early phase of this development programme, it became evident that the design team did not have a sufficiently clear overall view of the command, control and co-ordination processes operating within a command system. What was needed was a model to consolidate the fragmented ideas concerning command into a coherent, logical system, i.e. a model of command.

This contribution traces the development of such a model, starting with a simple three-process representation of the overall command system and culminating in the C-process. This is a generalised model of the command, control and co-ordination activities performed by any of the command centres (or cells) within a distributed command system.

During the development of the C-process model, many of the concepts were interpreted in terms of a world state space model. This interpretation is briefly described in Section 2.1.8. Section 2.1.9 uses a network of command cells to describe distributed command and discusses the role of co-ordination. A semantic model of the inter-/intra-cell command messages is summarised in Section 2.1.10. Finally, section 2.1.11 gives a description of the application of the model in the design of computerised command subsystems (called C-stations) and a summary of the scientific and engineering benefits gained by this attempt to model command.

2.1.2 The need for a model of command

Existing naval shipboard command dystems are primarily concerned with target acquisition and tracking, and weapon direction and control. They do not offer much assistance to higher-level decision processes such as planning. An evolutionary approach to the design of ASW command aids, involving the adaptation and refinement of existing systems, was therefore inappropriate.

Progress with the design was also frustrated by a lack of scientific understanding of the functions carried out within a naval command system. An attempt was made to discern these functions by examining naval doctine as described in tactical procedures. This provided an insight into the complexity of naval tactical command, but not an understanding of how the officer in tactical command and his subordinate commanders perform their command task. The list of requirements to plan and monitor the many aspects of a complex mission begs the following questions:

How is a plan to be formulated?
What does monitoring a mission involve?

Although the doctrine successfully identifies the products of a naval command system, such as the various kinds of policy, plans, guidance, etc., it does not explain the method by which these products are produced.

Obtaining an insight into how ASW command was performed, by observation and discussion with its practitioners, provided much information but not enough structure to discern any underlying method. There were too many command decisions to formulate a unique analysis of each decision problem.

We hypothesised the existence of a set of universal, general purpose command and control principles, collectively called *command*, and treated *ASW* command as simply a particular application of these principles to ASW. The analysis of ASW command was split into separate analyses of ASW and of command. Once an understanding of these separate disciplines was achieved, the application of command to ASW could be examined more easily, with their synthesis deferred until later in the design phase.

Preliminary investigation of command suggests a number of important aspects:

Interpretation of sensory data (perception)
Command
Control
Co-ordination
Communication.

The ensuing analysis and model building concentrated on command and control, their relationship with each other and with perception and co-ordination.

2.1.3 The overall command system

The development of the model started with an examination of the overall command system, which was considered to be a system for achieving objectives using assigned resources to interact with the physical world.

Initially a highly simplified model of the overall command system (Fig. 2.1.1) was used, which comprised three processes:

(a) Perception process   System feedback path

(b) Idealised command process ⎫
(c) Idealised control process   ⎬ System feed forward path
                                  ⎭

and the following signal flows:

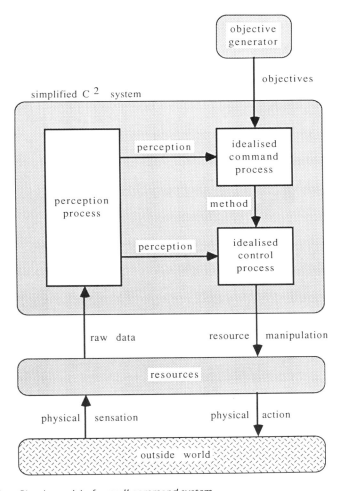

Fig. 2.1.1 *Simple model of overall command system*

Objectives   Goals (desirable situations) to be attained and 'penalties' (undesirable situations) to be avoided
Raw sensory data   Sensor resources' response to physical sensation
Perception   Updates to the state and structure of a world model
Method   Demanded resource activities, i.e. a required course of action
Resource manipulation   Orders given to resources to be translated into physical acts performed on the outside world or other resources.

It was originally assumed that objectives originate outside the system, enter the forward path of the system, and are transformed into method. 'What needs to be achieved' is transformed into 'How it is to be achieved'. The method is subsequently translated into the manipulation of resources in the physical world. Raw sensor data enters a perception process which completes the system's feedback path from the physical world.

2.1.3.1 The perception process: The perception process encapsulates that part of the overall command system which is concerned with understanding the world. This entails interpreting and integrating the whole spectrum of sensory data, ranging from sonar echoes to political factors. Consequently, this process encompasses a vast range of activities — from the detailed signal processing associated with target acquisition and tracking, to intelligence interpretation.

However, all forms of the perception process incorporate a world model which they endeavour to keep up to date. This world model provides structure to the received sensory data, and also expresses known dynamic relationships between past, present and future which provide a predictive capability. This is necessary because command and control can never influence the present — only the future. Knowledge of the current world state is only of value as a contribution to understanding the future.

Consequently, the act of perceiving involves not only updating this world model's current state (the current perceived world state), but also, when necessary, refining the world model itself by amending the parameters, structure and/or dynamics of the model, i.e. by learning. The resulting up-to-date model constitues a perception of the world. The role of the perception process is to provide and maintain this perception so that the command and control processes have an understanding of how the world is expected to behave in the future.

It was assumed that a perception interface exists, across which the updated state of the world model and any world model refinements are made available to the command and control processes.

2.1.3.2 The idealised command process: The idealised command process must transform objectives, together with a perception of the world, into a course of action which will achieve these objectives. This process operates in two modes, corresponding respectively to the preparatory and execution phases of a mission:

Mission planning   Searching for and selecting a method which is expected to achieve the objectives.

Effectiveness monitoring   Continually reassessing the suitability of the promulgated method.

2.1.3.3 The idealised control process: The idealised control process must transform the required course of action, together with a perception of the world, into resource manipulation orders such that actual resource activity complies with the required resource activity.

The C-process: a model of command 23

It should be noted that it is not sufficient just to think in terms of primitive error-driven control which can only correct after disturbance. A successful controller must be capable, to some extent, of anticipating change and taking preemptive avoiding or corrective action.

As for idealised command, the idealised control process operates in two modes corresponding to the preparatory and execution phases of resource manipulation:

Resource order generation Constructing resource orders that are expected to give rise to the demanded resource activity
Compliance monitoring Determining how closely the demanded activities are being, and will be achieved.

2.1.3.4 Functional similarities: The distinction between command and control may not always be obvious in a command and control system because the participants will often be performing both functions. Although the simple command system model had been constructed in an attempt to clarify, and to some extent emphasise, this distinction between command and control, there are some valuable similarities.

Since compliance with prescribed behaviour constitutes a particular form of objective, control can be regarded as a specialised form of command. Thus order generation is a specialised form of planning, and compliance monitoring is a specialised form of effectiveness monitoring. Furthermore, these similarities extended to both preparation and execution modes of operation. Both seem to necessitate assessing the extent to which the objectives or compliance can be achieved. Consequently, mission planning, effectiveness monitoring, resource order generation and compliance monitoring, all appear to share the same underlying functionality.

These similarities led to the speculation that it might be possible to find a functional model of the command and control processes which was generally applicable and which could describe the processes taking place in any of the command/ control cells within the overall command system. If all such cells could be considered as instances of such a general functional model (subsequently christened the C-process) then any particular command organisation could be represented by the interconnection of separate C-processes, where each C-process represents a cell with a declared responsibility within this structure.

2.1.4 Development of the C-process model

The functional model was developed in three stages. Initially, a 'proposal assessment' loop was suggested as the underlying paradigm for problem solving within the idealised command and control processes.

Consideration of the hierarchic nature and complexity of command led to the two idealised processes being abandoned in favour of a single generalised process. This necessitated augmenting the basic proposal assessment loop and gave rise to the C-process in its simple form.

Complexity arising from the need to consider simultaneously many interrelated

24 The C-process: a model of command

topics led to further development of the simple C-process structure to form a complex or multitopic version.

2.1.5 The basic proposal assessment loop

The idealised command and control processes are both required to handle difficult, possibly unforeseen or unique problems which, in general, cannot be solved by applying prescribed procedures and formulae. It is postulated that 'solution seeking' within the idealised processes is carried out by iterative application of a 'proposal assessment' loop (Fig. 2.1.2). This loop, which lies at the heart of the functional model, comprises three lower-level process functions:

PROPOSE A generator of putative resource activities (command) or putative resource orders (control).

WORLD MODEL Predicts the expected outcome of the putative resource activities (command) or the putative resource orders (control). This function is related to the world model in the *perception* process and is maintained by state updates and model refinements arriving via the perception interface.

MEASURE Applies effectiveness metrics to predicted outcomes.

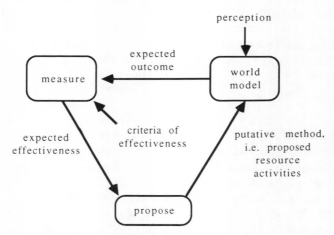

Fig. 2.1.2 *Basic-proposal assessment loop: heuristic interpretation*

This proposal assessment loop, combined with the supporting perception process, closely resembles the canonical decision model proposed by Pugh (1976).

The operation of the proposal assessment loop is described below in terms of mission planning and effectiveness monitoring, that is, the command preparation and execution modes. Similar arguments can be applied to represent the corresponding control activities, order generation and compliance monitoring.

2.1.5.1 The proposal assessment loop in planning mode: The PROPOSE function submits putative resource activities to the WORLD MODEL, which predicts

The C-process: a model of command

the future including the consequence of the putative activities. This predicted future is then subjected to quantitative assessment by the application of measures of effectiveness. The resulting estimate of effectiveness expected to ensue from following the putative activities is then returned to the PROPOSE function. This function then either terminates the search for an acceptable proposal or generates an alternative. Once an acceptable proposal has been selected it can be promulgated as the demanded method or plan of action.

2.1.5.2 The proposal assessment loop in monitor mode: The promulgated plan is resubmitted to the WORLD MODEL. As before, the expected future is predicted and subjected to the same quantitative assessment. As time progresses, and the plan is put into effect, the expected outcome is repeatedly updated by the WORLD MODEL, using the latest perception information. If all goes well, each predicted outcome will continue to match the desirable outcome, as originally predicted, and the expected effectiveness remains high.

In practice the expected outcome may drift away from the desired outcome, and the corresponding effectiveness will be inferior to that originally anticipated for several reasons:

(a) Own resources have not acted as expected, i.e. they have failed to implement the specified plan.
(b) Enemy and neutral objectives have not behaved as anticipated.
(c) The environment has not behaved as anticipated.

The expected future must be continually reassessed and, if insufficient effectiveness is predicted, then the planning mode must be re-entered.

The MEASURE function will also compare the current perceived world state against the desired world state. A match signifies successful competition of the allotted task.

2.1.5.3 Criteria of effectiveness: The introduction to the proposal assessment loop presupposes the existence of measures of effectiveness appropriate to the current objectives. An attempt was made to understand what is involved in translating objectives into effectiveness metrics. This required re-examining what is meant by the term *objective*.

The achievement of objectives involves not only the attainment of goals or desirable situations, but also the avoidance of undesirable situations, which for the want of a better term were called *penalties*. A statement of objectives must define both the goals and penalties and their respective degrees of desirability and undesirability.

Objectives can be interpreted as a distribution of desirability of various world situations — in other words, a value system. Measures of effectiveness which are intended to correspond to a set of objectives constitute a model of that value system. Constructing effectiveness metrics is an exercise in identifying and specifying the *value model*. This is a difficult task because the value model should make explicit:

(a) All relevant factors, and not just the immediate principal value of, for example, disabling an enemy submarine, or self-survival. Longer-term and more subtle issues must be encompassed, e.g. the value of remaining covert, retaining resources, and maintaining morale.

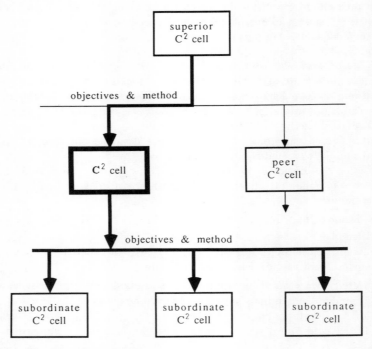

Fig. 2.1.3   *Hierarchy of command*

(b) The relative value of all relevant issues, e.g. the value of disabling an enemy submarine compared with the value of surviving the engagement.

In the case of control, the objective is simply that of compliance with the method imposed by command. Effectiveness would be measured by comparing expected resource behaviour against the required behaviour.

2.1.6 Hierarchic command

Command systems consist of a distributed, hierarchical organisation of cells, possibly performing a mixture of both command and control, and with all but the lowest level of cells generating objectives for their subordinates (Fig. 2.1.3). This results in a staggered transformation of objective into method. Command cells at intermediate levels within a command chain will receive and produce hybrid tasking comprising a mixture of objective and method. They are presented with high-level plans, comprising a mixture of high-level objective plus some method, which will be transformed into low-level plans, i.e. lower-level objectives plus much method.

The C-process: a model of command

This had several implications for the functional model. One of the more significant was the realisation that the transformation of objective into method in a single step, as implied by the original model, is oversimplistic.

Normally it will be too difficult to find a method to reach a remote goal in a single step. A common strategy is to propose intermediate goals or 'stepping stones' and then to seek methods for stepping between these. This implies an ability to decompose a high-level objective into a set of intermediate objectives. Successive applications of this top-down approach appear to cater for command problems of any magnitude, and have the advantage that at any stage of problem decomposition the subproblems can be delegated to subordinate commanders.

Thus we need a model that can not only describe the functionality of idealised command and idealised control, but also represent general hybrid tasking at any level.

Although in principle the model applies to control, the developments described below concentrated on the problem of command for the following reasons:

(a) The responsibility for achieving the objectives resides with command, which therefore plays the dominant role within the overall system.
(b) Control is believed to be a degenerate, and better understood, form of command.

2.1.6.1 Implications of hybrid tasking: Hybrid tasking has two major implications, each requiring augmentation of the basic proposal assessment loop:

Hybrid incoming command messages These must be analysed to discriminate between the different kinds of message statements (objective and method) so that they can be handled appropriately.
Hybrid outgoing messages These imply an ability to decompose a high-level objective into a set of intermediate objectives. This necessitates overcoming the following limitations of the basic assessment loop:

(a) The PROPOSE function in the basic assessment loop is confined to proposing activities. Clearly the PROPOSE function must also be free to propose activities and/or intermediate objectives.
(b) The WORLD MODEL function, as originally conceived, was only able to predict forward from the present. An enhanced WORLD MODEL would be required to predict starting from either the current situation or from hypothetical situations.
(c) Finally, the MEASURE function is now required to assess proposals with respect to objectives generated either internally or externally.

It is also necessary to introduce an extra function called *INTERPRET OBJECTIVES* which establishes the criteria of effectiveness appropriate to the current objectives.

These additional requirements result in the simple form of the C-process shown in Fig. 2.1.4.

28   The C-process: a model of command

2.1.6.2 Simple form of the C-process model: The simple form of the C-process shown in Fig. 2.1.4 incorporates three functions similar to those in the basic proposal assessment loop plus four additional functions:

PROPOSE   Generates method and objective proposals.

WORLD MODEL   Predicts the consequence of the proposed method, concentrating on those aspects appropriate to the relevance specified by INTERPRET OBJECTIVES, and starting from either the current situation or from a hypothetical start situation.

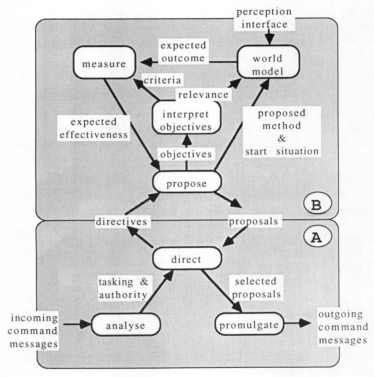

Fig. 2.1.4   *C-process: basic structure*

MEASURE   Performs quantitative assessment of predicted futures by applying effectiveness metrics corresponding to the objectives specified by the PROPOSE function.

INTERPRET OBJECTIVES   Translates objectives into criteria (effectiveness metrics) and relevance which indicates what kind of world modelling is appropriate.

ANALYSE   Examines and sorts incoming command messages to establish the tasking, authority and constraints prevalent at all times. This requires:

(a) Distinguishing between statements of authority and constraint, statements of requirement, and the object(s) of these statements

(b) Establishing the range of applicability of these statements

The C-process: a model of command

(c) Establishing whether recent statements supersede or augment previous statements.
(d) Distinguishing between mandatory and advisory aspects
(e) Announcing and identifying amendments.

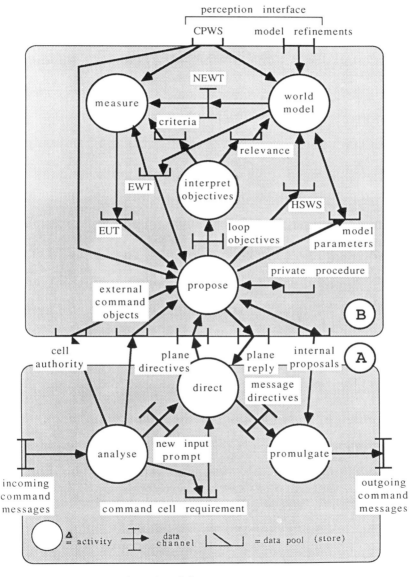

Fig. 2.1.5   *C-process activity-channel-pool diagram*
         CPWS   current perceived world state   HSWS   hypothetical start world state
         EUT   expected utility   NEWT   new EWT prompt
         EWT   expected world trajectory

30   The C-process: a model of command

DIRECT   Presents objectives to the proposal assessment loop (the role of this function is discussed in a later section.
PROMULGATE   Compiles and issues command messages as directed by the DIRECT function.

A more detailed definition of the structure of the C-process is presented in Fig. 2.1.5, which uses activity-channel-pool symbology (borrowed from MASCOT real-time systems design methodology) to capture the inter-process information flow more formally. A fuller description of the C-process components and data flow is provided in Galley (1983a).

2.1.7 Complex command
The final stage of this analysis of command and control resulted from consideration of the complexity of naval command, and the way in which total problems are decomposed into manageable subproblems. Complexity is ascribed to interactions between the multiplicity of objectives and resources typically associated with naval command:

Objective interaction   In general, pursuit of several objectives simultaneously will be more difficult than in isolation. Mutual influence between objectives prohibits the individual objectives being pursued independently. For example, attempting to find the enemy without betraying own position are examples of potentially conflicting objectives.

Resource interaction   There will usually be a large number and variety of resources within a naval force. Any interaction between these resources, which may be manifest as interference or resource interdependency, means that individual resources cannot be employed independently. For example, a platform may be capable of great speed and a sensor in isolation capable of great sensitivity. If however, platform motion degrades the sensor performance and/or sensor deployment degrades platform motion, then employing them in combination may demand difficult compromises.

Consider the following complex (multi-objective, multi-resource) planning task. Imagine that it is required to destroy a hostile submarine, at a known position, and to survive the engagement. The task will be simplified by assuming that it only involves weapon and platform resources. Even this simplified 'complex' problem serves to illustrate the wide range of factors which must be handled:

Influence of weapon on destruction
Choice of weapon(s) to cause sufficient damage with sufficient confidence
Extent to which firing weapon(s) warns the target to take evasive action

Influence of platform on destruction
Weapon's delivery dependency on platform position and motion
Extent to which the platform motion warns the target to take evasive action

Influence of weapon on survival
Possibility of causing self-inflicted damage

Extent to which firing weapon(s) betrays own position and alerts the target to mount a counter-attack

Influence of platform on survival
Extent to which platform motion betrays own position and alerts the target into mounting a counter-attack
Evasive steering to be performed in response to suspected counter-attack.

If this problem were approached *en bloc,* the compound objective, encompassing both 'destruction of target' and 'self-survival', would be presented to a proposal assessment loop which would be required to consider all objectives and resources simultaneously. The following section suggests an alternative approach based on problem decomposition.

2.1.7.1 Problem decomposition: It is postulated that large problems are tackled by iteratively:

(a) Dividing the factors involved into a number of topics, for example partitioning the total set of objectives/resources into individual objectives/smaller groups of resources
(b) Piecemeal solution of the subproblems associated with the individual topics
(c) Assessing these tentative subsolutions in the light of subsolutions to the other topics
(d) Modifying these tentative subsolutions accordingly, and returning to step (b).

For example, the afore-mentioned 'destroy and survive' task could be decomposed by objective, that is, treated as a target destruction problem plus a self-survival problem. The influence of both weapon and platform would have to be considered in each case. Alternatively, the task could be decomposed by resource type, i.e. treated as a weapon problem plus a platform problem, with both attempting to achieve the combined 'destroy and survive' objective. This approach can be taken further, i.e. the overall task decomposed by objective and by resource into four topics:

Weapon aspects of destruction
Platform aspects of destruction
Weapon aspects of survival
Platform aspects of survival.

The way in which a particular problem is 'sliced' will be a matter of convenience, convention or intuition. Whatever decomposition is used, it is essential to take account of all interactions between the various subproblem topics in order that cosistent and mutually acceptable total solutions are produced.

2.1.7.2 Complex form of the C-process: The simple C-process (Fig. 2.1.5) is divided internally into two subprocesses called the A-subprocess and the B-subprocess, the latter being that part based on the proposal assessment loop. As it stands, this structure can only address a single problem at a time.

32   The C-process: a model of command

To extend this simple structure to support parallel but co-ordinated consideration of tentative solutions to subproblems, there is a need for multiple, interconnected, proposal assessment loops. Figure 2.1.6 shows a stretched A-subprocess able to accept a number of B-subprocesses in much the same way that a computer bus backplane accepts electronic circuit boards. Each 'slot' provides the same

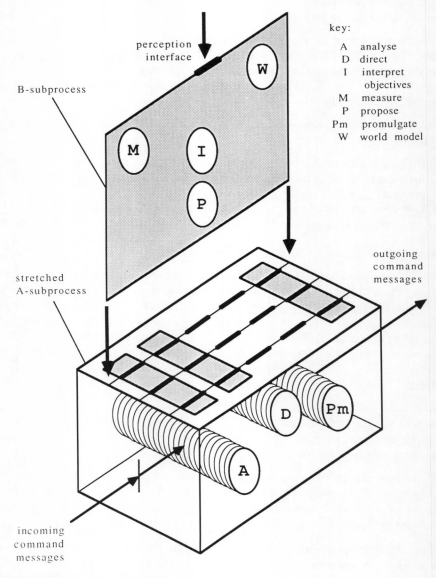

key:
A   analyse
D   direct
I   interpret objectives
M   measure
P   propose
Pm promulgate
W   world model

Fig. 2.1.6   *C-process: complex form*

interfacing between the A-subprocess and the B-subprocess as in the simple C-process (see Fig. 2.1.5).

The resulting functional model, called the complex (or multitopic) C-process,

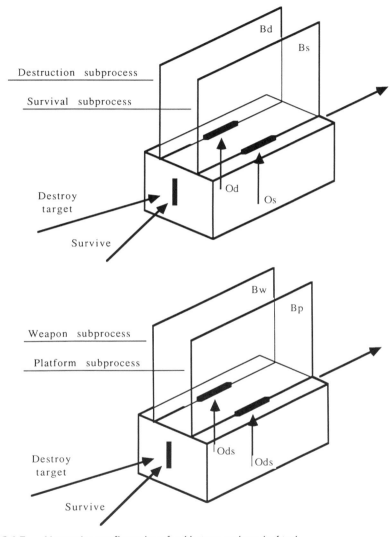

Fig. 2.1.7 *Alternative configurations for 'destroy and survive' task*

eases the task of the individual proposal assessment loops by allowing each to concentrate on simple, fundamental objectives and/or to specialise in particular types of resource. The difficulty shifts across to the A-subprocess, in particular the DIRECT function, which develops the total solution by the selection and mani-

34 The C-process: a model of command

pulation of promising interim proposals generated by the various proposal assessment loops.

Figure 2.1.7 show two different mappings of the example 'destroy and survive' problem on to a dual-topic (two B-subprocess) C-process. The upper diagram represents decomposition by objective whereby the A-subprocess presents the destruction objective (Od) to one B-subprocess and the survival objective (Os) to the other. Both must consider weapon and platform aspects. The lower diagram represents decomposition by resource, whereby the A-subprocess presents the composite 'destruction and survival' objective (Ods) to two B-subprocesses, each specialising in the use of a particular type of resource.

2.1.8 World state space interpretation of command

2.1.8.1 World state space: Past and present: It was assumed that the state of the world can be described by a n-dimensional vector and that the world state at a particular instant can be represented by a point in n-dimensional world state space. As time passes, the world's state will change. The locus of previous world states forms a trajectory through world state space representing history.

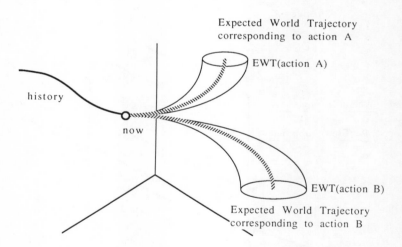

Fig. 2.1.8 *World state space representation of past, present and predicted future*

2.1.8.2 World state space: Future: Prediction of the world state at a future time will have an associated probability distribution in world state space, with the uncertainty of predicted state generally increased with time. A prediction of the future developing forward from the present corresponds to an extrapolation of the past world trajectory, i.e. an expected world trajectory (EWT), as illustrated in Fig. 2.1.8.

The C-process: a model of command 35

In general, the execution of different actions will give rise to different futures. Figure 2.1.8 illustrates alternative futures expected to result from following the alternative courses of action A and B.

2.1.8.3 Value model: A value model defines the relative desirability of different world situations or states, in which goals and penalties are desirable/undesirable world states (DWSs/UWSs) (Fig. 2.1.9).

2.1.8.4 Command decisions: If the objective of command is the attainment of goals (DWSs) combined with the avoidance of penalties (UWSs), the command function is the selection and application of actions to maximise intersection with DWSs and minimise intersection with UWSs. The modelling and measurement aspects of the proposal assessment loop can be regarded as the strategy of making decisions on the basis of maximising expected utility (Ackoff and Sasieniu, 1968).

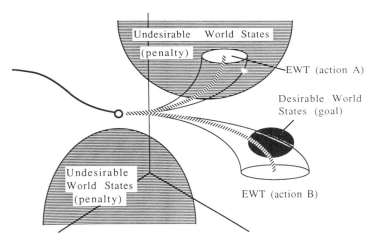

Fig. 2.1.9 *Distribution of value through world state space*

A course of action A_i is chosen so as to maximise its expected utility $E[U(A_i)]$, where

$$E[U(A_i)] = \sum_{\text{all } j} p[O_j|A_i] \, V(O_j)$$

where

A_i is the ith putative course of action
O_j is the jth outcome
$p[O_j|A_i]$ is the probability of outcome O_j given execution of activity A_i, i.e. the expected distribution of outcomes arising from action A_i
$V(O_j)$ is the value of outcome O_j

The role of the three components of the basic proposal assessment loop are as follows:

36 The C-process: a model of command

PROPOSE Proposes a course of action A_i.
WORLD MODEL Determines the probability distribution of outcomes $p[O_i|A_i]$ arising from action A_i.
MEASURE Applies the specified value system $V(O_i)$ to calculate the expected utility $E[U(A_i)]$ of pursuing action A_i.

This interpretation of the proposal assessment loop is shown in Fig. 2.1.10. The C-process model incorporates an augmented proposal assessment loop in which:

PROPOSE Proposes both action A_i and objectives;
INTERPRET OBJECTIVES Translates objectives into the value system $V(O_i)$ to be applied by MEASURE.

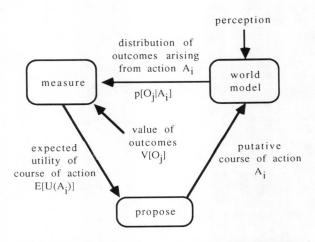

Fig. 2.1.10 *Basic proposal assessment loop: utility theory interpretation*

2.1.8.5 Planning and intermediate goals: During the preparation for a mission, command must seek a course of action which, starting from the current situation, is expected to achieve the current objectives. If an appropriate course of action is not immediately obvious, command may consider other situations from which the ultimate goal is attainable. If an intermediate situation is found which is itself attainable, albeit via other intermediate situations, then a viable plan has been found.

Figure 2.1.11 illustrates a plan involving two intermediate goals. Having failed to find a direct method for attaining the ultimate goal, command proposes initially to attain intermediate goal 1, from which it is planned to attain intermediate goal 2, after which the ultimate goal is realisable. The terms *forward chaining* and *back chaining* refer to the order in which a sequence of intermediate goals is sought, e.g.

Forward chaining Seeking method to attain:

(a) Goal 1 from start situation
(b) Goal 2 from goal 1
(c) Final goal from goal 2.

Back chaining Seeking method to attain:
(a) Final goal from goal 2
(b) Goal 2 from goal 1
(c) Goal 1 from start situation.

The approach used will depend on the nature of the problem. In general, a mixed strategy of forward and back chaining will be necessary.

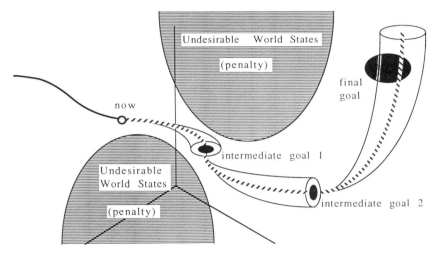

Fig. 2.1.11   *Intermediate goals*

2.1.8.6 Monitoring: The monitoring activity discussed in Section 2.1.5.2 is represented in Fig. 2.1.12. At time t_0 a plan of action was selected which was expected to give rise to a future (EWT_0) with a high probability of success. The plan is put into effect and is regularly reassessed during its execution. At time t_1 the revised expected future (EWT_1) still promises to attain the goal with satisfactory probability of success, i.e. there is still sufficient intersection between EWT_1 and the goal region. At time t_2, however, the expectation is that continuation of the promulgated plan is unlikely to provide success, i.e. EWT_2 does not intersect sufficiently with the goal region. At this stage it is necessary to re-enter the planning mode to find an alternative plan.

2.1.9 Distributed command and co-ordination

The following sections illustrate how the C-process can be used to describe the command organisation within a distributed command system. A discussion on the opportunities for mutual interference within a distributed force is followed by one on co-ordination.

38 The C-process: a model of command

2.1.9.1 Distributed command: There is a limit to the size of command task which can be performed by a single command process. There is a stage at which it is preferable to devolve some responsibilities, together with authority over some force resources, to subordinate commanders.

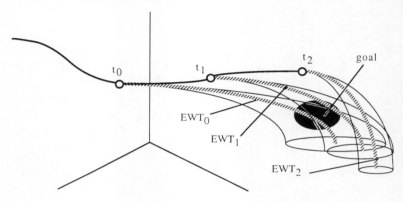

Fig. 2.1.12   *Monitoring mission effectiveness*

Command chains within a distributed command system can be represented by the inter connection of separate C-processes. Each C-process will have a declared responsibility within this structure. A simple example of a command chain is represented in Fig. 2.1.13. The flow of perception to each C-process is omitted from this diagram. It should be noted that, within a distributed command system, the perception must encompass the world both outside and inside the system. Therefore it includes knowledge of other Commanders plans.

The high-level commander C_0 has created two subordinate command posts, C_1 and C_2, and assigned to them a number of controllers ($C'_{11}, \ldots, C'_{22}$) and ASW resources ($R_{11}, \ldots, R_{22}$). Initially, the example command organisation will be discussed without reference to the co-ordination cell C_3.

The superior commander C_0 responds to the assigned tasks A...Z by producing a high-level plan (P_0) which, in general, will involve the participation of both groups 1 and 2. The specific task for each group, within the context of the overall plan P_0, will be defined by objectives 1 and 2 (O_1, O_2). The subordinate commanders C_1 and C_2 must in turn produce lower-level plans (P_{11}, P_{12} and P_{21}, P_{22}) aimed at fulfilling their respective objectives. Each lower-level plan will be assigned to a controller responsible for implementation of that plan.

2.1.9.2 Mutual interference in distributed command: The discussion of problem decomposition indicated the need for any plan to be consistent despite component parts of the overall plan being delegated separately.

Consider the example shown in Fig. 2.1.13. Controller C'_{11} will endeavour to avoid interference between items in its resource set R_{11}. This should be feasible assuming plan P_{11} to be internally consistent. Similarly, C'_{12} must avoid interference within R_{12}. Since plans P_{11} and P_{12} emanate from the same source they are likely to be compatible, in which case there should be little risk of interference between resources R_{11} and R_{12}. There is less confidence of non-interference

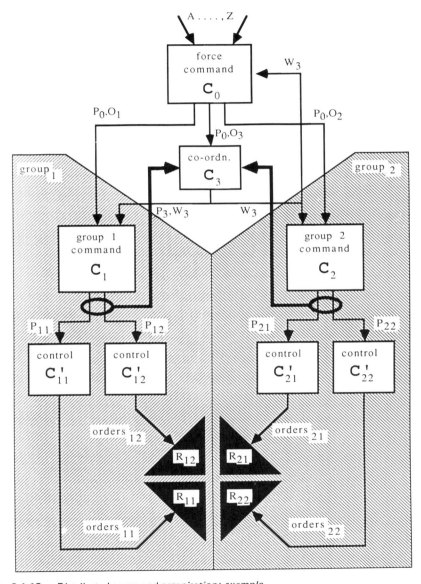

Fig. 2.1.13 *Distributed command organisation: example*

between resources R_{12} and R_{21} because plans P_{12} and P_{21} were produced independently, albeit within the context of a common high-level plan P_0.

2.1.9.3 Co-ordination: The purpose of co-ordination is to ensure co-operation between co-existing command and control processes in order that the force accomplish its total task with the minimum of internal conflict.

Consider the role of the cell C_3 in Fig. 2.1.13. The C-process is tasked, via objective O_3, with co-ordinating aspects of the activities of group 1 and group 2. The plans for C_1 and C_2 (P_{11}, P_{12} and P_{21}, P_{22}) are copied to C_3 and incorporated in C_3's world model, which can then predict the combined outcome of these plans. This prediction will be monitored against the objective of the required co-ordination. For example, co-ordination aimed at collision avoidance would include proximity measurement to predicted platform position.

When prediction of an undesirable situation is detected, the co-ordination process may be expected to submit proposals for solving the problem. The procedure for generating such proposals is identical to that for command.

Figure 2.1.13 shows the co-ordination process. Having predicted some undesirable aspects of group 1's and group 2's planned activities, C_3 has generated a proposed solution (P_3) and has submitted it to C_1 for consideration. If C_1 concurs with this modification to his plan, then C_3 may revert to the monitoring role. If, however, C_1 chooses to reject the proposal, then C_3 must issue warning (W_3) of the impending problem and refer to a higher authority for resolution.

There exist many forms of co-ordination; some are resource oriented, e.g. aircraft co-ordination, others are objective oriented, e.g. surface/subsurface curveillance co-ordination. Consider groups 1 and 2 in Fig. 2.1.13 to represent ASW and screen groups. In addition to the prime roles of conducting ASW and providing a defensive screen, these groups will normally be expected to contribute to the overall force surveillance. The responsibility for ensuring comphrehensive surface/subsurface surveillance is often given to a surveillance co-ordinator who, without resources under his direct command, must co-ordinate the surveillance activities of the divided force. In this case the criterion of effectiveness (objective O_3 in Fig. 2.1.13 would encompass overall surveillance.

2.1.10 Command message set
An integral part of the development of the C-process model has been understanding the information between command (and control) cells and similarly the information flow within cells. These information flows are obviously related since the intra-cell messages arise partly from the incoming messages and must ultimately be translated into outgoing messages.

In developing this model of command it has been assumed that the *command messages* of any level of complexity could be constructed by combining simple elemental statements, the simplest meaningful command statement being called a *command clause*.

Despite the wide spectrum of message types, e.g. plans, rules of engagement

The C-process: a model of command   41

(ROEs), resource allocation etc., attempts were made to find a command clause structure which was capable of expressing any type of command message. The proposal shown in Fig. 2.1.14 comprises a command instruction and latent commands. *Latent commands* are statements of authority or requirement concerning resources, actions and/or situations. A latent command does not, of itself, contain a command imperative. This is provided by the *command instruction,* which specifies its nature, scope and degree of compulsion.

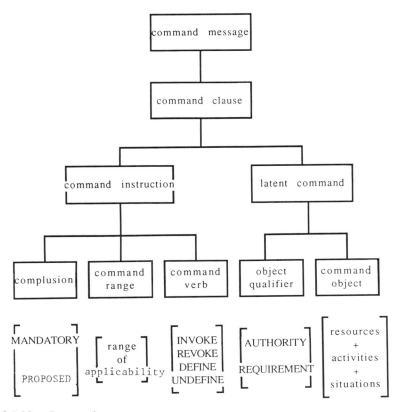

Fig. 2.1.14   *Command message structure*

2.1.10.1 Latent command: This contains the object (in a grammatical sense) of the command clause. The *command object* will define:

Resources, e.g. HMS *Ajax,* task force group 4.2,
Activities, e.g. search, attack; and/or
Situations, e.g. sector AB, sector AB devoid of submarines.

Qualifying these clause objects with either *authority* or *requirement* forms the basis for a command. For example,

(authority) (HMS *Ajax,* sector AB)
Refers to the authority for HMS *Ajax* to occupy sector AB.

(requirement) (HMS *Ajax,* sector AB)
Refers to the requirement for HMS *Ajax* to occupy sector AB.

2.1.10.2 Command instructions: The creation, activation, deactivation and removal of latent commands require an additional set of instructions called *command verbs*. The command verbs 'invoke' or 'revoke', in conjunction with the *object qualifier* within the latent command, determine the nature of the command imperative, i.e.

(invoke) (authority)   implies may ...
(revoke) (authority)   implies must not/may not ...
(invoke) (requirement)   implies must ...
(revoke) (requirement)   implies need not ...

For example,

(invoke) (authority) (HMS *Ajax,* sector AB)
Authorises HMS *Ajax* to enter sector AB.

(revoke) (authority) (HMS *Ajax,* sector AB)
Prohibits HMS *Ajax* from entering sector AB.

(invoke) (requirement) (HMS *Ajax,* sector AB)
Orders HMS *Ajax* to enter/occupy sector AB.

(revoke) (requirement) (HMS *Ajax,* sector AB)
Indicates HMS *Ajax* is not confined to sector AB.

An additional command verb 'define' provides the ability to specify commands to be subsequently invoked/revoked using abbreviated identifiers, e.g. a rule of engagement may initially be defined and later invoked by reference to its ROE number. A final command verb 'undefine' simply provides the ability to delete previously defined latent commands.

The application of command verbs to a latent command will, in general, be bounded, e.g. a latent command may be activated for a finite duration, within a finite region or subject to certain circumstances. Consequently, a command verb should be accompanied by a definition of its scope or *range of applicability*.

It is also necessary to distinguish between *mandatory* and *proposed* creation, activation, deactivation and removal of latent commands. This can be achieved with an additional qualifier called *compulsion.* Figure 2.1.14 indicates two grades of compulsion, but it may be necessary to allow 'shades of grey' between mandatory and proposed. Recommendations and requests are both considered as examples of proposals, e.g. a recommendation to authorise allocation of resources and a request for additional resources both need to be considered in a similar manner.

Table 2.1.1 gives examples of command clause construction.

Table 2.1.1 *Examples of command clause construction. Multiple entries under 'command verb' and 'object qualifier' indicate alternatives. Multiple entries under 'object qualifier' indicate any combination of resource, activity and/or situation definitions*

Command type	Construction			
	Compulsion	Command verb	Object qualifier	Command object
Task Goals to be attained and/or activities to be performed (possibly using specific resources)	mandatory	define invoke	requirement	resource activity situation
Authority Allocation of (and permission to use) resources, permission to perform certain activities and for certain situations to occur or be pursued	mandatory	define invoke	authority	resource activity situation
Constraints Prohibited situations to be avoided and proscribed activities and resources	mandatory	define revoke	authority	resource activity situation
Plan A related sequence/combination of goals and method performed by specified resources, i.e. in general, a plan is a combination of goals, method and resources	mandatory	define invoke	requirement	resource activity situation
Objectives A description of goals to be attained and prohibited situations ('penalties') to be avoided	mandatory	define invoke	requirement authority	situation

Note: an objective does not, of itself, address the method of achieving the goal or avoiding the prohibited situation.

The C-process: a model of command

Table 2.1.10 (contd.)

	Compulsion	Command verb	Object qualifier	Command object
Method Related combination/ sequence of activities	mandatory	define	requirement	activity
Procedure definition A prescribed method to be retained for future use	mandatory	define	requirement	activity
Order Activities to be executed (e.g. invocation of a prescribed procedure), possibly using specified resources	mandatory	invoke	requirement	resource activity
Goal Desirable world state (DWS) which is to be attained	mandatory	invoke	requirement	situation
Approved world states (AWS) which can be pursued or countenanced, e.g. travel within a protected lane, acceptable attrition etc.	mandatory	invoke	authority	situation
Penalty Undesirable world states (UWS) to be avoided, e.g. travel within a submarine patrol area	mandatory	revoke	authority	situation
Proposal A recommended, advisory or tentative plan submitted for consideration	proposed	invoke	requirement	resource activity situation
ROE definition A definition of objectives, method, resources and/or constraints to be retained for future application	mandatory	define	requirements authority	resource activity situation

Table 2.1.10 (contd.)

	Compulsion	Command verb	Onject qualifier	Command object
Invocation of an ROE Authority to pursue objectives or employ certain methods or use certain resources and to comply with certain constraints	mandatory	invoke revoke	requirement authority	resource activity situation
Request for assistance A request for assistance from co-operating forces; this may involve desired situations and/or activities	proposed	invoke	requirements authority	resource activity situation
Resource request A request for additional resources	proposed	invoke	authority	resource
Resource allocation Authority over warfare and/or resources	mandatory	invoke	authority	resource

2.1.11 The C-process as a basis for design

The development of the C-process was prompted by the need to design and build automated command subsystem called C-stations. The C-process model presents a network of function and information flow which is assumed to take place within any command situation. In other words, it is postulated that these processes apply whoever or whatever is performing them.

When applied to a purely *manual* system, the model helps to identify the essential functions carried out within the minds of the commander or command team, and the information flowing within the minds of individuals or between individuals, that is, man-man interfacing.

Alternatively, one could use the model to hypothesise a totally *automated* command system in which all functions are performed by computers and all information flows are within or between computers, that is, machine-machine interfacing. It should be said, however, that at present such a system is prohibited by a lack of understanding of some of the C-process functions, which is likely to persist for a long time.

2.1.11.1 Allocation of functions: In practice, only part of the task of the commander or command team can be augmented by computerised aids. In an applied design, each of the C-process functions were assessed in turn and a subset identified for which it was expected that computers could make a sensible contribution. The functions selected for automation, based on current capability, were:

WORLD MODEL
MEASURE
PROMULGATE
ANALYSE (part of)

A similar assessment for the purpose of intelligent knowledge-based computing suggested additional candidates for computer assistance, for example expert systems aiding proposal generation.

This categorisation of functions into those which were and those which were not machine supportable formed the basis for the design of the computerised system. The information flow between machine functions was identified as machine-machine interfacing, and the information flow between the machine and non-machine functions constituted the man-machine interface.

2.1.11.2 Functional decomposition: The 'machine' functions, together with their interfaces, constituted a high-level specification for the computerised system. It was then necessary to perform successive decomposition of these high-level descriptions until a logical design comprising sufficiently precise detailed specifications were achieved. In some areas, this involved four or five levels of decomposition before the functional descriptions reached a form which could be implemented.

It was during this decomposition phase that we took account of the specific operational requirements of the C-station. In our case we needed a C-station capable of supporting force-level anti-submarine warfare operations, e.g. logistics and replenishment at sea (RAS), defensive screen design, submarine search and weapon delivery.

This entailed designing a WORLD MODEL which included a capability for predicting:

(a) The position of platforms (including the probabilistic distribution of enemy submarines according to a variety of presumed hostile intentions)
(b) Fuel consumption
(c) The potential for platforms (including weapon delivery systems) to reach remote regions
(d) Acoustic and electromagnetic sensor detection performance.

This was accompanied by the consideration of candidate effectiveness criteria to be applied by the automated MEASURE function. Particular emphasis was applied to the development of an experimental metric for assessing the overall effectiveness of multisensor search strategies in the presence of uncertain submarine distribution/motion.

2.1.11.3 Physical design: The final stage was the mapping or translation of this logical design into an implementation design, taking account of the various machine constraints. Figure 2.1.15 illustrates the C-process as a basis for design.

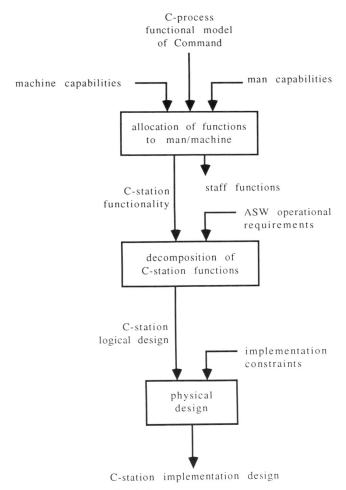

Fig. 2.1.15 *C-process as a basis for design*

2.1.12 Conclusions
This contribution has described an attempt to construct a formal description, called the C-process, of the activities colloquially known as command and control. Because of the immense complexity of this subject, it is recognised that the C-

48 The C-process: a model of command

process, like any model, is an incomplete description of what it seeks to describe. The emphasis of this model has been to describe the processes involved in the forward paths of an overall command system. It is felt that, despite its limitations, there are scientific and engineering benefits to be gained from this functional model of command and control.

2.1.12.1 Scientific benefits: The analysis and model building has generated a number of hypotheses which may form the basis for experimentation aimed at advancing a science of command. The main hypotheses are:

(a) That during both the preparation for and execution of a mission, the command, control and co-ordination all involve the prediction and evaluation of expected futures.
(b) That the principal role of perception is to provide and maintain models used to predict the future.
(c) That proper evaluation of situations necessitates a conscious awareness of all prevailing objectives, and that these objectives are translated into a value system.
(d) That non-trivial problems are solved by the application of two strategies:

Serial problem decomposition, i.e. a stepping-stone approach
Parallel problem decomposition, i.e. an amalgam of solutions to aspects of the overall problem.

2.1.12.2 Engineering benefits:

(a) A contribution is made towards a vocabulary of command and control which permits dialogue between system designers and system users.
(b) The model acts as a basis for logical and coherent design of command systems subject to the following provisos:

1 The design team have sufficient understanding of the PROPOSE, INTERPRET OBJECTIVES and DIRECT processes to carry out the necessary functional decomposition.
2 Automation to assist the predictive aspects of command (beyond simple deterministic endeavours) demands the development of probabilistic WORLD MODELs which can exploit uncertain perception and which generate probabilistic predictions (outcome probability distributions).
3 Automation can only assist the assessment aspects of command to the extent that objectives are declared explicitly and are faithfully translated into quantitative value models.
4 Automatic analysis of incoming command messages can only be applied to those messages expressed in an appropriate formal language, e.g. command message set (Galley, 1983b).

The model development described in this chapter was completed in 1983. Since then the C-process has been successfully employed in a functionalist approach to

The C-process: a model of command 49

the design of an experimental command subsystem called the C-station. A simulation of the C-station has been implemented which provides basic C-station features appropriate to ASW command. Current developments are aimed at enhancing this basic prototype to incorporate a fuller set of C-station facilities.

2.1.13 References

ACKOFF, R. L. and SASIENIU, M. W. (1968) *Fundamentals of Operations Research*, Wiley, New York

GALLEY, D. G. (1983) Redefinition of the C-process, ARE report A & D/XP20/AF.1

GALLEY, D. G. (1983) Command message set, ARE report A & D/XP20/AF.2

PUGH, G. E. (1976) Mathematical decision aids for the task-force commander and his staff, General Research Corporation, report 593W-01-CR

Chapter 2.2
MOSAIC: concepts for the future deployment of air power in European NATO

D.K. Hitchins
(Easams Ltd)

2.2.1 Introduction
Air power in Europe is conventionally deployed in support of the anticipated land battle. Since the World War II the many roles of air power have developed somewhat independently; NADGE (NATO air defence ground environment), for example, is the extant air defence (AD) element which draws the nations' AD forces together.

NATO is actively analysing the requirement for, and shape of, the future air command and control system (ACCS) which is required to ameliorate deficiencies in present command and control (C^2) of air power, particularly in the central region. These deficiencies are concerned with such fundamental issues as offensive/defensive integration, survivability, mobility and fratricide reduction.

Even as these early ACCS considerations are getting under way, technological advances are driving both the threat spectrum and threat counters to evolve at an ever-increasing rate. There is, therefore, a *prima facie* case for examining the underlying deployment of air power in these dynamic circumstances; otherwise, it may be that a future ACCS will be faced with the task of accommodating deployment deficiencies rather than providing the essential force multiplier.

It is timely, then, to consider alternatives to current force deployment in European NATO, to assess those alternatives and to consider the impact that they might have on air power (C^2). This paper introduces three interlocking concepts which together present a radical option to current air power deployment conventions, which is consistent with threat and technology trends and which at the same time addresses some present day shortcomings. These so-called MOSAIC concepts emphasise survivability and 'movability' as the key issues; integration of offence and defence is seen as a fundamental part of survivability, since neither perfect defence nor perfect offence is a viable military stance.

2.2.2 The changing European military scene

At the highest level, Western Europe could find itself in one of the following states: peace, tension, theatre conventional war, global conventional war, theatre nuclear war or global nuclear war. Transition from one state to another could be triggered by a number of factors; some are shown in Fig. 2.2.1, in which the six states are represented by boxes, and the causes of transition between states are at the commencement of arrows. For example, menace causes transition from peace to tension; invasion escalates tension into theatre conventional war; and so on. Of particular interest is the basis for transition from theatre conventional to theatre nuclear war; it is shown on Fig. 2.2.1 as rollback.

Rollback — the orderly withdrawal by NATO ground forces in the face of superior odds — is designed to provide sufficient time for a political solution to be set in place. Should rollback be far too fast, or the political solution fail to materialise, then it is possible that NATO would be obliged to use nuclear weapons to stem the tide. For rollback to be too fast implies that key land features or installations would be forfeit. Among those key installations may be included fixed air power facilities: radar, control and reporting centres (CRCs), sector operations centres (SOCs), runways, fuel and ammunition dumps. It therefore follows that forward siting of fixed or immobile air assets will contribute to the earlier need during rollback to resort to non-conventional weapons and hence to a lowering of the nuclear threshold.

With a swing in the tide of conflict, NATO forces might wish to pursue retreating forces into their own territory. Immobile air assets would limit pursuit effectiveness in this instance, by becoming progressively further behind the line of battle. Fixed air assets clearly have limitations when supporting a mobile war. On the other hand, the growth of such fixed assets has not seemed unreasonable over a period of thirty years of peace; static facilities have been developed, utilised, superseded and discarded with no recourse to mobility other than perhaps the occasional exercise.

The advent of new systems into the military inventories of both sides renders any complacency ill advised. Improvements in tactical weapons, delivery accuracies and the use of spaceborne sensors and weapons are two factors of relevance to the fabric of future air power in European NATO. Both President Reagan's strategic defence initiative and Chairman Gorbachev's move to eliminate nuclear weapons may result in fundamental changes in future NATO strategy. The only certainty seems likely to be one of change; for air power this implies flexibility, mobility and survivability as being key issues.

2.2.3 Air power in context

The concept of indivisibility of air power is writ large upon the minds of many air forces. Originally, the concept argued against naval air, army air and independent air forces such as those of the UK during World War II. Indivisibility also refers to the concept that offence, defence and support cannot sensibly be considered in mutual isolation. The original concept is not acknowledged worldwide; navies and

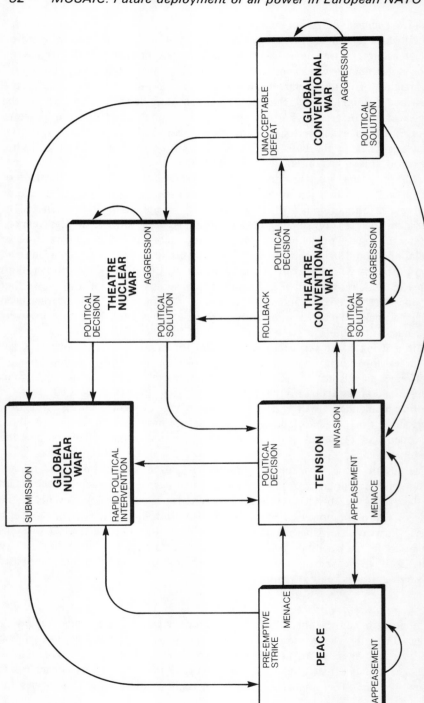

Fig. 2.2.1 Finite state diagram of European potential conflict transition/resolution
NB. Political solution generally implies victory/defeat

MOSAIC: Future deployment of air power in European NATO 53

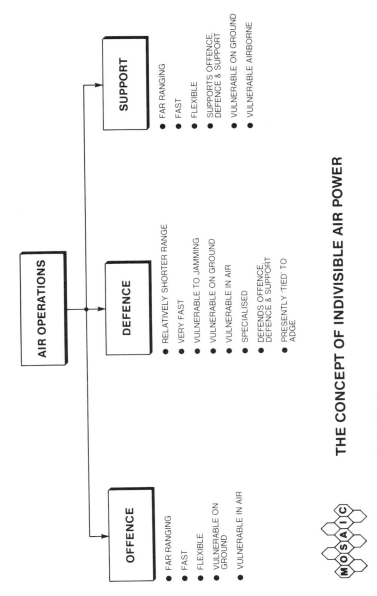

Fig. 2.2.2 *The concept of indivisible air power*

armies often have their own air forces, and there are cogent arguments in favour of such specialist and available air support. The second concept is more relevant to the discussion and will be expounded below.

Fig. 2.2.2 shows air operations as comprising three major arms: offence, defence and support. Some of the activities and characteristics of each arm are also shown. Examination of the characteristics shows:

(a) Air defence shields vulnerable offensive and support air as well as itself.
(b) Air defence is relatively short range and at present tied predominantly to fixed air defence ground environment (ADGE) facilities.
(c) Hence overall flexibility and high speed mobility are hampered by the relative immobility of defensive air, which is only partly overcome by the use of point defence weapons such as Rapier and Javelin during bridgehead operations.

There are, therefore, grounds for re-examining closer integration between the elements of air power. If the objective is to enhance the overall effectiveness of air power then, as Fig. 2.2.3 shows, a systems approach is needed. The figure shows the same three major air power divisions: (1) offensive (2) defensive and (3) support operations. Optimising each of these divisions individually need not result in an optimally effective air arm overall; an overall objective must be established for optimisation of air power as an entity.

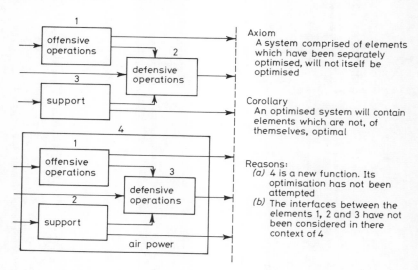

Fig. 2.2.3 *System design axiom and corollary*

Integrating all the elements of air power would be impracticable in most circumstances, however. Figure 2.2.4 shows four classifications of operations — offensive, defensive, maritime and air transport — with major activities peculiar to each class. Maritime operations, with its long-range maritime patrols (LRMP),

MOSAIC: Future deployment of air power in European NATO

anti-submarine warefare (ASW) and tactical aid in support of maritime operations (TASMO) is clearly set aside from the others in its roles, its tasks and its sphere of operations. Closer examination reveals that each of the others is also independent, and requires not only its own aircraft and weapons, but discrete procedures too; each attracts different personalities and characteristics among the air crew. It can also be seen that function grouping:

Concentrates and develops specialist skills
Affords a simple organisation
Enhances motivation and competition
Allows simple measures of effectiveness by function
Divides capital funding into defensive and offensive.

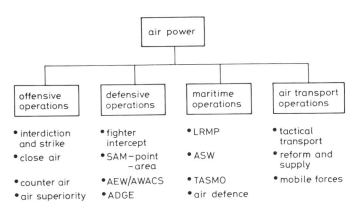

Fig. 2.2.4 *Air power — idealised organisation chart*

The last item proffers a two-edged sword. Since World War II, several periods of *détente* have arisen; during such periods it is politically more acceptable to spend money on defence than on offence. Spending on defensive facilities thus becomes continual and defence can be developed and optimised in isolation from its spasmodically funded offensive counterpart. Taken to its limit, this process lowers survivability of both air power and the land/sea battle it supports, by placing undue reliance on a defensive potential which cannot sustain itself in the face of continual attack.

The interrelation between the various elements of air power is illustrated in Fig. 2.2.5, which is a system dynamics view of air operations conducted in the face of an advancing enemy which is threatening to overrun forward air assets on the ground. In the so-called influence diagram technique, a plus or minus sign at the arrowhead signifies that the source of the arrow will enhance or diminish the focal point in question. Thus successful air defence engagements (top centre) reduce (minus) the advancing enemy ground threat (bottom centre) by improving conditions for air superiority operations. In the figure, enemy command and

56 MOSAIC: Future deployment of air power in European NATO

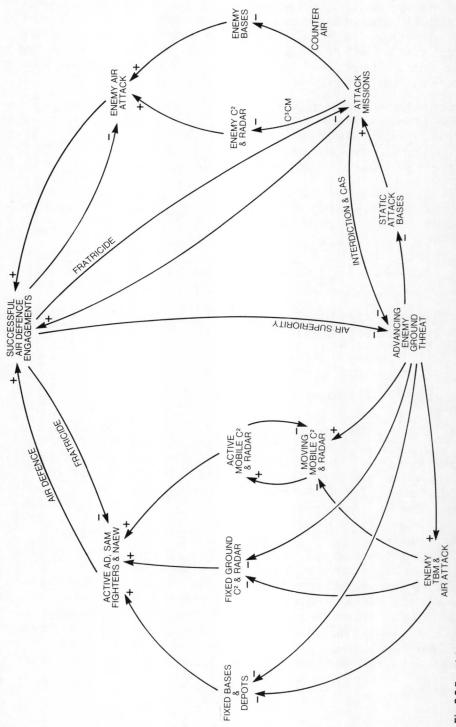

Fig. 2.2.5 Advancing ground threat and the air battle

MOSAIC: Future deployment of air power in European NATO   57

control (C^2), radar and bases are at the right. The very simplicity of this abstract representation of air operations highlights the importance of:

(a) Balanced offensive and defensive operations
(b) Common and dynamic C^2 where offensive and defensive effort and resources can be co-ordinated
(c) Mobility for C^2 and radar units, bases and depots
(d) Fratricide reduction.

Clearly any future deployment solution should recognise the close interrelation between areas of air power, while tempering the move to full integration at execution/operations level with the need to preserve specialisations.

2.2.4 The threat facing air command and control

Advances in C^2 are threat driven. The threat in the future air/land battle is developing swiftly; tactical threat developments can be categorised simply under three headings:

Launch vehicle
Delivery vehicle
Warhead, weapon and payload.

Launch vehicles include tactical ballistic missiles (TBMs) and cruise missiles which may be expected from any elevation. In the time frame of ACCS we must expect spaceborne weapons too; these could take many forms and arrive from any direction.

'Conventional' warheads will include chemical and possibly biological payloads as well as high explosive devices, and the variety of nuclear warheads is well catalogued.

Two very significant threat trends concern the detectability of threats — voluntary-emission control (V-EMCON) and Stealth. V-EMCON reduces or eliminates the detectable voluntary emission, including not only communications but also Doppler radars, radio altimeters and radiating DMEs (distance measuring equipments). Stealth seeks to reduce radar cross-sectional area, and encourages V-EMCON and the use of terrain to screen essential transmissions such as those from terrain following radars (TFRs). For entirely preplanned missions even TFR and radio altimeter emissions may be unnecessary in the foreseeable future.

There are clear pressures on the future ACCS and its associated subsystems, including sensors, communications and weapons, if the aims of ACCS are to be realised:

(a) Increased vulnerability to new all round threats.
(b) Need to operate in a degraded mode after sustaining damage, possibly from nuclear chemical or biological weapons, and in the future from beam energy weapons.
(c) Reduction in attack warning time, leading to increased risk of fratricide, and to reduce defensive weapon effective range.

58 MOSAIC: Future deployment of air power in European NATO

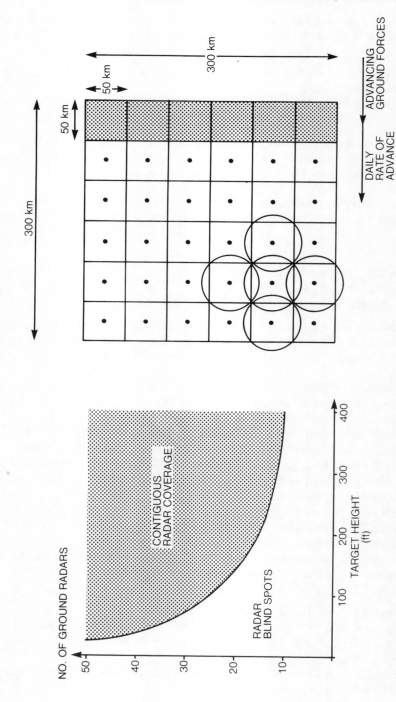

Fig. 2.2.6 Minimum ground radar density for continuous tracking of a 50' target (smooth earth)

MOSAIC: Future deployment of air power in European NATO

This last point is expanded at Fig. 2.2.6, which has two parts. At the left is a graph showing target height against an approximate number of ground radars needed to provide contiguous radar cover in a 300 km square, as shown in plan at the right. As target height reduces, the number of radars rises towards infinity, classically pursuing the law of diminishing returns. Advances in terrain following aircraft performance are continuing to reduce operational heights, and terrain screening can reduce pick-up range even more dramatically. The plan at the right of Fig. 2.2.6 shows a notional detection range of 25 km as requiring 36 radars to cover the 300 km square.

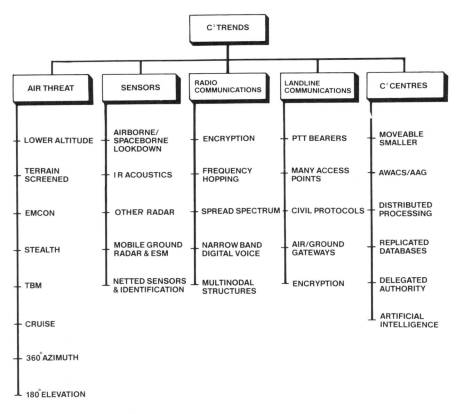

Fig. 2.2.7 *Balance in C^3 trends*

The resulting 50 km squares form columns and correspond to about the expected daily WP rate of ground force advance. Thus, in the given simple scenario, either six ground radars (one column) would be overrun per day, or (if mobile) they would move swiftly to the rear and redeploy. Simple though Fig. 2.2.6 may be, it amply illustrates two important points: the usefulness of ground radars is being restricted by enemy tactics, and those radars near the FEBA/FLOT should be mobile.

60   MOSAIC: Future deployment of air power in European NATO

Threat evolution is driven by the same technology that will provide the counters, although a time delay may arise between development of threat and of counter. Figure 2.2.7 illustrates the balance of C^3 trends, with the air threat at left. Remaining columns show selected sensor, communication and C^2 improvements fuelled by recent technological improvements. Taken together these improvements accumulate increased overall effectiveness. Command and control is of and by military personnel, however; major contributions to improved effectiveness will come from improvements in organisation, co-ordination and procedures which best employ and protect limited resources.

2.2.5 Air command and control deficiencies

Deficiencies in the current command and control of air power in European NATO have been identified, and may be summarised as follows:

(a) Lack of timely, secure and adequate information, leading to an inability to derive timely and high-quality command decisions
(b) Lack of co-ordination between all elements of air power, but especially between offensive and defensive air
(c) Vulnerability of sensors, communications, processing, HQs and C^2 personnel to physical and electronic attack
(d) An unacceptable fratricide rate due principally to the lack of high-integrity identification and weapons control.

In the light of threat developments, it can also be seen that the very nature of our air power deployment in European NATO is suspect, principally because:

(a) Facilities (sensors, communications, airfields) are almost entirely static
(b) Air defences assume that air threats come from the east and are oriented accordingly
(c) Current hierarchical C^2 organisations contain nodes and are not designed to tolerate damage.

The following sections introduce MOSAIC, which specifically aims to improve survivability and mobility but which also addresses the central issue of fratricide.

2.2.6 Introduction to MOSAIC

2.2.6.1 Air power deployment options: There are several ways in which air power may be tactically deployed. Figure 2.2.8 illustrates the terms and concepts employed here. Many forces employ layered air defence, in which offensive air power operates behind the relative safety of discrete defensive layers, comprising, for example, belts of surface to air missiles (SAMs), long-range interceptors and short-range interceptors. Layered air defence is perhaps at its most practical in naval use where the battle group presents a concentrated target. In the central region, the targets to be defended are widely dispersed and a 360° azimuth defensive layering

MOSAIC: Future deployment of air power in European NATO 61

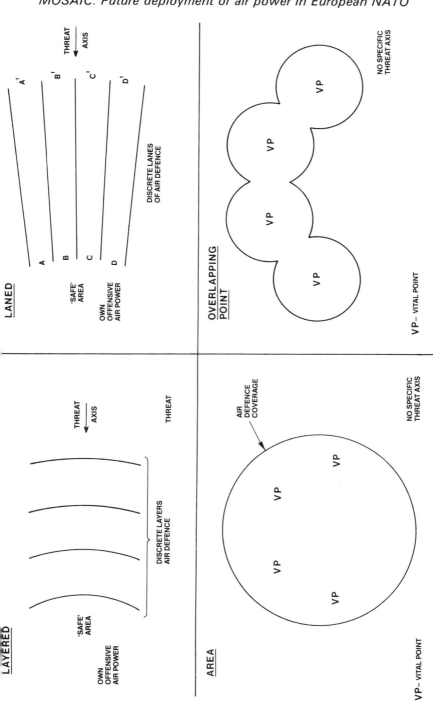

Fig. 2.2.8 Air power deployment options

would not be practicable due to size and cost. The layered system therefore covers only an arc, facing into the ground threat axis.

Laned defence (top right) presents lanes in the direction of the threat axis and is of use principally where there is no significant mix of weapon systems. Long-range interceptors, for example, might operate furthest from home, while shorter-range interceptors operated nearer the safe area from where offensive and support air must be deployed. Laned defence can be integrated with layered defence by making layers contain lanes.

Area and overlapping point defences do not presume a threat axis. Area defence covers vital points as sets within an area, while overlapping point, as the name implies, concentrates on point defence. In terms of concentration of force, point defence must be attractive when resources are limited and threat direction unknown, but of course it requires definition of what points are considered vital, which may prove to be far from trivial. MOSAIC concepts employ an area and point defence philosophy to protect vulnerable air assets on the ground from any direction.

2.2.6.2 The MOSAIC concepts: MOSAIC considers the evolving threat and the current C^3I shortfalls and takes these various influences to their natural conclusion in reaching a design for ACCS. MOSAIC is an acronym for movable semi-autonomous integrated cells.

Each of the words is separately important to the concepts, and the whole acronym is also appropriate. The aim of the MOSAIC concepts, of which there are three, is to greatly enhance survivability by selecting organisational structures which will continue to operate effectively in degraded mode. To achieve this, the structures must be:

Movable — either mobile or transportable — so that they may avoid targeting and overrun.

Semi-autonomous — so that loss of superior control can be accommodated by co-operation laterally with other groups in the structure and, *in extremis*, total isolation can be accommodated for a period.

Integrated — in two senses: first, to contain offensive, defensive and support elements in each group and so constitute a complete, highly survivable, fighting unit; second, so that the groups, although having the potential to operate semi-autonomously or even autonomously in degraded mode, normally act as one under superior control.

Cellular — so that individual cells, which by definition are movable, may be grouped together (interfaced) swiftly, easily and in various ways to reconfigure the super-structure.

2.2.6.3 The ground environment MOSAIC (GEM): Figure 2.2.9 illustrates the GEM concept, which addresses offensive/defensive integration and survivability directly by providing self-contained cells and by recognising that neither attack nor defence is, of itself, a tenable military role. Each **MOSAIC**, containing its indigenous

MOSAIC: Future deployment of air power in European NATO 63

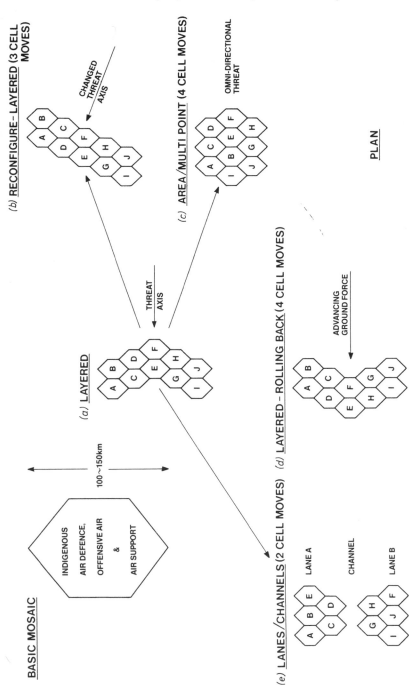

Fig. 2.2.9 The ground environment

air defence, offensive air and air support, can be redeployed. At Fig. 2.2.9(a) ten cells form a layered MOSAIC which corresponds to conventional layered air defence, except that offensive air and support air are not sheltering in a safe area but form part of the layer. Figure 2.2.9(b) moves three cells to accommodate changes in ground threat axis. Figure 2.2.9(c) shows area and overlapping point configurations, remembering that, since each cell has its own 360° air defence, the resulting MOSAIC also offers omnidirectional air defence. Figures 2.2.9(d) and (e) show the flexibility needed to fall back in the face of advancing ground forces and to configure killing zones if such were appropriate.

The concept of cell mobility to reconfigure tactical deployment clearly has potential for providing second-echelon support to front echelons experiencing undue pressure. Survivability too is potentially very greatly enhanced; any one cell remaining after an attack has the ability to move and to keep fighting. Sheltering of offensive air behind a defensive screen has been superseded for two reasons:

(a) The threat axis for air attack will not be known in the future, and certainly need not coincide with the main ground thrust, so that point/area defence of offensive assets is the only viable option.
(b) A rapid WP attack which broke through a conventional, layered defensive screen would find offensive assets both concentrated and vulnerable.

Other aspects of GEM and the two remaining concepts will be expanded after all three concepts have been introduced.

2.2.6.4 The sensor MOSAIC (SEM): The SEM concept is illustrated at Fig. 2.2.10 where it is compared with current philosophies at top left. The SEM concept concentrates on freeing air C^2 from its ground fetters and on integrating target data from airborne sensors.

Current C^2 for air power is strongly ground centred in (largely) static facilities. Ground sensors are linked together to form an integrated sensor picture, to which it is hoped to add the radar pictures from NAEW, AWACS or perhaps aerostats (tethered balloons) or airships. This ground-centred concept must be doubtful for the following reasons:

(a) The ground-based sensors will see less and less with the advent of V-EMCON, Stealth, TBM and cruise missiles.
(b) Static ground sensors and their equally static intercommunication links will be early targets.
(c) NAEW radar and the links with ground C^2 will be prime targets for electronic and physical attack (with ARM, for example).

The SEM concept recognises the introduction of track-while-scan radars and ECM resistant communications systems (ERCS) such as JTIDS into airborne service. Figure 2.2.10 also shows that the excellent lookdown capability of these radars forms a patchwork or MOSAIC of coverage, with some valuable system qualities when compared with ground-based static sensors:

MOSAIC: Future deployment of air power in European NATO 65

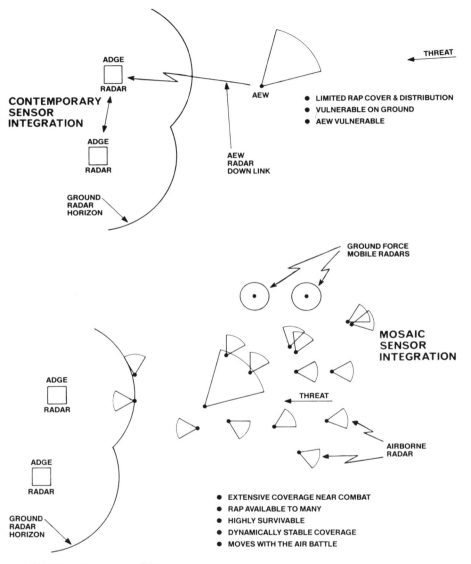

Fig. 2.2.10   The sensor MOSAIC

(a) There is a number of discrete sensors, each switching on and off as the crews pursue their operational roles.
(b) The density of such sensors increases near the air battle.
(c) Distances between sensors is relatively small near the battle, reducing the effects of jamming, especially on the ERCS.

(d) The sensor population is dynamically stable as fighters go off-station and are replaced.
(e) The sensors follow the air battle, unlike their ground-based counterparts.

For the SEM concept to be realised, the various sensor inputs must be brought together. The most likely candidate to undertake the integration is the NAEW/AWACS which, because it may be out of contact with the ground environment, assumes almost implicitly the role of air action group (AAG) co-ordinator. Taking this notion to its logical conclusion, we see that the sensor MOSAIC concept can free air power from the ground shackles which currently restrain it and limit its essential flexibility. The AAG is analogous to the naval surface action group, and in like manner it may have offensive as well as defensive elements, may move with the action, and may acquire temporary autonomy when conditions dictate.

2.2.6.5 The C^3 MOSAIC: The ground environment MOSAIC and the sensor MOSAIC may enhance flexibility, mobility and survivability but clearly present a challenging C^2 task, and the corresponding communications facilities will require careful analysis.

These aspects are the subject of the third concept, the C^3 MOSAIC, which is mostly directly relevant to current ACCS endeavours. The C^3 MOSAIC has the essential task of binding and controlling the movable GEMs and the AAGs.

Figure 2.2.11 introduces the concept, showing a three-tiered structure with cell C^2 controlling its own defence, offence and support and reporting to its respective cell group C^2. These group facilities control a varying number and population of cells during mobile operations, and are themselves capable of moving their defence, offence and support facilities. Above cell group level there may be either a further grouping or direct reference to the MSC (major subordinate commander).

Certain salient features become apparent from Fig. 2.2.11:

(a) The functional groups within each level have substantially similar divisions into operations, intelligence, plans, logistics and engineering.
(b) Within a base, cell or cell group control there will be continual interchange between the functional groups; this implies tight functional binding.
(c) Between cells and between levels, however, communication is specifically coupled by function; operations will speak externally to operations (as shown), intelligence to intelligence, but not intelligence to engineering or operations to logistics. (This and the previous features are common to all sophisticated C^2 organisations.)
(d) The sensor MOSAIC requires a number of line-of-sight air/ground communication gateways to accommodate AAG mobility.
(e) The communications structure allows one cell group to control another's cells, and also allows a cell to control an adjacent cell's base facilities should the need arise (battle damage, mobile operations etc.).
(f) In providing this cross-coupled communication system, the emphasis moves away from vertical communication in support of a simple command hierarchy

MOSAIC: Future deployment of air power in European NATO 67

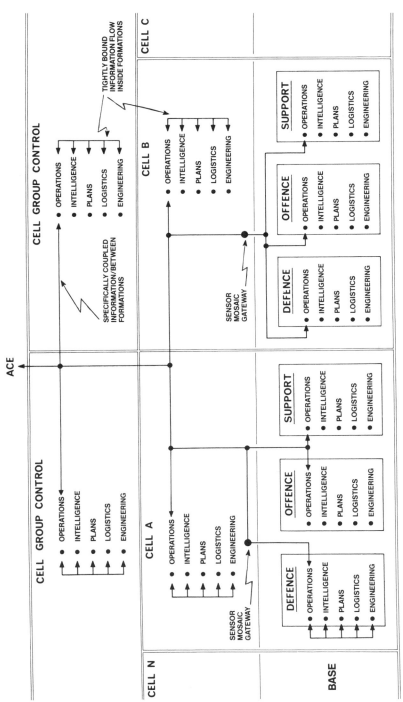

Fig. 2.2.11 The C^3 MOSAIC

towards a 'rectangular network' of vertical and lateral communications, with the communications channels grouped by function (operations, intelligence, etc.).

2.2.7 Expanding the MOSAIC concepts

The conceptual issues raised by MOSAIC are fundamental and, if fully embraced, would impinge on almost every aspect of air power. Before assessing MOSAIC, it may be useful to consider some of the key implications of the three interlocking MOSAICs:

(a) Movability of cells, cell controls, weapon systems, resources and support infrastructure.
(b) Interoperability between cells and their controls.
(c) Communications and interfaces to permit and enable movability and interoperability even after nuclear exchange if necessary.

2.2.7.1 Movability: Movability (i.e. mobility and/or transportability) of ground-based sensors is not a new concept, but it does have design implications. The need for enhanced performance, for example from antennae, must be reconciled with ease of breakdown and transportation. Similarly the size of CRC/SOCs in terms of personnel facilities, power supplies, hardening, air filtration etc. must be limited by movability dictates. These problems are generally understood. Less well understood perhaps are the implications of moving complete squadrons of aircraft with their supporting infrastructure, air traffic control, aircraft protection, recovery facilities, fuel, weapons, avionics maintenance etc. There are, however, ample precedents in the several methods employed by armies and particularly navies.

Navies carry their total air power infrastructure on board ship, including runways, surveillance, air traffic control radar, recovery and support. Their aircraft are adapted to the purpose but, as aircraft such as the Sea Harrier, F14, F18 and E2C Hawkeye illustrate, the process of adaptation need not inhibit performance. In a MOSAIC ACCS, aircraft might be similarly adapted.

One approach to air power movability, therefore, is to emulate the navy approach, at least in part. The transporting of runways is not out of the question, using special track laying vehicles and adapted aircraft, but there are many extant airfields which could be used, and autobahns have already been considered for some types of aircraft. There are also VSTOL aircraft designs for the future which could ease the problem further. Alternatively, the armies of the world have also been mobile since warfare began, and their approach is not dissimilar to MOSAIC in many respects. The use of step-up headquarters – so that one HQ may operate statically while the second is on the move, in leapfrog fashion – is tried and trusted and could be considered for some air force applications, although it does carry penalties of increased manning.

Looking at navy and army movability in this way, it becomes apparent that MOSAIC seeks to bring air power movability into line with the capabilities already

MOSAIC: Future deployment of air power in European NATO 69

present in navy and army air power, and in so doing impinges on the design of ground sensors, HQs and aircraft. Paradoxically, mobility may reduce manning levels to accommodate transport requirements. Air power composition seems likely to rebalance to meet the movability requirement: more short/medium-range STOL transports would be necessary to move cells rapidly during rollback or advance, when road/rail links might be choked or severed. MOSAIC also encourages the development of multirole fighter aircraft and aircrew; at present such aircraft change between roles with relative difficulty, and the aircrew often find the change even more difficult. In MOSAIC, the emphasis on combined offence and defence within each cell would be sensibly accompanied by a rationalisation of spares and weapons to reduce the logistics transportation burden, and by an ability of each fighter type/aircrew to perform a second role. This follows logically from the fundamental objective of survivability; if *in extremis* the cell must be able to function then too, as crews and aircraft are lost, remaining crews and aircraft must have the flexibility to attack, defend and redeploy almost at will.

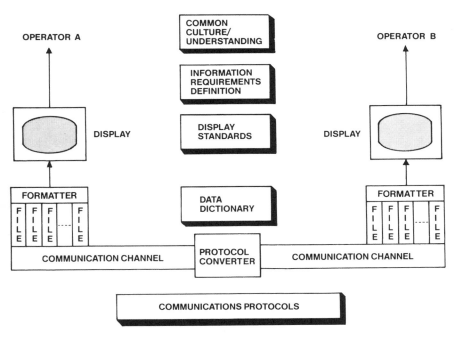

Fig. 2.2.12 *Interoperability*

2.2.7.2 Interoperability: Interoperability has many meanings in practice in addition to the interoperation between facilities in different nations. The MOSAIC concepts highlight a current problem in interoperability, which is illustrated by Fig. 2.2.12. Ideally, and subject to security protocols, operator A should be able to access not only his own parochial information but also that of his counterpart operator B in another cell, perhaps in another region. One operator could be in an aircraft, another in a mobile HQ.

It might be possible in principle to oblige both operators to store and display information identically, but in practice, due to language, operational and personal differences, each operator will tend to prefer his own display formats. Thus operator A, in viewing a file from operator B's database, will expect the display to appear as operator A's other files appear. For example, an aircraft readiness tote from a remote region should be formatted identically with a local aircraft readiness tote.

As Fig. 2.2.12 shows, this objective requires many factors to be in place:

(a) An ACCS-wide data dictionary
(b) ACCS-wide communications protocols, including file transfer protocols, such as the ISO open systems interconnection (OSI) seven-layer protocol.
(c) Agreed information requirements for operators on an operational function-by-function basis.
(d) Common or compatible display standards.
(e) A common understanding of the words and symbols transferred between systems.

Even then difficulties can be foreseen, since operators A and B may wish to discuss the different displays of identical information which they respectively view. This problem is likely to be resolved only by common training and practice.

The need for standard protocols has been mentioned and is as obvious as it is difficult to achieve. One concept which would bind the various communications bearers together in underpinning a common theme is that of an ACCS-wide tactical data interchange language, which may be called **TADIL-ACCS**. We already have **TADIL A, TADIL C** and **TADIL J** in association with **NATO LINKS 4, 11** and **16**. If the valuable **TADIL** concept is carried to its logical limit, then the problems of interoperability and movability would be considerably eased. The **TADIL-ACCS** concept is not closely coupled with the communications bearers; Fig. 2.2.13 illustrates the point, with a **TADIL-ACCS** track message being formulated by an airborne radar and being borne over a variety of bearers via **NAEW**, air/ground gateways and ground-to-ground links before initiating a track display.

The **TADIL-ACCS** philosophy thus engenders:

(a) **TADIL-ACCS**-based subsystem designs in aircraft and ships such that messages are formulated at source (button press) in **TADIL-ACCS**
(b) **TADIL-ACCS**-based processing and display systems on the ground
(c) **TADIL-ACCS** bearer systems, such as **MIDS, LINK 11, LINK 1** etc.

For full mobility in the **MOSAIC** sense, **TADIL-ACCS** addresses only part of the problem -- the real-time operational task. There will be a need for a parallel data interchange for intelligence, logistics, engineering and plans as well as for operations; each function will require an information interchange language. There may be, therefore, a need for LOGDIL for logistics data, ENGDIL for engineering data and INTDIL for intelligence data to provide interoperability at all C^2 infrastructure levels. With such a set of languages and with compatible databases for each cell, the prospect of moving a cell and 'plugging in' at a new site becomes realisable.

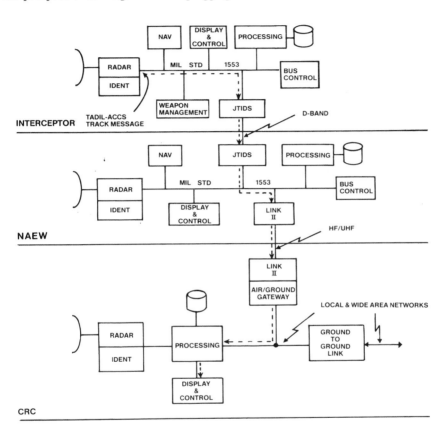

Fig. 2.2.13   *TADIL-ACCS track message routing (example only)*

2.2.7.3 Communications: Communications to support the MOSAIC concept must be battle damage tolerant and yet have sufficient capacity and flexibility to accommodate cell movability and the full bandwidth of communication in peacetime.

At present voice communications, although moving towards digital modulation, are none the less diverse (as Fig. 2.2.14 illustrates), and the consequent need for translation between systems is real now and would be exacerbated by MOSAIC.

Communications survivability can be tackled in a variety of ways, including:

Hardening
Non-nodal organisation
Multinodal organisation

For MOSAIC, hardening must be reconciled with mobility. Non-nodal communi
cations such as the TDMA variety of JTIDS would prove essential to MOSAIC for
air-to-air and air-to-ground communications survivability. The situation on the
ground is the most interesting, however. At present, the commercial pressures for
communications are causing PSTN to burgeon and to adopt fibre optics for cost

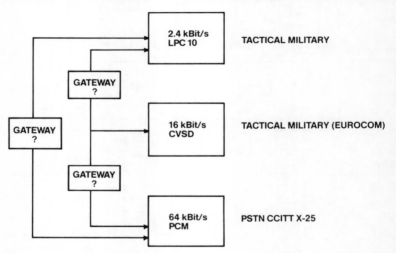

Fig. 2.2.14 *Voice communication gateways*

ease of maintenance and capacity reasons. This closely woven communication
backcloth will, in the near future, offer a ready-made solution for ground-to-ground
bearers, provided of course that the military can either adopt or adapt the CCITT
protocols. Last, but not least, HF radio communications would be a vital link in the
communications chain, especially after nuclear exchange. In particular, ground
wave may be the only effective communication in emergency, and as such would be
central to the MOSAIC concept for survivability.

To summarise, Fig. 2.2.15 shows the communications interface complexity of
MOSAIC and emphasises the need for systematic control on an ACCS-wide basis.
As the figure shows, there are sets of interconnected functional blocks which
appear as 'tiles' in a MOSAIC:

(a) PSC and cell controls form a command group.
(b) Cell control, cells (and the SOC/CRCs inside the cells) form a CIS functional
    block for near-real-time control.
(c) SOC/CRCs and NAEW form a real-time control.
(d) Finally, NAEW, fighters and tankers form the air action group (AAG).

MOSAIC: Future deployment of air power in European NATO 73

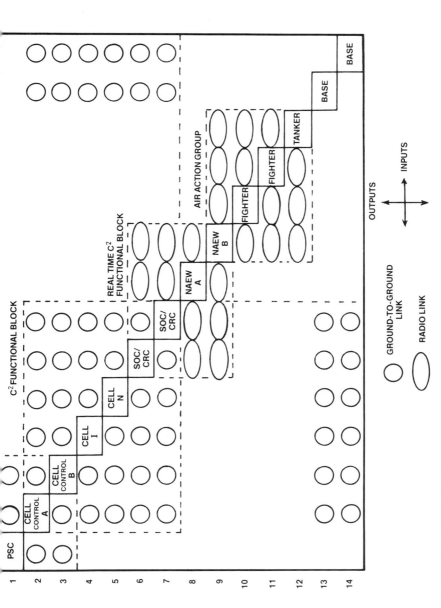

Fig. 2.2.15 Simplified MOSAIC interface structure

74 MOSAIC: Future deployment of air power in European NATO

The figure also shows two NAEWs to avoid one NAEW becoming a node, and to allow data relay when the AAG is out of communication range. Replication of NAEW will be an essential plank of survivability. implying that future NAEW will need to exist in greater numbers and cost less than at present; the US Navy's E2C sets an example, with its small size and cost, at least relative to AWACS/NAEW. Although Fig. 2.2.15 is MOSAIC based, it is noteworthy that a very similar diagram could be drawn for any ACCS. The interface problem is significant in any design MOSAIC simply highlights the needs.

2.2.8 Implementing MOSAIC

2.2.8.1 Military and political obstacles: European NATO is a patchwork of different nations, organisations and infrastructure mapped on to widely differing terrains from Norway to Greece. No one part of the organisation is perhaps any more complex than the funding and acquisition of equipments on a NATO/national basis. Thus MOSAIC faces two immediate implementation hurdles: it seeks to move cells according to overall military needs rather than political boundaries, and those cells are unlikely to be equipped with the same aircraft, radars and communication.

MOSAIC also prejudices the current organisation of power within individual national air forces. For MOSAIC to be implemented requires a greater degree of delegated authority than is presently accepted within NATO air forces, in order to allow for MOSAIC cell and AAG potential autonomy. (This mirrors the relative authorities of, say, an air base commander and a ship's captain; the latter generally has much greater local authority consistent with his relative independence.)

Land forces in the central region have their organic air defence SAMs. Naval forces in the Baltic and Mediterranean also form part of the AD fabric. For MOSAIC in particular, with its dynamic mobility, these land and naval elements of air power present a special need for co-ordinated battle management.

Last, there are certain features of air power employment which lend themselves to central rather than distributed control and management. These include reconnaissance, interdiction, recovery, search and rescue, electronic warfare, in-flight refuelling, contingency planning, resource management, joint service co-ordination and many others. The degree to which delegation of these presently unified facilities can, or need be, effected is uncertain. Instead it may be that some functions should stay centralised and be kept 'above' the MOSAIC tactical level.

2.2.8.2 Army, navy and national cells: MOSAIC lends itself to the introduction of national cells, with defined operational capabilities and interfaces, but with few specific requirements within the cell. So we can envisage a number of established Norwegian, Greek, Italian, UK and US cells, say, capable of being integrated into a MOSAIC layer using standard interfaces to communicate laterally and hierarchically. Further, the connection of land and naval MOSAIC cells can be envisaged, and this concept reveals one of MOSAIC's underlying strengths; such land force and naval cells already exist in current practice, are already mobile and already integrate offence, defence and air transport support.

MOSAIC: Future deployment of air power in European NATO 75

2.2.8.3 Applicability: These simple extensions of the MOSAIC concepts can overcome many of the basic military and political obstacles to funding, preserving national interests within NATO and co-ordinating land, sea and air power. A full embracing of MOSAIC in place of our current evolutionary approach might be ill advised, however. MOSAIC is unlikely to find universal applicability across European NATO, with its widely differing threats, terrains and political objectives. Rather MOSAIC can be seen as most relevant in areas where mobility of the ground battle will be greater, notably in the central region and its bordering areas. In other areas, a move towards some mobility might introduce a mixture of fixed and MOSAIC-like features.

2.2.8.4 Transition and cost: Since MOSAIC is different from current systems, it is reasonable to suspect that transition to MOSAIC would alter time scales and cost *vis-à-vis* the present plans.

The MOSAIC concepts outlined in this paper describe a complete system which might not be fully in place for several decades. Having set the end target, it is possible to backtrack to our present baseline and to define transition stages.

For example, MOSAIC would contain few static radars other than perhaps OTH-B and satellite trackers. Since most radars would be mobile, we can establish mobile radars, CRCs and SOCs as constituting part of phase 1. Similarly, new aircraft designs such as the European fighter would be affected by the need to operate from prepared strips, or to use arrester gear on landing. Such requirements might also form part of phase 1.

The establishment of a communications master plan would constitute part of phase 1, with the definition of TADIL-ACCS, the definition of protocols, interfaces and gateways, of data dictionaries, display standards and information requirements and the location of nodes for plug-in-points during deployed operations.

A mobility plan would interlock with the communication plan, but would also address resource dumps, runway availability, maintenance facilities, power supplies and the administration inherent in mobile operations, and already well known to navy and army personnel in their respective spheres of operation.

The air action group concept, which mirrors its naval and army counterparts, is not an essential feature; it requires considerable study by experts to establish credibility. The underlying sensor MOSAIC, however, is a technical enterprise offering air picture survivability which does not depend fot its value on the AAG. Until recently, the SEM would not have been practical; now it is, given the appropriate airborne communications and NAEW processing. Specification of requirements could form an early part of phase 1, since the SEM can operate in isolation from the other MOSAICs.

Cost is difficult to estimate but could conceivably be reduced in a MOSAIC ACCS. On the one hand, mobile facilities are of themselves often simpler; on the other hand, more such facilities may be needed and the communications infrastructure must be provided. It is reasonable to suppose that, overall, more money will be spent at the execution level, and less will be spent on the higher levels of command and control due to the integration of offence, defence and support. Thus the burden of expenditure will shift towards the cutting edge of air power.

2.2.9 MOSAIC strengths and weaknesses

So far MOSAIC has been introduced and discussed in relative isolation; no comparisons have been undertaken with alternatives, and indeed no evaluation process has been considered. One structured approach to comparing options is shown in Fig. 2.2.16, which *inter alia* identifies the principles of air weapons employment and option assessment, both of which will be enlarged upon below. At the highest level, then, it may be asked whether MOSAIC is consistent with the principles of air weapons employment; that question is addressed in part at Table 2.2.1, from where it can be seen that the three MOSAIC concepts find their strengths chiefly in:

Flexibility
Concentration of force
Unity of effort
Survivability
Maintenance of the aim.

Potential MOSAIC weaknesses are also apparent, and are addressed below.

2.2.9.1 Economy of effort: MOSAIC could require some increase in manpower in particular cases owing to the need for mobility. If support personnel contribute to economy of weapons employment, which in one sense they do, then MOSAIC would not score in those particular cases. The term usually applies, however, to accuracy of weapon delivery and sufficiency of damage; in this context MOSAIC is no worse or better than current systems.

2.2.9.2 C^2 at highest practicable level: The C^3 MOSAIC is based on the assumption that operation may be necessary in a degraded mode. MOSAIC would therefore compromise between C^2 at the highest level and sufficient delegation to allow effective succession of command (SUCOC). The key issue centres on the word 'practicable' and requires further assessment.

2.2.9.3 Comparative assessment: Since the basis of MOSAIC is to restructure the deployment and grouping of the elements of air power, it is essential to compare the MOSAIC approach with that which would be achieved by evolution from the present systems, and indeed to compare MOSAIC with other concepts, as previously suggested by Fig. 2.2.16.

To effect a sensible comparison it is necessary to establish a set of criteria against which to judge, and a methodology for trading advantage under one criterion with disadvantage under another. Fig. 2.2.17 divides assessment into two areas: performance and survivability (of performance). Performance is concerned both with the effectiveness of forces and the ease of command and control. Survival too is sub-

MOSAIC: Future deployment of air power in European NATO 77

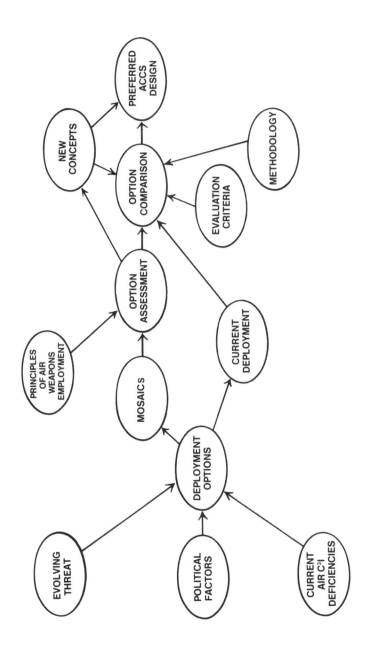

Fig. 2.2.16   Structured approach to option comparison

Table 2.2.1 MOSAIC and the principles of weapons employment

Principle of air weapons employment	Ground environment MOSAIC	Sensor MOSAIC	C³ MOSAIC	Comments
1 Attainment of a favourable air situation	●	●	N/A	MOSAIC contributes by surviving
2 Flexibility	●●	●●	●●	Principal objective
3 Concentration of force	●●	●●	●	Principal objective
4 Economy of effort	?	?	?	Mobility means more personnel
5 Unity of effort	●●	●●	●●	Combines offence, defence and support
6 C² at highest practicable level	●	●●	?	C³ MOSAIC presumes degraded operation as a potential need
7 Offensive action	N/A	N/A	N/A	Principal objective
8 Survivability	●●	●●	●●	Reconfigurability and the AAG contribute to surprise
9 Surprise	●	●	N/A	
10 Preplanning	N/A	N/A	N/A	
11 Maintenance of the aim	●●	●●	●●	Principal objective

MOSAIC: Future deployment of air power in European NATO

divided into survival of facilities and personnel on the one hand, and the ability to fight on the other.

Each of the broad headings of Fig. 2.2.17 is capable of considerable elaboration. Certain key issues are discussed below.

2.2.9.4 MOSAIC fratricide reduction: Fratricide presents one of the most intractable problems facing the employment of air power. The MOSAIC concepts offer surprising and unexpected potential to reduce fratricide levels significantly, for a number of reasons:

(a) Own offensive forces would be mixed in with defensive forces, and would not therefore traverse a complete layered air defence when going to and returning from a target. On average, and viewed simplistically, they would overfly only half of the friendly defensive positions from which they are at risk, offering at a stroke the potential for a major fratricide reduction.
(b) The close co-ordination of offence and defence essential for fratricide reduction is made possible at cell and cell group level in MOSAIC.
(c) The lateral communications and the possible inclusion of land and naval forces in the MOSAIC concept lend themselves to effective development of the indirect subsystem (ISS) of NIS, the NATO identification system, which will contribute directly to fratricide reduction.

MOSAIC offers sufficient scope for fratricide reduction that for this reason alone it may be worth pursuing MOSAIC concepts further.

2.2.9.5 Availability: Operational availability of weapon systems and sensors is likely to be affected by movement of bases, and generally the effects may be adverse. This aspect pales into insignificance, however, when set against the rise in survivability offered by MOSAIC which, in any conflict, would rapidly overcompensate for loss in availability, and in the employment of less sophisticated ground systems which tend to be more reliable.

2.2.9.6 Force management: Because MOSAIC presents many more options, and because it tends to operate with smaller groups in many cells, force management will be more complex than for static conventional systems. It also has the potential to be immeasurably more effective by rapidly redeploying mobile units – the essence of force multiplication.

2.2.9.7 Airspace management: Centralisation of military air movement plans and 'actuals' within MOSAIC cell groups will assist with airspace management, which will none the less remain a largely intractable problem owing to the separate interests of civil and military air traffic.

2.2.9.8 Avoidance of detection: Movability as the keynote of the GEM is concerned with concealment of battle order and avoidance of detection, in the

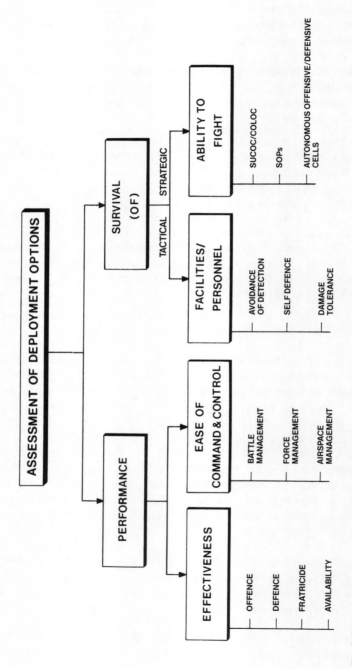

Fig. 2.2.17 Assessment of deployment options

MOSAIC: Future deployment of air power in European NATO

sense that cells should be able to set up, operate, break down and move in the time it would take an enemy to detect, locate, target and attack. The use of PSTN communications, passive sensors, LPI radars and LPI communications are all contributions to avoidance of detection.

2.2.9.9 Self-defence: The MOSAIC concept engenders complete fighting units as cells, including self-defence, which is therefore one of the key benefits of MOSAIC.

2.2.9.10 Damage tolerance: Within each MOSAIC cell, damage tolerance may be prejudiced by lessening the degree of hardening consistent with the need for movability. However, the multinodal nature of the MOSAIC cell pattern provides overall structural damage tolerance in another way, since the separate cells can operate, in the final analysis, autonomously. MOSAIC thus enhances damage tolerance of the air power overall.

2.2.9.11 SUCOC/COLOC: The C^3 MOSAIC provides a natural ease for both SUCOC and change of location of command (COLOC), since the lateral communication facilities will provide the basis for alternative command locations and the multilayer cellular structure presumes SUCOC as a means of degraded mode operation.

2.2.9.12 Standard operating procedures: Standard operating procedures (SOPs) are the inertia which will keep the flywheel of operations rotating when the command and control driver is absent. For MOSAIC, SOPs would be doubly important to promote semi-autonomous and autonomous operation as the C^2 fabric became progressively disrupted by battle damage.

2.2.10 MOSAIC in action

Before drawing to a close, it is interesting to look into the future and envisage an engagement in which MOSAIC concepts are in action under a future threat. Figure 2.2.18 shows a collection of systems, some real, some conceptual. At bottom centre is a ground environment MOSAIC (GEM) being threatened by a ground force at right and by a 360° air attack. The GEM contains all of the essentials to attack, defend and redeploy autonomously if need be. One GEM cell has been vacated at right as part of rollback and is reforming at left. Immediately overhead are NATO early warning air command aircraft, acting as focal points for air action groups, long-range radar sources and communications relays. At left a NATO emergency air command post (NEACP) aircraft is postulated, containing a senior battle commander to co-ordinate *inter alia* MOSAIC activities.

In space is a set of satellites for navigation, communications, nuclear flash detection (IONDS is the US integrated operational nuclear detection system), surveillance, intelligence, missile IR flash launch detection (IMEWS is the US integrated missile early warning system) and missile/warhead engagement. Fighting

82 MOSAIC: Future deployment of air power in European NATO

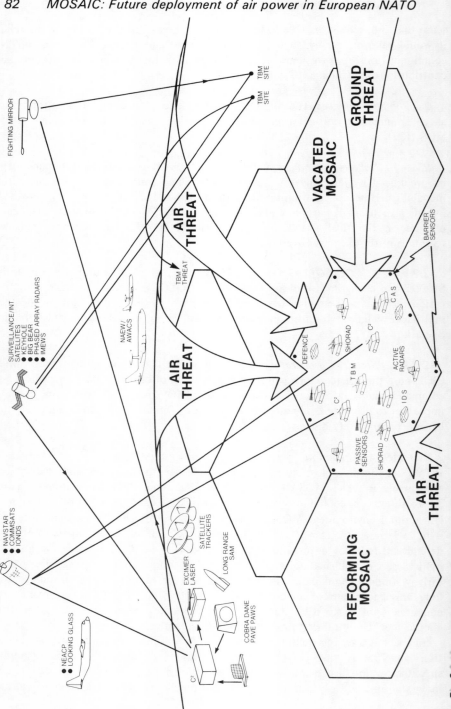

MOSAIC: Future deployment of air power in European NATO

mirror is one of the SDI concepts for such engagements, but is modified in this representation. In the full SDI concept a ground-based excimer laser directs energy at a mirror in geosynchronous orbit, which reflects that energy to a second, lower-orbit mirror and thence on to a moving ballistic missile or warhead target, delivering up to 50 megajoules per square metre and vaporising the metallic target covering. In Fig. 2.2.18 only one mirror is shown for simplicity.

At mid-left, in addition to the excimer laser, are satellite trackers; C^2 HQ employs US military satellites and two long-range radars, COBRA DANE and PAVE PAWS, the latter being a phased array warning radar said to be able to detect a grapefruit-sized target at 1000 miles. PAVE PAWS is a so-called JANUS radar because it can comprise two back-to-back phased array antennae.

The scenario of Fig. 2.2.18 is interesting in several respects:

(a) Only fighting mirror is truly futuristic; with the exception of MOSAIC, which is not technologically innovative, all the other systems are operational now.
(b) The space sensors cover *inter alia* areas of direct interest to a future ACCS, MOSAIC or otherwise.
(c) Although the space sensors and weapons may have a strategic label, the speed and accuracy of responses implicit in their implementations renders them of great tactical importance too.

2.2.11 Conclusions

The presentation and assessment of MOSAIC have:

(a) Provided a conceptual ACCS design solution which is consistent with the evolving threat and political factors, yet which addresses some current air C^3I deficiencies
(b) Highlighted the implications of full mobility, and in so doing illustrated the implications of any significant degree of mobility in any future ACCS concept
(c) Introduced new concepts which arise out of MOSAIC, but which may have much wider application:

The Air Action Group
TADIL-ACCS and the other interchange languages

(d) Identified the potential for very significant reductions in fratricide using MOSAIC
(e) Presented a set of high-level criteria by which MOSAIC and other ACCS concepts may be judged
(f) Assessed MOSAIC against the principle of air weapons employment and found it to be substantial
(g) Indicated that transition to MOSAIC is practicable and offers a highly survivable, effective and flexible ACCS.

One major conclusion to be drawn from analyses such as MOSAIC is that, in driving system designs towards their logical limits, existing concepts need to be reassessed and new concepts may be introduced which, like TADIL-ACCS and SDI interfacing, are of value to any future ACCS design.

Chapter 2.3

C^3 effectiveness studies

G.H. Lammers
(Plessey Defence Systems)

2.3.1 Introduction

The design of large-scale C^3 systems has in the past relied heavily on expert judgement and intuition with considerable emphasis on improving individual component elements in the existing system. This contribution reviews a flexible and effective method for the study of such systems. The method allows significant achievements to be made in a relatively short time, primarily by maintaining a system-wide overview while taking account of the interrelationships of component elements.

C^3 systems are invariably large and complex. Any discussion of their overall effectiveness will be equally complex and can consume vast amounts of time and energy. Through a qualitative and structured analysis, the method given here provides a tool for the identification of all significant system attributes, their mutual interrelationships and the key system management issues. When appropriate, this initial exercise can readily be followed by a more substantial quantitative analysis, normally involving the use of a computer simulation.

In this work a review of the system dynamics modelling technique illustrates the general applicability of the technique to the study of C^3 systems. Specific aspects of effectiveness are then discussed, leading to the application of system dynamics the study of information flows and the analysis of transitioning problems.

It is shown that, although the subject is indeed complex, it is possible to adopt a structured approach that provides a system-wide overview while maintaining an appropriate level of detail. There is a heavy emphasis on finding a practical way to tackle the complex issues, and use is made of examples wherever possible. By the end of Section 2.3 the reader should have grasped the relevance of the analysis technique to system effectiveness and have a clear view of the benefits of thinking in terms of the dynamics of such systems, including analysis of the ever-changing environment in which the system is embedded.

2.3.1.1 Overall approach: The term 'effectiveness' embraces many attributes of a system. The attributes most likely to come to mind immediately include surviv-

C^3 *effectiveness studies* 85

ability, security and performance, but there are many more. The method described here allows examination of each of these attributes in a structured way and leads to a logical mapping of an operational requirement on to current (and future) techniques and technologies. It is this structured analysis that paves the way for a subsequent quantification process which is necessary for meaningful trade-offs to be achieved.

A particular application of the analysis technique is to examine the flow of information in a system through the use of a model. A top-down approach to model building is described, with the emphasis not on low-level detail but on gaining a system overview, taking into account all the functional areas of a command and control system. Thus consideration is given to all data processing, communications and human elements in the system. Such an analysis gives the scope necessary to identify the key issues relevant to the C^3 system, and this provides the basis for the development of a test bed for the evaluation of specific architectures. The subsequent trade-off process takes account of the relative contributions to overall effectiveness from all attributes of the candidate architectures. Specific designs can be considered to aid the trade-off process.

Having studied and assessed the options, it is necessary to consider the transition from the existing system to the preferred option. This in the extreme could show up unexpected problems, leading to a change in the preferred option. The application of the analysis to transition planning is considered at the end of this contribution and, as summarised in Fig. 2.3.1, this completes an analysis of the important issues in C^3 effectiveness.

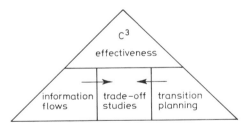

Fig. 2.3.1 *Overall approach to effectiveness analysis showing the application areas*

2.3.1.2 Techniques for the analysis: The system dynamics modelling technique forms the basis of the method discussed in this chapter. The technique was developed in the 1950s as a method of tackling the problems of industrial managed systems (Forrester, 1961).

It has been used extensively since then to examine a wide range of problems, and has more recently been applied to the study of command and control. The technique offers a rigorous approach that proceeds through a number of phases. A system dynamics analysis seeks first to understand the system behaviour in terms of its structure and controlling policies, and second to discover ways of improving its behaviour by suggesting changes to the structure or policies (Coyle, 1977).

The initial process in a system dynamics analysis is the production of

86 C³ effectiveness studies

influence diagrams. These represent the main factors in the problem, their interrelationships and how they interact with the environment. The influence diagrams are progressively refined by constructive discussion until a reasonable representation of the system is arrived at. The production of influence diagrams is the most important part of the analysis and is an end in itself. These diagrams provide valuable insight into the system operations, and this insight facilitates the identification of potential problem areas. The notation used in the drawing of influence diagrams is relatively simple and is illustrated in Fig. 2.3.2. The format of the level equation is

level now = level before + (elapse time × rate of change)

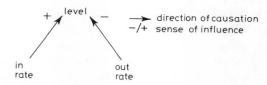

Fig. 2.3.2 *The influence diagram notation*

The format of the rate equation is

rate of change = input rate − output rate

Influence diagrams form the basis for the second phase in system dynamics modelling — the production of a quantified simulation. This involves establishing quantitative relationships for the identified influences leading to *level* and *rate* equations. There are tailor-made simulation languages (e.g. DYSMAP and DYNAMO) that facilitate the production of such computer-based models. Level equations correspond to the measurables in the system and they are determined by the integration of flows. These flows (i.e. the rates) themselves represent system actions.

Application of the technique provides a number of benefits. The quality of the system can be judged by its ability to withstand shocks from the environment in which it is to operate and still be able to perform satisfactorily. The system dynamics approach is particularly good for assessing this ability and the effectiveness or relative merits of different system options. Further, it is possible to develop a high-level model in a relatively short time. The model can then be used to highlight potential problem areas and allow investigation of possible improvement in design procedures, manning levels and management. It is also possible to incorporate the results from other analysis activities, and thus provide a framework for more detailed specific analysis to be correlated into the overall system analysis.

2.3.2 Effectiveness

One of the biggest problems in designing modern C³ systems is to know which of the important areas to concentrate on in order to achieve a system that is both

C^3 effectiveness studies 87

more operationally effective and more cost effective. It is possible to think of many factors that contribute to the overall effectiveness, but it is not necessarily clear what the relative merits of each of these contributing factors are. For instance, is it more important to make the system survivable, or can a bigger improvement be achieved by making the system more secure? In principle both should occur but in practice, with limited resources, it is necessary to trade off one benefit against another, aiming for some compromise whereby both are achieved to a lesser extent.

To further complicate the decisions that the designer faces, the present situation in the world is one of rapid technology change. It is important to realise that such changes affect both the system and the threat to the system. The effectiveness of a C^3 system must, however, be maintained as far as possible, in the face of all such changes. Failure to achieve this has in the past led to some systems becoming obsolete almost as soon as installed (if not before!) because of unexpected developments in technology.

2.3.2.1 Use of influence diagrams: In order to be able to improve this situation and ensure that system designs take these issues into account, it is crucial to have a clear understanding of what effectiveness actually means. The term 'effectiveness' as applied to a C^3 system is all embracing; it breaks down not into well defined parameters but rather into a large set of often highly correlated attributes. This is where the techniques described in this work first become useful.

At this point, it is essential to appreciate that the modelling analyst can never hope, in isolation, to generate a realistic model of C^3 effectiveness unless he is particularly gifted and endowed with a vast knowledge of C^3 systems and their problems. To achieve a well balanced model, it is important to involve a number of experts who each have specialised knowledge of the system. These experts could be communications, automatic data processing and computer specialists, operational people having direct experience of the systems operation, system designers, reliability engineers and so on.

One of the most powerful features of the approach is the use of the influence diagram to communicate the thoughts of such a wide spectrum of people in a common language. Influence diagrams mirror very closely the human thought process in establishing causal links. Specialists often claim that they have a 'feel' for their subject and can predict system behaviour quite accurately without recourse to detailed calculation. However, this 'feel' is derived from long experience where the causal relationships have been embedded into the subconscious by the human learning process. If nothing else, using influence diagrams allows the thought process to be structured and made visible.

2.3.2.2 What is understood by C^3 effectiveness?: In order to understand what is meant by the term 'effectiveness', it is necessary to break it down into a set of parameters representing the attributes of a system. These attributes can then be related through effectiveness to the operational requirement and thus the threat (Fig. 2.3.3). The effectiveness parameters have been identified as:

88 C^3 effectiveness studies

Interoperability
Security
Survivability
Flexibility
Availability
Adaptability
Operational efficiency.

All of these terms refer to specialist areas, but already they cannot be considered entirely in isolation. In fact these attributes are interrelated through highly complex interactions. Provision of certain security features contributes to an improved survivability. Availability, or lack of it, could also have a large impact on survivability. Nor is it obvious which of these factors are the most important to overall effectiveness. The influence of each could vary enormously from one geographical region to another, affected by such differences as the terrain, the climate, the skill level of the human resources available and the perceived threat from the enemy. In some organisations, particularly those of a static nature well away from front-line fighting, survivability of the headquarters may be paramount in maintaining overall

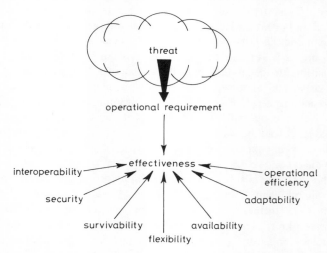

Fig. 2.3.3 C^3 effectiveness, showing the connectivity between the total threat and the operational characteristics through the operational requirements

system effectiveness, and physical hardening may be the means of achieving it. However, for other organisations, operational flexibility through mobility and reconfigurability may be more important.

To appreciate the extent of the interactions, it is necessary to resolve the original terms further until a more extensive set of parameters is arrived at. One of the main problems in such a exercise is knowing when to stop, as the breakdown can be continued almost indefinitely. The answer is usually subjective and depends on whether the level of breakdown achieved is sufficient to say something useful about the system. Furthermore, a C^3 system is not static; it changes dynamically as it responds to changes in the threat, technology, the political situation and so on. An effective system today may cease to be so in five years' time owing to changes in one of the areas mentioned above. In order to have a more realistic appreciation of effectiveness, over the time scale of interest, these dynamics have to be taken into account in addition to any static analysis of the effectiveness.

2.3.2.3 Modelling of effectiveness: Figure 2.3.4 provides an example of the application of influence diagrams to effectiveness modelling. The effectiveness attributes identified previously are each further resolved into lower-level attributes until a set is achieved that is considered to be fully representative of the system qualities. In the example, survivability can be considered in terms of electronic and physical survivability. The figure shows some of the factors that can be considered as influencing each of these. Thus, for example, one of the key threats to physical survivability is sabotage; in terms of overall system effectiveness this aspect of

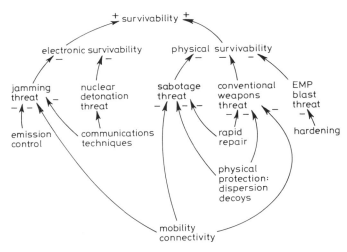

Fig. 2.3.4 *Breakdown of survivability, showing the technique of identifying low-level factors while maintaining an overview*

survivability is determined by the rapid repair ability, the degree of physical protection, the use of decoys and dispersion, together with overall mobility and connectivity within the organisation. This breakdown is an example of the principle of subdividing all the attributes through the addition of lower-level parameters until the parameters at the lowest level are easily grasped. Meaningful discussion at the lower levels can then proceed while remaining within the context of the overall effectiveness. This achieves the stated objective of maintaining a system-wide overview while relating this to lower-level attributes.

The expected period of evolution of these attributes has to be determined in order to complete the picture. This would be derived by considering likely developments in technology, the perceived threat, and any long-term projects

90 C^3 effectiveness studies

currently under way. With the evaluation criteria thus derived, using the influence diagram technique it will be possible to map the characteristics of the various system design options on to the attributes. Thus the applicability of any specific design to the C^3 system can be evaluated.

2.3.2.4 Method of presentation: The set of effectiveness parameters can be viewed in a number of ways, but one of the simplest techniques for rapid assimilation and comparison of relative merits is for the parameter to be plotted on a *histogram of merit*. This histogram (an example is shown in Fig. 2.3.5) is a means by which the contribution of each of the test parameters can be shown. Since these

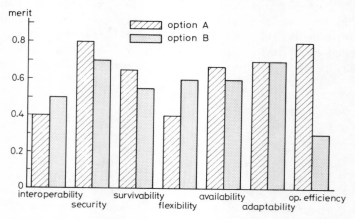

Fig. 2.3.5 *Histogram of merit, showing the relative contribution to system effectiveness of each of two architecture options*

test parameters are common to all systems, only their individual values will change between systems, and these histograms therefore provide an easy means for comparison. Further, by use of online graphics terminals, the histograms can be made time varying to represent dynamic changes in the effectiveness measures.

The histogram of merit is scaled from 0 to 1, where 1 corresponds to the maximum value and 0 corresponds to the minimum value. All the test parameters will be scaled in the same manner to ensure that the histogram is easy to use. However, by changing the value of the threat of today to a value anticipated for the future, the differences between the test parameters due to this threat change can also be shown on the histogram.

2.3.3 Information flows

The complexity of modern warfare has highlighted the commander's desire for greater availability of information. This information may relate to numerous different aspects of the state of the commander's own forces or of the enemy, as well as other factors such as the weather and terrain. Advances in technology have met this desire by generating a vast amount of information. This information is collected by sensor and action units, and passed by various means through the chain of command until it reaches the commander. The sheer volume of information may render it unusable, without regard to the problems of interpretation imposed by delays and distortion.

In order to understand this problem, and to look at the impact that this has on overall effectiveness, it is desirable to have available a flexible tool that facilitates the study of the behaviour of these information flows. Use of such a model will allow exploration of different aspects of information handling and management as well as being useful in assessing the possible impact of new technology.

The ability to handle information quickly and efficiently is one of the most important attributes of any command and control system. In order that a commander is able to control his forces effectively, he must have timely and accurate information. The system dynamics modelling technique can be used to construct the sort of tool necessary to analyse the information flows in such a system. This analysis will lead to a greater understanding of how information propagates through the system. The nature of the tool is such that it also gives an indication of the impact of sudden changes in the information traffic as a result of, say, the transition between peace, tension and war.

Before describing such a model, it is appropriate to obtain a clear view of what purpose the model is to serve. There are in fact many possible purposes for such an overall system mode, including:

(a) To assess the process of deriving measures of effectiveness
(b) To investigate the bottlenecks in the handling of information
(c) To investigate the impact of accuracy and timeliness of information
(d) To derive an optimum system configuration including issues such as the allocation of functions to entities
(e) To assess the effectiveness of operating and control procedures.

The model described here investigates the timeliness of the information flows, determines the most critical factors affecting this timeliness, and provides a measure of the impact on timeliness of these critical factors.

In order to produce such a model of the information flows, it is necessary to start from a definition of the functions performed by each entity in the organisation, the connectivity between these entities and the information load throughout the system. Information exchange does in fact take place at very many different levels. It could be from one major headquarters to another, or from a headquarters to an action unit whose task is to execute an order. Each entity in a command and control system has one or more well defined roles. In order to perform these roles, it has to operate a sequence of functions each of which will involve information processing. The basic feature of the model is to represent through key parameters both the rate at which these functions are performed for a given load and the amount of information that is generated by these functions.

2.3.3.1 Analysis steps: As with any analysis exercise, it is essential that some structure is built into the process. For a study of information flows, the analysis

92 C^3 effectiveness studies

typically proceeds through a number of steps, as summarised in Fig. 2.3.6. This structured approach to model building is seen as an essential ingredient and has a number of benefits:

(a) Confidence in the model is increased.
(b) The analysis process is fully traceable.
(c) The impact of any amendment is easily assessed.
(d) The whole analysis is repeatable.

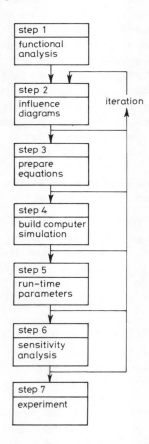

Fig. 2.3.6 *Steps in the analysis process, showing the iterative nature of the model development*

An outline of each of the steps in the analysis process is now provided. The remainder of this discussion of information flow modelling is concerned with the construction of a particular model in order to demonstrate the process. The steps are as follows:

1   Produce a functional analysis leading to identification of the information flows between functions.

C^3 effectiveness studies 93

2   Derive influence diagrams summarising characteristic information levels (backlogs), workrates and delays.
3   Construct a set of equations to be used as a basis for a computer-based simulation.
4   Use a special simulation package to construct a computer-based program.
5   Define specific run parameters and establish an accepted baseline model.
6   Subject the model to a sensitivity analysis to identify the most critical assumptions and parameters.
7   Define and perform a series of experiments, leading to a full, quantified analysis.

2.3.3.2 Generic entity module: To model information flows for a large C^3 system is a formidable task, but is made easier by the fact that in any military structure there are many entities which perform basically the same functions, although at different rates. It is possible therefore to create a generic model for such entities. The basic differences are then encapsulated in the values assigned to the model parameters.

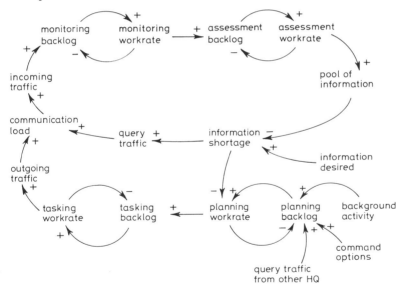

Fig. 2.3.7 *Generic headquarters model, illustrating the use of influence diagram and backlog/workrate pairs*

Figure 2.3.7 shows a generic model that might be used to represent a typical headquarters. In this model there are basically four functional areas — monitoring, assessment, planning and tasking. The workrates for each of these functions are variables in the model to reflect the fact that different headquarters may operate at different rates.

The messages that arrive at the headquarters (incoming traffic) will typically include commands and directives, status and intelligence reports, and query traffic. These filter through the headquarters until they end up as a pool of assessed information. Comparison between this and the required pool gives a measure of the shortfall of information. The greater the shortfall, the greater the effect on planning. With less information available, the commander is less able to make decisions, or will be more likely to make decisions based on his own intuition rather than on hard facts. Attempts will be made to rectify the shortfall by generating query traffic to other headquarters, and in turn the first headquarters will be subject to similar query traffic from other headquarters. Feedback into the planning functions will result and will increase the load in the planning area. A possible disastrous situation could result if the query traffic is allowed to build up to too high a level.

2.3.3.3 Entity hierarchy: In order to examine the timeliness of the information flows in the entire command and control structure, it is necessary to construct an overall model of the hierarchy. In doing this, it is equally necessary to strike a balance between the level of detail in the model and the accuracy of the model outputs. The overriding objective is initially to identify the most critical parts of the system, and then to decompose these parts to a greater level of detail. In this way the model size is kept under control and attention is focused on the key issues.

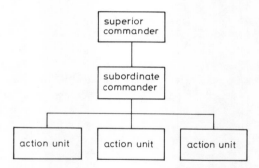

Fig. 2.3.8 *Hierarchy considered in the model*

In the light of this, it is appropriate to represent the command and control hierarchy in terms of a number of entities, each of these undertaking a number of functions. The information flows between functions (both inter- and intra-entities) can then be discussed. A typical hierarchy of entities is shown in Fig. 2.3.8. The hierarchy includes:

Superior commander
Subordinate commander
Action units.

Here the same functions are being performed in each entity, although of course they will be at different rates and at different levels of command. These differences will be captured at a later stage during the quantification process through the assignment of parameter values.

The entities shown in the figure are linked by various communications facilities which could be provided through a number of techniques (such as satellite or land lines). The current model will represent these techniques as appropriate.

2.3.3.4 Command response time: The behaviour of the system as represented by the information flows is clearly complex. Many different phenomena can be observed and many different measures can be proposed. However, the skill in using such a model is to select carefully the measures to observe in order that an accurate picture of the model behaviour can be gained without the need to study an overwhelming wealth of computer printouts. A number of measures suggest themselves as candidates for a global measure of system behaviour. However, the so-called command response time is often the most useful and is the measure that will be used here.

The measure reflects the time it takes for a directive to be sent by a high-level commander and to be conveyed to an action unit through the subordinate commander, and for some form of acknowledgement to be returned to the originating commander. This measure gives an indication of the efficiency of the communication links, as well as the information handling and decision making processes at the subordinate and superior entitites. It does not include any indication of the success or failure of specific military actions. The measure will typically be monitored throughout the simulation, and will be summarised over the whole run.

2.3.3.5 Assumptions: The functional analysis identifies the functions to be considered in the model, and the information flows between the functions. It now remains to make some assumptions in order to construct a dynamic model of these flows. The influence diagrams can be used to identifiy the assumptions that are needed for the model.

A number of distinct assumption categories are found in the simulation:

External triggers   Information arriving from outside the modelled system.
Function behaviour   The internal behaviour of the functions.
Parameter values   Values assigned to each of the identified parameters.

These assumptions can be considered as model boundaries. Typically, they will be derived from expert opinions, operational experience (e.g. trials) or even other models. This illustrates one of the strengths of the technique in that results from other analysis activities can readily be incorporated.

The final set of assumptions needed to complete a baseline model relates to the specific values for the identified parameters used in the simulation. Values are given to each of the parameters identified at run time.

2.3.3.6 Validation: Validation of a model is a key part of the model building process. Where a modelled system is in service, this is a straightforward task. Validation of a predesign model is, however, a more imprecise art and the strength of the system dynamics technique lies in the fact that it uses the collective judgement of a number of experts, together with any relevant available data. Validation of such a model is an iterative process, testing the model behaviour until all concerned are reasonably confident that the model produces sensible results. This will include checking results with other methods.

In this type of model, validation is initially a bottom-up process; only when the nuts and bolts of the model are seen to be working correctly can one step back and begin to consider the model as a whole. The nuts and bolts in this case are the backlog/workrate pairs shown in the influence diagrams that make up the separate entity functions within the model.

To this end the model is run with all the inputs set to zero so that no information enters the system. In this state all the backlogs remain empty and the workrates are zero. A single pulse can then be introduced to one of the functions of one of the entities — the equivalent of dumping a quantity of information into the backlog of the function. Using this technique, the progress of this pulse of information around the model can be monitored and analysed. Any information within the system at any time after the pulse was introduced can be attributed only to the pulse, as there is no other data entering the system. This allows information paths to be easily traced through the backlogs and workrates of the model.

2.3.3.7 Action units: Action units are both recipients and generators of information. They could well be the subject of a submodel. The basic role of an action units is to receive information and act on it, reporting back when the action is completed. When the connectivity of the system is established it is possible to bring together all the participating units into an overall information flow model. It must be noted that this model attempts not to measure the success or failure of actions, but merely to examine the ability of the system to handle the information flows. It is possible to extend the scope to include the likely impact of the availability or otherwise of information on the course of the battle itself, and this idea is currently being researched.

2.3.3.8 Baseline model: Once the correct mode of operation of the individual backlog/workrate pairs is verified, it is possible to move on to use the model. This initial use is primarily aimed at testing how the system copes with an expected normal level of activity. This will establish the basic behaviour of the command and control organisation as modelled and its general responsiveness. This in turn will be the first step in the process of gaining full confidences in the accuracy of the model.

The graphs shown in Figs. 2.3.9 and 2.3.10 are typical of the outputs seen at this stage. Figure 2.3.9 shows the variation of the backlog at the correlation function as the simulation steps through time. Surges of work come and go, but quite clearly the backlog is manageable and does not grow out of control. A similar view of all other backlogs of interest can be taken.

C³ effectiveness studies 97

Figure 2.3.10 is a more interesting result and shows graphically the command response time and how it varies through the simulated period. For completeness, the maximum time and average time are shown superimposed on the continuous plot. It will be clear to the reader that this type of output provides an insight into the overall C^3 performance. Further, a breakdown of the results shown here can provide information on bottlenecks in the system, and can identify the critical areas and the key management issues. However, before attaching too much significance to these results it is necessary to undertake a sensitivity analysis.

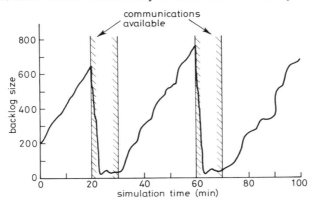

Fig. 2.3.9 *Information transmission backlog at a naval headquarters*
Response time   Average response   Maximum response

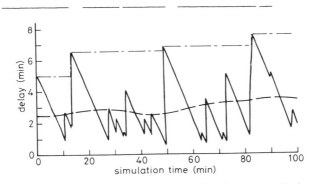

Fig. 2.3.10 *Command response time between a naval headquarters and subordinate*
Response time   Average response   Maximum response

2.3.3.9 Sensitivity analysis: As with any model, it is necessary to establish a degree of confidence with regard to the dependence of the output on the accuracy of the inputs. There are typically two types of input parameters used in the model — those reflecting assumptions made, and those reflecting design factors. Provided that the model is reasonably robust with respect to inaccuracies in parameters reflecting the assumptions, it is then possible to go on and test sensitivity to the design parameters, which is where the main interest lies.

98 C^3 effectiveness studies

The objectives of the sensitivity analysis are therefore threefold:

(a) To give confidence in the model by showing that it behaves in a predictable and acceptable fashion
(b) To show how the model can be used to predict the variation in system behaviour with changes in the parameters
(c) To show to which such parameter changes the model is most sensitive.

The parameters that are varied in the sensitivity analysis during the study of a C^3 system might well be:

(a) Transmission capacity of the communications links
(b) Availability of the communications links
(c) Capacity within the functions at the entities
(d) The amount of information entering the system
(e) The rate of generation of query traffic.

On completion of the sensitivity analysis it is appropriate to move on to a set of experiments — the main purpose of building the model. In these experiments a baseline set of system parameters is fixed and the behaviour of the system is studied under representative scenarios. Some typical experiments are now discussed which show how the model can be used to test the robustness of different systems under various external conditions.

2.3.3.10 Information shock experiments: A good system must be able to cope with shocks generated either externally or internally. What happens, for example, if headquarters is disabled by enemy attack and another has to take on the extra load? What is the implication of a commander exercising an option that generates a large increase in the planning backlog? Such experiments should provide valuable information on the system's capability and highlight potential problem areas which can then be investigated in greater detail.

To illustrate this further, consider the following scenario. Routine status and intelligence information might be arriving continuously at the headquarters, while information from higher commands and national agencies might arrive in short bursts. A general pattern of behaviour can thus be established by examining the dynamically changing situation. This would represent the baseline model and is in itself valuable. However, the strength of such a dynamic model is the ability to observe the effects of a disturbance to the system. By injecting into the simulation some specific events (e.g. removal of links, reduction of planning capacity, surge of new information), the overall effect (and particularly the time spread) of such effects through the system can be observed.

Figure 2.3.11 shows how a peak of activity at one command location passes through the system and how the corresponding peak at another location occurs some time later. This delay is a significant contributory factor to the assessment of the effectiveness of the system, since a command and control system that is not responsive enough will certainly feature unacceptable delays. Another assessment criterion is the amount of information available. Figure 2.3.12 shows how this

C^3 effectiveness studies   99

might be derived from the simulation described in this section. The assessed current information at a headquarters is seen to drop dramatically after the major communication failure, and the minimum level reached determines whether the system remains effective. Further, the time that it takes to recover from the failure, together with the extent to which full recovery is achieved, are also valuable characteristics of the command and control organisation — in this case reflecting the effectiveness of the control procedures designed to cope with the specific situation considered.

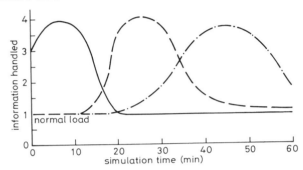

Fig. 2.3.11 *Activity peaks within the organisation, showing the magnitude of inherent delays*

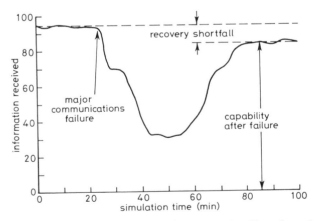

Fig. 2.3.12 *Assessed current information at HQ, showing the effect of a major failure*

It is clear from the above discussion that the type and number of experiments that can be performed is very large. The intention here is to convey the flavour of the experimentation process and to leave the reader with a clear view of how useful such an assessment tool is. Its particular value is in answering 'What if . . .?' questions, where such questions range in scope from the systemic ('What if we maximise automation of all functions at all entities?') to the specific ('What if we improve the efficiency of the man—machine interface at the review function of the subordinate commander?').

At this point it is appropriate to move on to discussion of another crucial issue in C^3 architecture — the transition from today's system.

2.3.4 Transition planning

Once the effectiveness modelling has demonstrated the desirability of a certain design, it remains for an implementation plan to be derived. There are many factors to be taken into account in the transition from the old to the new system. Most important is the requirement to maintain continuity of operational effectiveness. It would be totally unacceptable, for example, to remove major components of the old system while the new one is being installed. Other important factors in the transition from the old to the new system include the compatibility between the new and existing systems, their protocols and standards, the availability of spares and the growth potentials of the system. The interrelationship of the many factors involved can be explored using the influence diagram technique.

Figure 2.3.13 is a simplified influence diagram of some of the factors involved in transition planning of a static subordinate headquarters. This influence diagram looks at some of the more basic elements of transition planning as related to a static system. The parameters of interest have been identified as:

The operational capability of the site
The levels of trained manpower
The availability of support facilities
The availability of space.

Equipment and associated facilities can take up considerable amounts of space — a scarce and expensive resource. It would be desirable to make do with space already available, but this could mean dismantling existing equipment. The command function would have to decide what is an acceptable level of disruption to the operational task for this approach to proceed. In any case, some building expansion will almost certainly be necessary owing to the overlap of the two systems. The extra building space will become less of a problem as the old system is progressively removed. In fact, at the end of the transition there could be large surplus of building space. Careful planning will be required to minimise this problem.

Another most important consideration is manpower. There will no doubt be little manpower availability as most of the personnel will be occupied with the old system. This will leave only a relatively small proportion available to undergo training in the use of the new system. How much training is required is governed by the complexity of the system. The recruitment of new staff might appear to be a solution but in practice would be undesirable, because at the end of the transition there would be a large surplus of staff which would be an expensive luxury. It is particularly acute in this sort of problem as the new system would probably be more automated with less reliance on manual effort. The commander may have to endure some reduction in the operational capability to release space and service facilities as well as manpower. One can identify a number of feedback loops in this system. The situation is more complicated than the diagram suggests, however, as only a few influences have been included; those arising from the external environment have been suppressed.

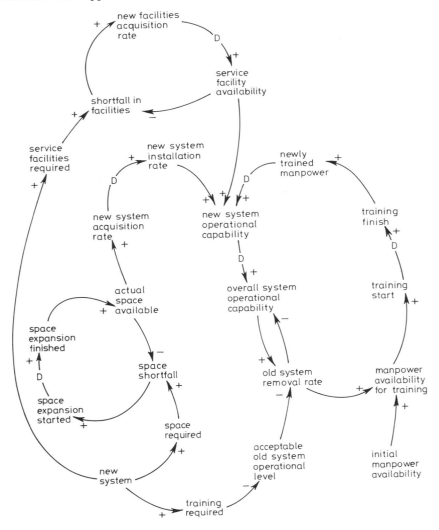

Fig. 2.3.13 *A simple transition plan influence diagram (D = delay)*

By way of example, consider the situation summarised in Fig. 2.3.14. This represents the application of the model to a simple case where the command aims to achieve 85% operational capabilities of the final system before releasing any capability in the old system. As the new system is installed, the operational capability, which was at 60%, gradually increases to 85%. In order to achieve this additional capability, investment in new facilities, building space and manpower has to be made, with the appropriate cost penalties. At about week 25, with the outlined 85% operational capability, the old system is run down and overall

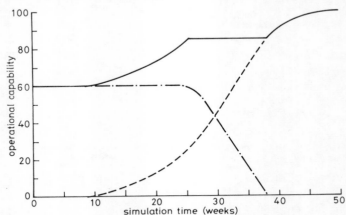

Fig. 2.3.14 *Transition plan modelling — 85% operational capacity before release of resources*
Total operational capability   Old system capability   New system capability

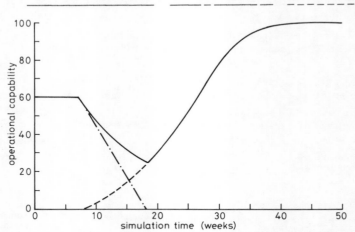

Fig. 2.3.15 *Transition plan modelling — 25% minimum acceptable operational capability*
Total operational capability   Old system capability   New system capability

capability remains at around 85% until all the old system has been removed. The final phase in the installation of the new system brings the operational capabilities to near 100% by around week 50.

An alternative to this is where the command can accept a loss in the operational capabilities of the old system to facilitate the installation of the new (Fig. 2.3.15). A minimum acceptable capability has been set at 25% of final capability. Here the old system is removed more rapidly and the overall capability rapidly increases until by around week 35 the system is at its 100% operational capability. So a much more rapid and cheaper transition can be achieved, but of course subject to some loss in operational capability for a period.

The transition plan for a large C^3 system would be very complex indeed. It is governed principally by the concern of maintaining overall system effectiveness during the transition. System dynamics offers a powerful approach to study of the problem. It can analyse fully the implications of the introduction of changes in the system. The construction of a quantitative simulation model would enable an extensive sensitivity analysis to be performed to ensure that the key factors in the problem stay within acceptable bounds.

2.3.5 Concluding remarks

In any large and complex system there are many factors to consider with complex interrelationships. Change one factor and you could, unknowingly, affect a number of others through a complex chain of relationships. Worse than this, many of the factors and the relationships change with time. Today's rules will not necessarily apply tomorrow.

In seeking to improve existing command and control systems or design new ones, it is essential to consider the global implications of any specific changes to the system and to compare the relative impact of alternative designs on the effectiveness of the overall system. The system dynamics modelling technique, through the medium of the influence diagram, provides the means of discussing and analysing such problems in a structural and logical way, ensuring as far as possible that the final system is able to perform its intended task effectively through careful planning.

2.3.5.1 Summary of the benefits: Although the application of such modelling techniques is still at a relatively early stage, it is proving a very valuable aid to the process of understanding complex C^3 systems. Influence diagrams in particular facilitate communication between specialists and enable a wide spectrum of people to understand the purpose of the system.

A system dynamics analysis provides a more scientific approach to system design. One can never replace good system designers, but a technique such as system dynamics can make the design process visible and allow a structured assessment of the relative merits of alternatives.

The visibility of the design process enables C^3 systems to be tuned to the specific needs of the user so that a more cost effective system results. It facilitates the design process by allowing a means of investigating the interplay of factors in complex systems and in identifying dynamic behaviour.

The result of the analysis is a better understood system whose weaknesses and strengths are highlighted, so that the design can be amended or enhanced to ensure that the eventual system is as effective and robust as possible.

2.3.6 References

COYLE, R. G. (1977) *Management System Dynamics,* Wiley.
FORRESTER, J. W. (1961) *Industrial Dynamics,* MIT Press

3: Databases for C³ systems

Chapter 3.1
Spatial database management for command and control

C. A. McCann, M. M. Taylor, M. I. Tuori
(Defence and Civil Institute of Environmental Medicine)

3.1 Spatial database management for command and control

3.1.1 Introduction

The interactive spatial information system (ISIS) project at the Defence and Civil Institute of Environmental Medicine, Downsview, Canada has a double purpose: to study how people use spatial data for decision aiding, and to provide models on which operational spatial information systems can be based. The models take the form of working systems, whose design derives from the conceptual understanding achieved at some particular stage. One such working prototype — spatial database management system (version 1) (SDBMS-1) — is described in this contribution. The primary application area is military command and control, though the potential applications cover a much wider field.

We begin with a discussion of the background to this project, and its implications for command and control (Section 3.1.2). In Section 3.1.3 the goals of an ideal ISIS are outlined, with emphasis on the need for both linguistic and spatial forms of human—computer interaction (HCI). Section 3.1.4 offers a summary of the implementation and capabilities of SDBMS-1, while Section 3.1.5 summarises ongoing and future work.

3.1.2 Background

The original motivation behind the ISIS project was a concern that the presentation and manipulation of spatial information had not been given enough emphasis in the development of current command and control systems. Recent systems based on digital information handling can handle more data more accurately than their predecessors, but typically this information is offered to the user as alphanumeric text rather than a graphical presentation. Psychological studies reviewed by Taylor and Taylor (1983) have shown that humans manipulate information both logically (linguistically) and associatively (spatially). The importance of the spatial mode of thought for planning and decision making is clearly evident in any command post, where the situation map is usually the focus of attention. Spatial thinking enables humans to connect different aspects of patterns into a coherent whole, to notice consistencies, to recognise analogies, to 'see the big picture'. It is the intuitive half of thought. The logical mode, however, permits the construction of sequences, fine discriminations, the interrelation of multiple contingencies, and the extraction of correct interpretations from a multitude of possibilities. The logical mode is much better than the spatial in analysing the details of complex situations. Spatial display helps the human to take advantage of the intuitive mode of thought, whereas language (including the language of mathematics) gives preference to the logical mode. The two modes are complementary, and together provide a power that neither alone can achieve.

How does spatial information differ from non-spatial? Spatial data are those whose relationships depend on their location as well as on their function. Spatial data can be geometric or topological. Geometric data have measurable distances and directions between objects; both distance and bearing can be measured on a continuous scale. The space containing the data is 'dense'. An object could be placed at any location, though in practice the resolution of measurement of object position is limited. Topological data have no measurable distances or directions but they are related in a spatial way: one object may be *between* two others, or to the *left* or *right* of some arc, or *inside* a closed curve.

Spatial arrangements, whether of topography or of military units, have no inherent significance. Their significance derives from the association of their parts into meaningful structures. Skilled commanders, whether they be military tacticians or chess masters, can remember meaningful arrangements of units much more easily than random arrangements. They remember groupings suited for functions such as attack, mutual support, area defence and so forth, and can reproduce these structures as parts of the overall situation. This kind of pattern recognition is much easier in a pictorial display than in an alphanumeric list.

Since space is continuous, spatial displays permit presentation of dense information structures in which neighbouring information (such as terrain elevation or trafficability) is highly correlated. A picture of an area in which manoeuvres are anticipated can often be much more informative than a description. The combination of picture, map and description can together give a much better idea of the possibilities and problems than can any single mode of display.

It is important to recognise that the spatial display of data and the display of spatial data are not the same. Spatial data are those whose relationships depend on their location rather than on their function. The control of a military operation is a good example of a situation where the display of spatial information (positions of friendly and enemy military units and resources with respect to the terrain) is very important. Spatial display of non-spatial information is also possible and often useful. For example, the 'media room' developed at the Massachusetts Institute of Technology (MIT) displays all sorts of information laid out spatially so that the user can 'move around' in it, plunging deeper into items of interest, or scanning the display space for useful material (Bolt, 1984). Some of the information is spatial;

Spatial database management for command and control   109

the user may dive into a map of MIT to focus on a single building, of which an aerial picture may be displayed, and then into a single office in that building, then into a book lying on a desk in that office. Other information, such as a calendar of events, is non-spatial, but it can be found from spatial memory because it is visually located at a particular place in the large display space.

Since planning and problem-solving activities are carried out by a combination of deliberate logical reasoning and of less deliberate associative action, an ISIS must support both. Logical reasoning can be pursued, for example, via well formed requests and statements through a formal query language, while less focused spatial reasoning can be done through more casual means like browsing. Logical reasoning can, moreover, be performed by a computer, perhaps through the medium of 'expert systems', whereas spatial (intuitive) reasoning can be done only by a human. The desirability of having command and control systems that simultaneously support both logical and spatial styles of interaction is clear. To our knowledge, no such command and control system exists.

3.1.3 ISIS: the goal

The development of our ideas for an interactive spatial information system has been coloured by familiarity with military land operations, and in particular by a concern for the problems that arise in tactical planning. However, the concepts described in the following sections can be applied to all command and control environments.

At a minimum, ISIS should provide the user with a highly interactive environment in which to move freely over a large virtual map, zoom in on particular areas of interest, call up and remove spatial information on a display, modify stored information, invoke spatial analysis tools, and query the system for answers to particular questions. The requirements include that:

(a) ISIS should serve as a repository of detailed spatial and non-spatial information on relevant aspects of the command and control situation. It should include information on geographic regions, topographic and tactical entities within those regions and their interrelationships, and user annotations or detailed military plans.

(b) ISIS should respond to direct questions, such as 'What enemy units are within *this* region?' and 'How far is it from *this* point to *that* point?'

(c) The user should not be restricted in the form (linguistic or spatial) used to query the system. For example, a user might phrase the same question as 'Display the symbol and characteristics of unit number X', or as 'Tell me about *this*' (pointing at the symbol of the unit). In the same way, ISIS should be able to portray the answers either as a spatial representation on a graphical display or as an alphanumeric description in the form of text output.

(d) ISIS should be able to select or synthesise visual imagery showing a battlefield situation from various viewpoints.

(e) ISIS must embody algorithms for doing routine computations, such as finding the total number of weapons owned by a military unit, or the distance between a weapon and a potential target. Eventually, it should provide tools for solving more complex problems such as the computation of mobility conditions for areas selected by the user, taking into account vegetation, ground conditions, slope and recent weather.

More sophisticated interaction and response by the system becomes possible with the inclusion of techniques from artificial intelligence, including natural language interaction, expertise in logistical requirements and military doctrine, and advanced methods of human—computer dialogue:

(f) Rather than insisting on precise explicit queries, ISIS should be able to infer the *intent* of a user's query. It should react intelligently to user requests, interpreting them on the basis of the user's task and expertise and on the basis of the current military situation.
(g) The system should be able to display information in a form and to a depth appropriate to the user's context and military role.
(h) If an item of information creates an apparent inconsistency (e.g. a position for a tactical unit is incompatible with previous information), the system should be able to point this out and request confirmation.
(i) ISIS should have knowledge of the principles of military engagement and be able to project the likely outcomes of proposed operations.

3.1.4 SDBMS-1: a demonstration prototype
The spatial database management system (version 1) is a demonstration prototype in which some of the basic capabilities of ISIS have been implemented. It is intended both as a vehicle to demonstrate concepts and as an environment for conducting experiments on the human factors of interaction with spatial information. Ultimately, ISIS will provide assistance for a wide variety of spatial problem-solving tasks. Practical considerations dictated that we concentrate initially on a specific application encompassing a typical range of map tasks. The application chosen was tactical planning at the brigade level.

3.1.4.1 Tactical planning study: We assessed current map usage in tactical planning by observing experienced military planners while they made tactical estimates of a military scenario using a standard paper map (McCann and Moogk, 1983). Verbal protocols presented by the subjects were analysed to identify the spatial information and the level of detail used by the planner, and to identify the geographic problem-solving tools that would assist the commander to estimate the military situation more completely or efficiently.

The study produced a list of spatial objects required by the planner, and some of their attributes and relationships. It also determined that planning could be assisted by: a more complete set of terrain features and characteristics than is presented on standard paper maps; better representation of some terrain characteristics, including elevation and gradient; computational and planning aids such as distance, area and intervisibility calculators; and aids for the assessment of options, such as the selection of the best crossing point or the best approach.

Spatial database management for command and control 111

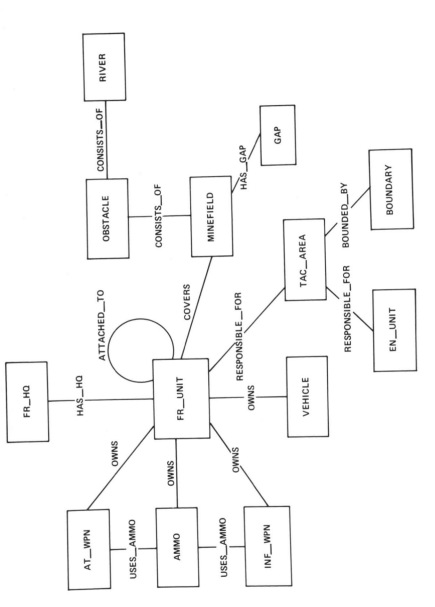

Fig. 3.1.1 Entity-relationship diagram for tactical information

112 Spatial database management for command and control

3.1.4.2 Information modelling: A preliminary investigation of the field of spatial data handling (Moon *et al.*, 1984) concluded that, of currently available data management techniques, the relational data model was the most appropriate approach to spatial information modelling and management. To reduce implementation time, a commercially available relational database management system (DBMS) called Empress (Rhodnius, 1986) was used for the core of SDBMS-1.

Two extensions to Empress were required to permit spatial data to be handled. The first was the provision of a variable length bulk data type of arbitrary format. Bulk data in SDBMS-1 include X, Y co-ordinate streams, raster imagery (e.g. Landsat) and other representations. A mechanism was also required to allow application-specific operators (e.g. distance, areas, containment) to be used in high-level queries to Empress. Empress was therefore modified to allow 'external operators', which are defined for a specific application. This modification was essential for the efficient incorporation of geographic calculations such as distance, or of more complex tactical planning aids such as intervisibility between points in the terrain. The actual operators are not known in advance, and may change with circumstances. In some cases, they may even involve expert systems rather than mathematical algorithms.

High-level conceptual information modelling was done using the entity-relationship (ER) approach. In the ER approach (Chen, 1976, 1981), similar objects are grouped as entity classes, as are significant relationships between classes. The entities and their relationships are conveniently depicted in the form of an ER diagram, in which a rectangle represents each entity class, and a labelled arc between entities represents a relationship between them. Other diagrammatic forms have been suggested (Lenzerini, 1985; Tamassia, 1985), but the principles are the same.

For tactical planning, object classes in the database of SDBMS-1 need to include topographic features such as roads, rivers, towns and lakes, as well as tactical features such as friendly or enemy units, equipment, boundaries, control measures and obstacles. The topographic component of the SDBMS-1 conceptual data model is described by Tuori and Moon (1984); a subset of a possible ER diagram for tactical information is given in Fig. 3.1.1. The relationships among the object classes link them together in a network-like structure. For example, the relationship *owns* might be used to describe the ownership of military equipment by friendly units. Weapons can be further linked to the class *ammo* by a relationship called *uses_ammo*.

Once the objects and their relationships have been at least partially defined, each can be examined to determine the characteristics or attributes that will be needed to describe each instance of that object. The object *road* might require attributes that describe its type (e.g. expressway, city street), width and name, together with co-ordinate data to define its shape. Attributes for *inf_wpn* (infantry weapon) might include name, type, weight and range. For a relationship, the attributes must include those that provide the linkage between the objects

Spatial database management for command and control

Table 3.1.1 Friendly unit (fr_unit) table for SDBMS-1

name	parent	role	size	no_pers	dtg	x	y
10LLAD	1CDN	arty	regt	547	240530	300000	4884000
11CMB	1CDN	mechinf	bde	5496	232350	308000	4892000
11RIFLE	1RRC	inf	coy	129	240600	324300	4886500
12CMB	1CDN	mechinf	bde	5185			
12RIFLE	1RRC	inf	coy	827	240600	331000	4891000
13CLH	13CMB	armd	regt	669	240530	321300	4882500
13RECCE	13CMB	recce	sqn	140	240200	322500	4880500
162MOR	16CBTSP	mor	plt	31	232400	331900	4888300
16CBTSP	1RRC	sp	coy	159	232000	317000	4885500
17RIFLE	1RHLI	inf	coy	121	240600	325300	4876600
1RRC	13CMB	inf	bn	827	240630	325500	4889000
951LLGUN	95LLGUN	arty	tp	28	240400	302300	4885300

they join; in the case of *owns* the attributes must include the name of the owner and the equipment name and class. Equipment class is necessary in this case, since *owns* links units with several classes of equipment (e.g. infantry weapon, vehicle).

Upon completion of the conceptual data model, the next stage — database design and construction — is straightforward. Each object or relationship class becomes a table in the relational database, and the attributes become the column headings in the table. (Part of the friendly unit table, *fr_unit*, is in Table 3.1.1.) When the database is actually filled with data, each row in the table contains one unique instance of the object or relationship. In addition to the co-ordinates of the object's position or shape, we include the co-ordinates of the minimum horizontally oriented rectangle that encloses the graphical representation of an object (the minimum enclosing rectangle or MER). The addition of the MER improves the speed of spatially determined retrievals by allowing a fast search by MER, followed by a slower (but less voluminous) search by actual co-ordinates.

Data modelling and design is not a trivial job. The efficient design of the ER network has a strong bearing on the later efficiency of the whole system, in terms of both how easy it is to ask questions and how fast the system can answer them. The use of the relational model facilitates extension of the database by the addition of attributes to existing tables, and by the addition of new tables. However, it is hard to modify or regroup inappropriate design elements once tables are defined.

3.1.4.3 The user interface: The user interface to SDBMS-1 allows for multimodal interaction; the user can choose keyboard, voice or gesture for input, while information is output on either an alphanumeric or a graphics display screen. The user formulates system requests linguistically by typing them on a keyboard or by speaking them through a speech recognition system (a Verbex Model 3000). In many cases, query formulation requires that the user refer to graphical information already displayed (e.g. to indicate an area of interest, or to point to a displayed feature) or input new graphical information (e.g. draw a new feature). Graphical interaction of this sort is done using a pointing device (a mouse or digitising tablet). (The Verbex Model 3000 is a continuous speech recogniser, with a maximum vocabulary of 360 words. Only about 100 words were used in the SDBMS-1 vocabulary. A discrete word recogniser can understand single words, spoken one at a time, by contrast, a continuous speech input system listens to sentences spoken as a whole, and then determines the individual words.)

Interactions with SDBMS-1 are conducted in a simple formal language that translates into a subset of the query language of the underlying Empress DBMS. There are 14 commands. Each is initiated by a keyword that may need to be followed by further inputs to complete the command:

show Extracts and displays information about one class of object or relationship in the database.
unshow Erases information from the display.
join Extracts information that demands access to more than one class in the database.
window Changes the geographic area covered by the display.
insert into Allows a new object or relationship to be added to the database.
update Changes the values of attributes of an existing object in the database.
copy Duplicates an entity in the database (normally followed by update of the duplicate).
remove from db Removes objects from the database.
draw Allows the user to annotate the graphic display.
history edit Allows the user to change the history.
save/restore base pic Saves or displays a background picture obtained from some sequence of queries.
run query Executes a script file of queries.
quit Ends the interactive session.

Conceptually, the commands can be divided into two groups: those affecting only the information currently displayed on the graphics and text output screens; and those affecting the information stored in the database. The commands *show* and *unshow* control the displayed information, in contrast to commands like *update* which modify the database itself.

Formulation of any request to access the database requires the appropriate command to be specified, followed by the class(es) of object concerned, the aspects or attributes of the object and any constraints on the selection of instances from the class (given as a Boolean expression). The syntax is:

⟨command⟩ ⟨class⟩ aspects ⟨attribute⟩ such that ⟨Boolean expression⟩;

The query to retrieve the role and size of a particular friendly military unit would be:

show fr_unit aspects role, size such that name match '1 RRC';

Spatial database management for command and control 115

In a spoken query, the final semicolon is replaced by 'please' and commas are spoken as 'comma'. The same command in spoken form is:

show friendly unit aspects role comma size such that name match 1 RRC please

We have attempted to design the interface so that the user is permitted to mix modalities, even within a single query. The user could have elected to pose the previous query using voice and keyboard modalities as:

show keyboard aspects role comma size such that keyboard please

SDBMS-1 permits the user to complete the portions of the query flagged by the term 'keyboard' (namely, class and the Boolean expression constraining the selection of tuples from that class) by typing. This feature is especially useful in circumventing the limited vocabulary of the voice recogniser, by allowing terms not contained in its vocabulary to be typed. Only complete logical phrases within a query can be selected for typing in the middle of a spoken query.

SDBMS-1 allows the user to integrate graphical gesture into a query. If the friendly unit named 1 RRC was already displayed on the graphics screen, the user could refer to it (using the keyword 'this') as an example of the class of object required in the following query:

show this aspects role, size;

To complete the query, the user picks, by graphical gesture, the unit named 1 RRC from the graphics screen. The system deduces that the selected object is of class fr_unit, substitutes that fact in the query and responds, giving the user the role and size of the other friendly units. It might have been better to use the key phrase 'this class' rather than 'this', to permit easy display of attributes for the particular entity 'this item'.

The graphic description of areas can also be used as a constraint in a query. The following query will show only those friendly reconnaissance units within a geographic area described graphically by the user:

show fr_unit aspects x,y such that role match 'recce' and x,y within here;

Graphical interaction is essential when the graphical attributes of an object (such as its location) must be inserted or modified in the database. To change the boundary associated with a friendly brigade, the user would pose this command:

update boundary such that this fill in aspects coord please

The last part of the query tells the system which attribute is to be updated (i.e. coord, or shape). The user is then placed in a graphical editor which provides an environment for drawing the new boundary.

A more complete functional description of the user interface to SDBMS-1 can be found in McCann et al. (1984).

3.1.4.4 The design of SDBMS-1 using MASCOT: The construction of the SDBMS-1 prototype demanded a design and development environment that would

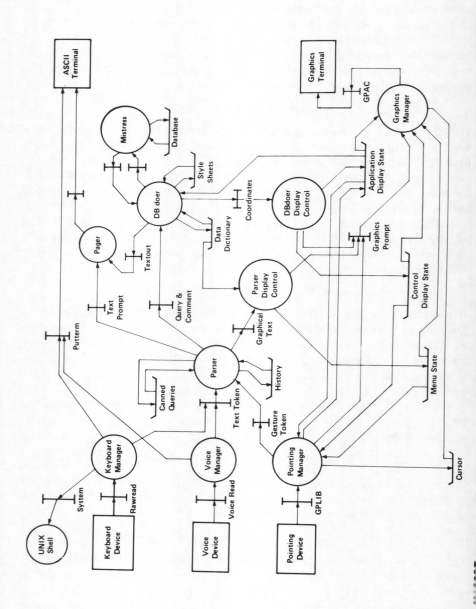

allow easy ongoing modifications. The MASCOT approach (modular approach to software construction, operation and test) was chosen; it was developed at the Royal Signals and Radar Establishment (UK) in the early 1970s (Jackson, 1977; JIMCOM, 1983). MASCOT designs can be operated on single computer systems or on distributed networks of computers. At the moment, SDBMS-1 is implemented on a single processor whose UNIX operating system has been modified to support MASCOT.

In order to discuss the design of SDBMS-1, it is necessary to define briefly the MASCOT notation and conventions. A MASCOT design can contain four kinds of objects — activities, channels, pools and devices. An activity is a sequential computer program. Individual activities communicate with each other through inter-process data areas (IDAs) called channels and pools. Channels pass messages among activities; once a message has been read by an activity, it is gone, and cannot be read again. Pools, in contrast, hold information that needs to be read by several different activities or several times by the same activity. The information in a pool is stable until it is deliberately changed by an activity. The fourth class of object is the device; a device connects a MASCOT net with the outside world, either as an input or as an output. The MASCOT diagram for SDBMS-1 is given in Fig. 3.1.2. Activities in the diagram are represented by circles, channels by I-beams, pools by U-shaped buckets, and devices by squares. The diagram shows the only legal means by which information can flow in the system. The general flow of information in this diagram is from left to right; input devices are represented on the left side of the diagram, and output devices on the right. Activities in the upper part of the diagram are concerned with the processing of linguistic or textual information, while those in the lower part deal with the graphical.

All input into SDBMS is passed to the parser activity through the text token and gesture token channels. Text streams are sent by the keyboard and voice devices, and gestures by the pointing device. Input for a single query can be intermixed in the three input streams as desired by the user. Furthermore, the input streams are asynchronous from the point of view of the human. A graphical gesture can be input at any time using the pointing device, even if syntactically it should be placed at the end of a query being formulated on the voice device. It is the parser's job to find the right place for it in the query and to transform the text and gesture tokens into a standard format for passage to the DBdoer. Provision for this multimodal interaction is one of the important features of SDBMS-1.

The DBdoer activity interprets requests to the DBMS, which in turn accesses the database pool for retrieval and storage of information in the relational tables. DBdoer is also responsible for determining where the retrieval information will go — whether to the ASCII terminal device or to the graphics terminal. It does this on the basis of information on the *style* of display, contained in the style sheet pool. Style sheets specify the default display mode (text or graphics) for all attributes of all objects, and also describe the coding method to be used (for graphical attributes, coding methods include colour, size shape, texture etc.). The user can, to some

118 Spatial database management for command and control

degree, specify the style of the displayed information. For example, by inserting the phrase 'as graphic' in a query, as in

show fr_unit aspects name as graphic;

the user can force the name of the unit, normally displayed on the alphanumeric screen, to be displayed as a graphic label. Certain dependencies are incorporated in the descriptions of style. For example, it makes no sense to display unit names on a map unless the unit symbol is displayed. Hence a request to show a name as a graphic label will also result in display of the associated symbol.

Further control of text output is provided by the pager activity. Management of graphics output is somewhat more complex, as reflected by the several pools at the bottom of the diagram. The system must keep track of the objects displayed (via the application display state pool) and provide a tight coupling to a graphics cursor for fast gesture feedback to the user.

Our experience in using MASCOT has been positive. A small part of the network (the text handling part) was built first, and other aspects added later. Although work initially proceeded more slowly than expected, later addition of other modules was fast, and had little ill effect on existing code.

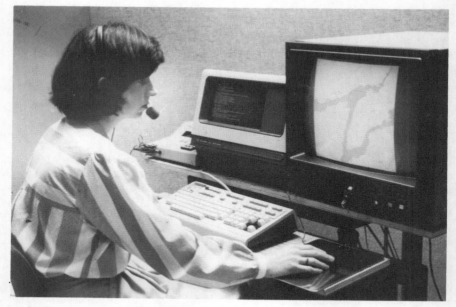

Fig. 3.1.3 *The user workstation for SDBMS-1*

3.1.4.5 Implementation and performance results: SDBMS-1 has been partially implemented on a Perkin Elmer 3242 computer under UNIX V7. (The user workstation, consisting of graphics and alphanumeric displays, an optical mouse and Verbex voice recognition device, is shown in Fig. 3.1.3.) Digitised terrain data

Spatial database management for command and control 119

for a small part of southern Ontario (an area roughly 80 km by 30 km), corresponding to that available on Canadian 1:50 000 National Topographic Series maps, has been loaded into the database. A small amount of fictitious tactical data has also been stored. The current size of the database is about 31 Mbtye. The area contains complex shorelines recorded at a resolution of 5 m. The request to draw all shorelines for this area, 'show shorelines aspects coord;', takes six minutes to complete. Object classes containing less visual detail, like military unit symbols, are displayed within seconds.

At the moment, neither the data model nor the database itself is extensive enough to allow **SDBMS-1** to be used for any realistic military task. It lacks the standard information about military units and their configuration. Neither does it have the topographic information that would be required for computations concerning, for example, mobility (e.g. soil type, tree type and spacing). From this point of view, the database for **SDBMS** can be expected to expand considerably; this will affect the response time, an important characteristic of the user interface. In other respects the database is much bigger than it need be, in the sense that the resolution of the topographic information (typically about 10 m) is greater than would be required to satisfy, for example, most 'show' queries in the application. For certain computations which depend on having precise spatial information, however, the full resolution might be needed. We anticipate the requirement to keep different versions of the database which are customised to suit the spatial information and precision needed by individual users. Eventually the customising might be done dynamically by the system, either through explicit questioning of the user or through inferencing based on current and previous dialogue.

3.1.4.6 Software tools for intervisibility analysis: The tactical planning study demonstrated that military planners are consistently concerned with intervisibility across the terrain during their assessment of the military situation. This factor arose in the determination of arcs of fire, key terrain, ground dominating an approach, and the possibility of mutual support among positions. Accordingly, one aspect of the **ISIS** project has been to explore computation and presentation of intervisibility.

A suite of software programs (Collins *et al.*, 1985) was developed to generate a raster digital elevation model (DEM) for a given geographic area and from the model to calculate and display depth below line of sight or a wire frame perspective of the terrain. The DEM is a raster elevation model in which each cell in a grid (of arbitrary resolution) contains the height of a small square patch of land. The DEM is obtained by painting the elevation information (stored as polygonal contour lines or as control points in **SDBMS-1**) on to the grid and then interpolating among these values to compute the elevation of intervening cells. The visibility of a particular cell from a given point on or above the terrain, or its depth below line of sight, is determined by extending lines of vision outward from that point. Cells become invisible when the line is lower than that to some intervening cell. The results of this calculation are displayed as a coloured map, in which visible regions of water and land are shaded in solid colours, and deepening shades are used to colour regions of increasing depth below line of sight.

3.1.4.7 Topographic information content and integrity: In attempting to use available digital cartographic data, we have had to deal with problems in the detail and form of information (Lehan and Collins, 1985). Errors of position, connectivity and duplication were few, but demanded attention. Of greater concern was the lack of topological information about geographic features. The data, when drawn, produce a relatively complete *picture* of the region; however, for planning and problem solving, additional information is needed about the deeper structure of geography. For example, it is not sufficient that when all road segments are drawn they form a visual road network (e.g. as would be displayed on a graphic output device); we must also know which road segments should be combined to form a complete named entity (for example, highway 401). Moreover, for computation of routes between points, the intersection of entities must be known. In a traditional map, this information is implicit; in a digital database, it must be explicit. Lehan and Collins propose a descriptive model of the forms that spatial information can take (Table 3.1.2). The least explicit, the point, keeps only the co-ordinates of individual geographic points and a set of attributes relating to each point. Higher forms hold increasingly more explicit information concerning the relationships between cartographic elements. For example, the spatial entity form contains complete information about a feature and its association with other features (e.g. a lake with name, size, adjoining road access, and connections with other water bodies). Special software can restructure and enhance topographic information captured in a lower form by computing and adding the topological

Table 3.1.2 *Forms of spatial information*

Form	Definition	Example	Query by
Point	Set of co-ordinates + non-spatial attributes	X, Y labelled 'water'	Point elements
Graphic segment	Chain of discrete points + non-spatial attributes	X, Y chain labelled 'road'	Line segments
Cartographic segment	Graphic segment + non-spatial attributes about surfaces on either side	Road with attribute 'forest-cropland'	Boundary segments
Cartographic entity	Ordered set of cartographic segments forming a feature + features attributes	Building	Cartographic feature
Spatial entity	Cartographic entity + attributes on relationship to nearby entities	Airport (including buildings, runways, fuel tanks, towers)	Relationships between features
Temporal entity	Set of spatial entities giving representations over time	Airport in 1978, 1985	Relationships over over time

Spatial database management for command and control   121

characteristics of higher levels, but this processing is difficult and expensive. It was concluded that deep geographic information should be captured as early as possible in the digitising process.

The topographic data contained in SDBMS-1 were cleaned and enhanced using a suite of software tools developed to permit both automatic and manual correction. The removal of visual detail beyond that needed for SDBMS-1 resulted in data reductions of 15%. Coalescing graphic fragments into larger entities reduced the number of distinct tuples stored in the database, and thereby reduced the time to display (for example) shorelines from 14 minutes to 6. Complete polygonal representations for such features as shorelines, buildings and wooded areas were formed. These features can now be shaded in solid textures, improving the quality of the visual display.

3.1.5 Ongoing and future work

3.1.5.1 Improved workstation architecture: The SDBMS-1 system was conceived and implemented in the context of a passive colour frame-buffer display system. Although it is very fast as drawing vector and raster graphical output, it provides no capability for hardware-assisted scroll or zoom. Moreover, the display subsystem maintains no information on the objects that are displayed; picking an object by pointing at it or encircling it must be done through the host computer, at a cost in time, space and responsivity. By replacing the colour display subsystem with a more powerful workstation (Silicon Graphics IRIS-2400), connected to the database system over an Ethernet local area network, we expect several benefits.

The workstation will perform the dialogue interface functions of the IFIP user interface model (Dzida, 1983), thus relieving the host UNIX system of much of the work of user interaction and graphics display. Furthermore, the workstation can perform some of the tools interface functions of the IFIP model without reference to the host system. Because the IRIS is based on very fast custom VLSI graphics hardware, it can support rapid scroll and zoom across the surface of a map, or around and within three-dimensional displays. The user can operate multiple independent windows on the workstation for text and spatial views of different information, or of the same information from different points of view. Hardware depth coding in the IRIS will allow the investigation of techniques for controlling clutter in spatial displays. For example, depth could be used to represent how recently a feature was drawn, with older features gradually fading out.

3.1.5.2 Layered protocol dialogue design: The design of the human–computer interface in SDBMS-1 was arrived at in a relatively *ad hoc* manner. We believe that a system of this complexity should be designed using more formal methods. One such method, called the layered protocol framework, has been developed under the ISIS project (Taylor, 1987a, 1987b).

The layered protocol methodology rests on two separate bases – layers and protocols. The idea of layering is fashionable in all aspects of computer design, and has been used in several descriptions of user interfaces (e.g. Buxton, 1983; Foley and van Dam, 1982; Moran, 1981). The underlying concept is that each layer communicates with layers above and below it, and each provides some services that can be used and take for granted in the design of higher layers. In computer—computer communication, one layer may handle the electrical details of message passing, another the routing of packets between computers, a third the correction of errors in the passing of packets, a fourth the connection of processes within two computers, and so on. Such layering has been systematised in various *de facto* standards, such as the Arpanet or ISO-OSI standards (Tanenbaum, 1981).

The second important concept in the layered protocol methodology is that of a protocol. Communications between a human and a computer is seen as carried by messages. Each layer carries messages in both directions between the partners. As in the ISO-OSI model for computer—computer model, the messages within any layer are 'virtual' in that they are actually carried by messages at the next lower level. Only at the lowest layer (whose definition is arbitrary) are the messages considered to be 'real'. Within a layer, each message conveys some context which is to be passed to the next higher layer (or acted upon in some other way). The content of the message is clothed in some form, such as a constrained string of characters for a word, or a selection from a menu of options. The form a message takes may demand a particular kind of response. For example, a menu selection usually demands that the selection be highlighted. The forms of messages and the kinds of responses they demand together make up the protocol for the layer. The agreed protocol is the means through which messages sent by one partner can be correctly interpreted by the other.

The framework allows for mixing of protocols using different modalities within and across layers, as shown in Fig. 3.1.4. The highest layer (layer 3) is supported by both linguistic and spatial protocols at lower levels.

It is impossible in this short space to completely detail the layered protocol framework, but it may be illustrated by example. A user of a conventional computer system wants it to perform some action (e.g. move a file into a save area). This intention is the content of the message that must be communicated at layer 3. To send the message, the user must inform the computer as to what action is to be performed (move), on what objects (*from* where and *to* where), and with what subsidiary conditions (none in this example). The computer can determine from the form of the message whether all elements are present, and can request clarification if they are not (e.g. 'destination area must be supplied'). These are aspects of the command protocol that are supported by various possible protocols at layer 2. At the next lower layer, the user can determine how to transmit the message. Using a keyboard, the characters in a command may be typed and interpreted by a command line interpreter — a linguistic protocol — but in a more advanced computer it may be possible to send the message by gesturing at a spatial display or by mixing voice with gesture and typing (as in SDBMS-1). The spatial display may be a visual metaphor for objects and places or it may be a series of menus showing commands and legal arguments for those commands. All of these different methods can support the same messages at the command level (layer 3) without changing the protocol at that level.

The layered protocol methodology offers promise for analysis of proposed or existing interfaces, for the design of portable interfaces that can allow users to change hardware freely or that can allow easy upgrading of system hardware, for the experimental analysis of interfaces at different conceptual levels, and for the development of interfaces suited to people of different psychological or cultural makeup.

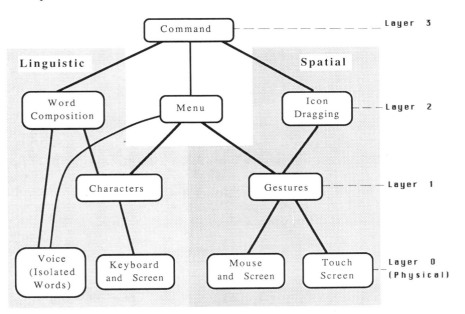

Fig. 3.1.4 *Layered protocol framework showing multimodal protocols at different layers*

3.1.5.3 ER diagram navigation: In common with other databases query environments, SDBMS-1 assumes that the user has a good conceptual model of the way in which information is organised in the database. It provides relatively little assistance concerning the names and attributes of features, and no assistance to the user on the linkages between features (i.e. the *relationships* in the ER model). There is no way to tell the end user about the data model used by the designer to structure the information set, although knowledge of this model is essential in constructing effective queries. For example, to pose a query about the number of anti-tank weapons of a given type held by friendly units, the user must know the names of the tables (*fr_unit, owns, at_wpn*), the attributes of the tables that are pertinent to the query, and how to link the three tables using the 'such that' clause. The query is posed primarily by textual means. Furthermore, there is no way for the user to change the model by extending it, renaming features or attributes, or collapsing parts of the model into a single structure.

124 Spatial database management for command and control

To address this problem, we are developing an environment that will allow a database designer/user to structure the information model using graphical ER editing tools; to create and populate the database; and then to visually navigate the resulting database by graphically exploring the active ER network representation. This workstation for active network design and navigation is being implemented on an Apple Macintosh, using a prototyping approach and taking advantage of the tools contained in the Macintosh user interface toolbox. The network creation and editing tools will allow the designer to create and describe network elements (e.g. entities and relationships) by manipulating graphical icons; to link them and state the characteristics of those links; to adjust the layout of the diagram in the editing space by moving objects whilst retaining the network topology; and to change the representation parts of the network by grouping and compressing the elements therein. Although the emphasis is on manipulation of the graphical representation, provision will be made for text views of the network and its elements.

The graphical representation of the data model can then act as an 'intelligent' vehicle for user exploration of the database, along the lines suggested by Zhang and Mendelzon (1983). We envision an interface that allows the user to point at (or speak the name of) an entity on the diagram, get the attributes associated with the entity, and find out what values occur in the table for a given attribute. By tracing a route through the network diagram, and making selections from the attributes along the way, the user will be able to find data values in one entity (table) that are linked to particular values of attributes in another (e.g. to answer the question 'How much ammunition for 105 mm guns does the 23rd artillery regiment have?'). This style of interface permits a query to be developed incrementally, rather than requiring that the user input it in a single chunk. Of course, it will be necessary to provide access to a spatial representation of the *content* of the database (probably through a sketch map) as well as the textual representation.

The development is adopting the concepts of Taylor's layered protocol model, and in fact will result in an environment for the design of other kinds of active networks, including MASCOT diagrams and augmented transition networks.

3.1.5.4 Intelligent dialogue: SDBMS-1 does not provide an environment for real dialogue between the user and the computer system; queries, cast in an explicit form, are initiated by the user. If the query is syntactically correct, it is processed against the database. If the query does not satisfy the user's intent, being either syntactically or semantically incorrect, the onus is on the user to make the appropriate corrections; except for pointing out simple syntax errors, SDBMS-1 does not provide help.

An important aspect of true dialogue is *connectivity* – the ability of each partner to create coherence between messages in a dialogue. Mechanisms for local connectivity will allow the system to make backward reference to other parts of the dialogue to clarify the intent of an interaction, thus reducing the need for the user to be explicit in his requests. For example, intelligent scoping may derive from earlier elements in the dialogue. In SDBMS-1, some connectivity is provided

Spatial database management for command and control 125

through the accumulated output on the graphical display. In constructing a query, the user can allude to features already displayed on the screen (e.g. give more information on 'this' feature already displayed). In ISIS, global connectivity might be provided by a model of the user's task and goals developed from military doctrine.

An intelligent interface would try to take account of the user's intent at each stage in the dialogue. It would interpret the user's interactions, whether linguistic or spatial or some combination, on the basis of previous interactions in the same or other dialogue sessions. The system must therefore hold *models* of the user (his capabilities and style of interaction) and also of the user's task. These models must be dynamic, and capable of being updated implicitly through the dialogue and explicitly through direct command (as when the commander announces 'I've just been up for 24 hours, so please give more detailed explanations than normal'). The availability of models to guide the dialogue between a human and a computer has been identified by Edwards and Mason (1987) as one of the key requisites for intelligence in dialogue.

The *control* of intelligent dialogue is another key aspect, having to do with the amount of initiative that is available to the dialogue partners. Traditionally, computer systems have been designed for trained users who, by virtue of knowing their goals and the system's capabilities, can control the execution of those goals from the interface. As systems like the proposed ISIS will be used more for iterative problem solving and decision support by relatively untrained users, the computer must have the ability to direct the dialogue, perhaps to aspects of the problem that the user has not considered. This will be especially true when the decision maker is operating under stress or fatigue.

A layered protocol analysis allows the initiative to be held by one partner at one level and the other at a different level. For example, in the HAM-ANS research system for hotel reservation (Marberger and Nebel, 1983), the computer holds the top-level initiative, asking the user about his requirements for a room and obtaining the necessary personal information, but at a lower level the user can direct the dialogue, providing information and asking questions. The problem of initiative control within the layered protocol model requires further research to determine the protocols that humans can comfortably use to control initiative at different levels, as well as the effect that initiative at high levels has on initiative at lower levels.

Military commanders often plan operations and consider options in discussions with human staff officers. Dialogue between humans takes place across several channels using different modalities (visual, tactile and acoustic). This gives the participants the opportunity to express the same message in many ways using well developed lower-level protocols. It also allows for redundancy in communication. An intelligent interface for military decision support must also allow interaction through different *modalities* (sometimes simultaneously) and provide ways of representing messages in different *forms* (e.g. graphical, acoustic). SDBMS-1 provides some support for these aspects.

To support an intelligent user interface, ISIS must have appropriate methods for structuring knowledge and for acquiring new knowledge, either by explicit update of the knowledge base or by learning. In a command and control application, it will also have access to external information sources that may or may not be accessible to the military decision maker. These aspects of *knowledge representation*, *knowledge acquisition* and *information sources and targets* are the three remaining aspects of intelligent dialogue identified by Edwards and Mason (1987). Methods of knowledge representation in SDBMS-1 are limited to the use of a relational database management system for storing both spatial and non-spatial information. SDBMS-1 treats space as an attribute of objects. All access to data is through the objects themselves, and only by accessing all objects in a given area can the contents of the area be determined. Objects can, however, be indexed in many ways for rapid access. Location and extent are among these attributes, but space is not natural to SDBMS-1, as it must be in a fully functioning ISIS. In order to answer questions such as 'What area can be seen from anywhere along *this* road?', SDBMS must extract from the database a digital terrain model covering the whole area that might be relevant. The work is then done on this terrain model. A purely spatial database, however, would be able to access information using location as a primary index. Its canonical question would be 'What is near here?' rather than 'Where is this thing?' The full ISIS would be based on both kinds of canonical question.

3.1.5.5 Spatial problem-solving tools: ISIS must be able to compute many different kinds of spatial relationships between features if it is to be useful for geographic problem solving. Two major categories for spatial relationships can be distinguished. Metrical relationships are those that depend on distance or relative distance. The following specific metrical relationships might be the basis of useful spatial operators in ISIS:

(a) Distance between two points, e.g. range, path distance
(b) Proximity, e.g. shortest distance from a point to a curve, distance between the edges of two areas
(c) Area, e.g. true surface area, projected area
(d) Plane angle, e.g. field of view.

Topological relationships are those based on connectivity, including:

(e) Connectivity of lines (networks)
(f) Adjacency of surfaces and solids
(g) Intersection of lines, surfaces
(h) Containment.

These too, must be provided as basic spatial operators in a fully functioning ISIS.

The intervisibility tools discussed in Section 3.1.4.6 currently operate in a stand-alone mode. They will need to be integrated as a functional component of SDBMS. Additional tools for mobility, range of fire and other complex spatial analyses related to command and control should be developed and integrated. The usefulness of these tools, and the form of presentation at the user interface, should be evaluated through human factors analysis and experimentation.

3.1.5.6 The representation of time: A fundamental characteristic of information in a command and control environment is its dynamic nature. This has been recognised in most systems by provision for a date/time group attached to tactical information that reflects the time at which the information was updated. A more comprehensive capability to handle time-dependent data is desirable, however. It should be possible to store historical information about enemy movements or status that could be replayed in an animated form; such a presentation might provide intelligence staff with clues about the enemy's future intentions. Similarly, a commander might find it desirable to play out possible future tactical consequences of one or more operational plans. A capability to handle time-dependent data will be necessary to allow weather predictions to be incorporated into mobility calculations, and to allow simple projection, for example, of the current position of an enemy tank sighted a few hours earlier. Future systems should be designed to accommodate time-dependent information.

3.1.6 Concluding remarks

The overall goals of the ISIS project remain well beyond our capability. Nevertheless SDBMS-1 demonstrates that some of these goals are attainable with today's technology, and provides us with an experimental environment in which to study their impact on military planning for command and control. We are convinced that robust, usable spatial information systems for command and control are possible only if their designers and developers are aware of the evolving principles of human–computer interaction. Although improvements in hardware are essential, they must be coupled with a better understanding of the problem-solving skills and requirements of the planner.

3.1.7 References

BOLT, R. A. (1984) *The Human Interface – Where People and Computers Meet*, Lifetime Learning Publications

BUXTON, W. (1983) 'Lexical and pragmatic considerations of input structures', *Computer Graphics*, 17, pp. 31–7

CHEN, P. P. (1976) 'The entity-relationship model – toward a unified view of data', *ACM Trans. Database Systems*, 1(1), pp. 9–36

CHEN, P. P. (1981) 'Entity-relationship approach to information modelling and analysis', ER Institute, PO Box 617, Saugus, CA 91350

COLLINS, S. H., LEHAN, T. J., TEASELL, S. and COLLINS, J. (1985) 'Development of software tools for intervisibility analysis, using the DCIEM spatial database management system', contract report, contract #8SE84-00244, DCIEM, Downsview, Ontario

DZIDA, W. (1983) 'Das IFIP-Modell fuer Benutzerschnittstellen', *Office Management* special issue on man–machine communication, pp. 6–8

EDWARDS, J. L. and MASON, J. A. (1987) 'Evaluating the intelligence in dialogue systems', *International Journal of Man–Machine Studies*, to be published

FOLEY, J. D. and VAN DAM, A. (1982) *Fundamentals of Interactive Computer Graphics*, Addison-Wesley, Reading, Mass.

JACKSON, K. (1977) 'Language design for modular software construction', *Information Processing 77*, IFIP

JIMCOM (Joint IECCA and MUF Committee on MASCOT) (1981) *The Official Handbook of MASCOT*, Computing Policy and Standards Section, RSRE, Malvern, UK

LEHAN, T. J. and COLLINS, S. H. (1985), 'Problem report: Development of methods and tools for correcting large volumes of spatial data', contract report, contract #8SE84-00278, DCIEM, Downsview, Ontario

LENZERINI, M. (1985) 'SERM: semantic entity-relationship model', Proceedings 4th international conference on entity-relationship approach, pp. 270–8

MARBERGER, H. and NEBEL, B. (1983) 'Natuersprachlicher Datenbankzugang mit HAM-ANS: syntaktische Korrespondenz, natuersprachliche Quantifizierung and semantisches Modell des Diskursbereichs, report ANS-22, Research Unit for Information Science and Artificial Intelligence, University of Hamburg

MCCANN, C. and MOOGK, C. (1983) 'Spatial information in tactical planning', DCIEM report 83-R-60 (restricted)

MCCANN, C., TUORI, M., TAYLOR, M., MOON, G., TEASELL, S. and COTE, D. (1984) 'Spatial database management system-1: user interface design', DCIEM Technical Communication 84-C-13

MOON, G. C., COLLINS, S. H. and BLAIR, G. (1984) 'Considerations for the design of an interactive spatial information system (ISIS)', DCIEM contract report 8SE80-00052

MORAN, T. P. (1981) 'The command language grammar: a representation for the user interface of interactive computer systems', *International Journal of Man–Machine Studies*, **15**, pp. 3–50

RHODNIUS (1986) *Empress: Relational Database Management System*, Rhodnius Inc., Toronto

TAMASSIA, R. (1985) 'New layout techniques for entity-relationship diagrams', Proceedings, 4th international conference on entity-relationship approach, pp. 304–11

TANENBAUM, A. S. (1981) 'Network protocols', *Computing Surveys*, **13**, pp. 453–89

TAYLOR, I. and TAYLOR, M. M. (1983) *The Psychology of Reading*, Academic Press, New York

TAYLOR, M. M. (1987a) 'Layered protocols for computer–human dialogue I: Principles', *International Journal of Man–Machine Studies*, to be published

TAYLOR, M. M. (1987b) 'Layered protocols for computer–human dialogue II: Some practical issues', *International Journal of Man–Machine Studies*, to be published

TUORI, M. and MOON, G. (1984) 'A topographic map conceptual data model', Proceedings of the international symposium on spatial data handling, Zurich, pp. 28–37

ZHANG, Z. and MENDELZON, A. O. (1983) 'A graphical query language for entity-relationship databases', in *Entity-Relationship Approach to Software Engineering*, Elsevier, North Holland, pp. 441–8

Chapter 3.2
Systems design and data management problems in the utilisation of local area network architectures

A. S. Cheeseman and R.H.L. Catt
(Software Sciences Ltd)

3.2.1 Introduction
The local area network has brought enormous opportunities to the C^3I designer. It allows him to design flexible and modular systems which can be more accommodating of change and enhancement than systems of the past, and which can also exhibit a much greater reliability than was previously possible. By its very nature, a system incorporating a local area network can continue to function even though some of its nodes have failed.

What is not generally appreciated is that, to realise the potential benefits of such an architecture, a much greater degree of preliminary systems design is required than was necessary in the past. The procurement agency will therefore have to adapt its organisation to accommodate this additional task.

This contribution describes two aspects of the preliminary design process:

(a) The need to design the overall C^3 system, the functions of the component system and the message flows between them, before moving to procure the individual components themselves (weapon systems, sensor systems, command systems, communication systems etc.)

(b) The need to anticipate particular technical problems, such as the management of distributed data, and to identify the potential solutions and costs prior to the procurement process.

The authors have been working for a number of years as consultants to the UK Ministry of Defence, and have been involved in the development of communications standards for shipwide combat system networks, the management of data exchange between the component systems of a combat system, and research into distributed database management techniques for combat systems. In particular the authors have been associated with three projects which, although managed and funded separately, have a great deal in common. They are:

130 Systems design and data management problems

(a) The combat system data interchange study (**CSDIS**) which was originally sponsored by the XCC division of the Admiralty Research Establishment (**ARE**) but which is now funded by the Chief Naval Weapons System Engineer (**CNWSE**).

(b) The type 23 data management support project which is funded by the Director-General Surface Ships (**DGSS**).

(c) The distributed information architecture for ships (**DIAS**) project run by the XCC and latterly the AXT divisions of the Admiralty Research Establishment.

All of these projects are based at the Portsdown site of **ARE**.

3.2.2 Historical development

Digital computing was first introduced into Royal Naval systems in the 1960s. The early computerised systems consisted of relatively unsophisticated weapons and sensors coupled to a centralised digital computer system. Figure 3.2.1 illustrates the architecture. The computer system supports both command and control. The command aspects are picture compilation, threat evaluation, weapon assignment and command aids; the control aspects are sensor data processing, fire control and aircraft control.

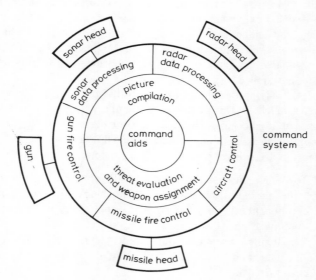

Fig. 3.2.1 *Functions within early centralised systems*

The computers were typical of their generation, being physically large and having limited memory. Because digital processing capacity was at a premium, the central computing system was expected to perform all the necessary digital computation for the weapons and sensors and also to provide the analogue drive for those equipments.

Systems design and data management problems 131

The last twenty years have seen increasing demands made upon this digital processing capacity, with those functions already computerised requiring greater sophistication, additional functions requiring computerisation, and more advanced threats requiring faster and even automated responses. The addition of the new weapon and sensor systems developed during the last decade has seen a migration of some of the digital processing to these systems, thus relieving the central computers of some of their original tasks and enabling them to undertake new tasks previously performed manually.

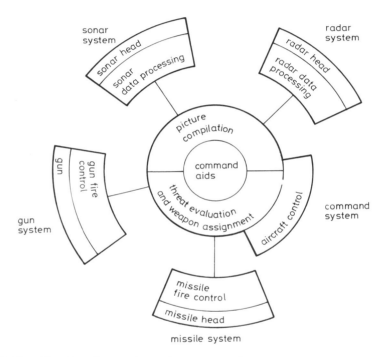

Fig. 3.2.2 *Migration of processing to the weapons and sensors*

Figure 3.2.2 illustrates this trend, and typifies the situation for today's designers working on systems for the immediate future. The centre of the now distributed processing system is the command system, which retains the functions of picture compilation, threat evaluation, weapon assignment, aircraft control and command aids. The processing pertinent to each weapon and sensor system is now located in computers associated with each system. The overall system has come in navy circles to be termed the combat system, but is in effect what is more widely called the C^3 system.

As processing power moved out to the weapon and sensor systems, it became necessary to define digital interfaces between them and the central processing system. The interfaces that were developed were normally with the central processing system. Any need for a weapon or sensor system to communicate with another weapon or sensor system was achieved either by the central processing system providing a store and forward message switch service or by the provision of dedicated point-to-point links.

This gradual migration has resulted in each weapon and sensor system having a unique privately agreed interface with the central system, each interface independently defining an electrical interface, a communication protocol and local data standards. Thus, although the digital processing power of the combat system has become distributed, the combat system data exchange architecture has remained essentially centralised, with the individual weapon and sensor systems having point-to-point star connections to the command system and with the command system controlling the data flow.

3.2.3 Introduction of a local area network

The extent of the distribution of processing power is now such that it is quite natural to consider the introduction of a local area network to achieve the interconnection of the various elements of the combat system. The computers internal to each weapon and sensor system perform the most time-critical processes, and consequently the time requirements of the interface between the command system and the individual systems are less stringent. A shared communications service has become more feasible.

At the simplest level the local area network interface is a replacement for the current special purpose interface, but in addition its introduction can have the following beneficial effects:

(a) Provide total connectivity between all the systems comprising the combat system, for each system is connected to all the other systems
(b) Provide a common interfacing policy for all the individual systems
(c) Provide considerable physical simplification at the command system, which will no longer have to support many unique interfaces
(d) Considerably reduce the number of cables in the ship and allow early specification of cabling to the shipbuilder
(e) Considerably simplify the introduction of new systems during refits.

Figure 3.2.3 shows two different views of the introduction of the local area network, first as a development of the previous figures, and secondly as the combat system might be drawn without considering current implementations.

The presence of the local area network requires the combat system designer to make a fundamental decision about the type of data exchange architecture to be used. The choice lies between the centralised architecture and the distributed architecture.

The centralised data exchange architecture is a perpetuation of the past. Each weapon or sensor system uses the network for point-to-point data exchange with the command system, and the local area network merely replaces the former point-to-point links. This approach realises the hardware benefits of using the local area network, but the combat system architecture is no more flexible than before the introduction of the network.

The distributed data exchange architecture enables the combat system designer to take full advantage of the potential benefits of using the local area network. It uses the total connectivity available to allow data to flow freely within the combat system. Data is no longer constrained to flow via the command system. The command system is no longer required to set aside resources to provide a centralised message store and forward service. Data output by broadcast from sensors is immediately available to all users, and this obviously improves reaction times. Users of sensor data have direct access to the output of all sensors and can readily switch from main to alternate sources when necessary.

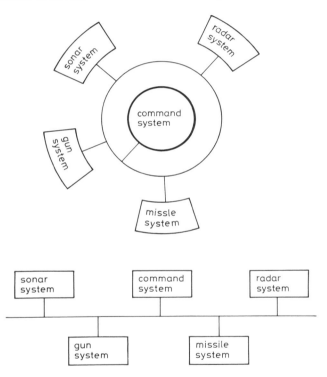

Fig. 3.2.3 *Alternative views of the network system*

The command system remains the normal controlling system of the combat system, but its unavailability through equipment failure or action damage is not as catastrophic as with a centralised architecture. In the distributed architecture, the weapon and sensor system can still communicate in the absence of the command system, thus enabling reduced modes of operation to be achieved.

However, the realisation of the benefits of a distributed data exchange architecture requires an explicit design of the combat system, with the command system

134 Systems design and data management problems

being considered as one of the constituent systems. The combat system architecture is the framework in which the weapon, sensor and command systems must be designed if they are to co-operate successfully and if the potential attributes of a distributed system are to be achieved. In the past the total system design was often subsumed into the design of the command system, but this cannot be the case for a distributed architecture.

3.2.4 Distributed system architecture

It is now appropriate to consider the basic aspects of distributed systems and what is required in their development from a purely technical viewpoint. A distributed system is a set of computer systems linked by some form of digital communication. In this section the complete system is often referred to as the *total* system and the individual constituent systems as *member* systems.

Today's distributed systems have arrived from one of two historical paths. The first path is where computer systems have been designed to run independently but have subsequently been linked to achieve more efficient operation. The second path is where a centralised processing system has become dispersed, with processing power migrating away from the centralised computer to peripherals of increasing intelligence.

In either case, if the necessary co-operation between the constituent parts of the distributed system is not contained within their original requirement specification then the degree of co-operation achievable on interconnection is unpredictable indeed, if there is any co-operation at all, it is purely fortuitous.

When connecting independently designed systems to make a distributed system the potential incompatibilities are manifold:

(a) There can be functional incompatibilities, where functions may be missing or duplicated.
(b) There may be data incompatibilities, where definitions or classifications of data may overlap or be inconsistent.
(c) There may be communications incompatibilities, where the protocols and/or the electrical signalling employed by one system may not be supported by another system.

These incompatibilities may sometimes be overcome by modifying each member system until the required level of co-operation is achieved, or alternatively by adding further computers to the total system to act as intermediaries for the established systems. In some cases the incompatibilities may be so severe that neither of these solutions is cost effective and so the connections cannot be made.

The difficulties of connecting different computers has been realised for many years, and is the *raison d'être* for the pursuit of the open system interconnection standards. Indeed, these difficulties pose the major obstacle to the achievement of information technology.

For the purposes of this section, a distributed system's architecture is split into three separate sub-architectures:

Systems design and data management problems 135

A functional architecture
A data exchange architecture
A communications architecture.

The functional architecture is the way in which the various functions to be performed are to be distributed among the individual systems — the member systems.

The functional architecture has to encompass all the required functions and attributes of the whole distributed system. It must take into account the nature of the distribution and any constraints imposed by the underlying communications network. In the specific case of a combat system, it is important to realise that these individual functions, when put together and combined with human resources, form the warfare functions which are only available from the total system, and it is these higher-level functions that have to satisfy the requirement for the combat system itself.

Consideration of the way in which functions are to be assigned to each member system and the consequent coupling between them is essential where real-time systems are concerned. For a combat system, great care must be taken if it is to be a quick-reaction system capable of responding to threats in a co-ordinated manner, and if it is to display good resilience to action damage.

The data exchange architecture is derived from the functional architecture. The scope of the data exchange architecture is not solely confined to the definition of the data itself, for it includes the definition of the context of the data as well. This combination of data and its context is often regarded as constituting information.

The data exchange architecture is concerned with the data flow between member systems, that is between the functions contained within one member system and the functions contained within another. Data exchange is concerned in particular with application data context, application data structures, application message sequencing and data representation (i.e. frames of reference, reference origins and units of measurement), all of which must be established before any two systems can exchange data meaningfully. The data exchange architecture is not concerned with the message structures of the underlying communications system.

The communications architecture is concerned with the conveyance of data between the individual member systems of the distributed system. From a communications viewpoint, all systems are viewed as member systems and their functionality is of little importance except where there may be a requirement for a priority service. The communications architecture is concerned solely with the provision of the communications service and not with the data that is actually carried.

Figure 3.2.4 illustrates the three sub-architectures, showing functional assignment to systems, the data exchange between the functions, and finally the communications view of the system. The three sub-architectures, taken in the order

Functional architecture

136 Systems design and data management problems

Data exchange architecture
Communications architecture

form, in fact, a top-down design sequence for a distributed system. Unfortunately human experience develops in a bottom-up way, such that:

(a) Communications architecture are well understood.
(b) Data exchange architectures are now being better understood, but are some way off being resolved.
(c) Functional architectures are the least well understood.

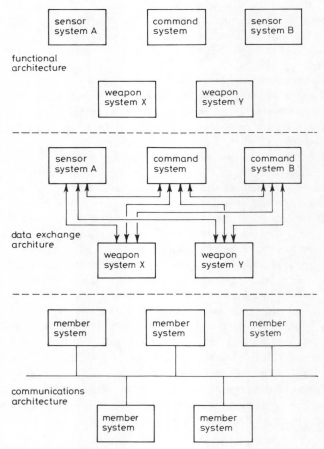

Fig. 3.2.4 *The three sub-architectures*

The consideration of the functional or application area standards has just been commenced by the open system interconnection committees, but only for simple commercial applications. It will be a considerable time before any commercial

Systems design and data management problems 137

functional standards are available which are equivalent to the complexity of those needed for military command and control systems.

The generation of an unambiguous and complete requirement specification for a distributed system such as a combat system seems to be beyond our current capability. The history of C^3 systems is one of a considerable shortfall between user expectation and actual achievement. This can normally be attributed either to an ambiguous requirement, or to a continuously changing requirement, or sometimes to no formally documented requirement at all. The writing of a sound requirement specification for a weapon or sensor system performing mechanistic functions is difficult, but it is a simple task in comparison with that of writing a specification for a C^3 system such as a command system or a combat system. Such a requirement would have to ensure:

(a) That all the functions to be provided are compatible with current operational procedures
(b) That human factors and human organisation is taken into account
(c) That all the required attributes are incorporated, such as resilience and accommodation of evolution.

This particular requirement specification problem has to be solved in a pragmatic way to the best of our ability until a sound formal approach has been proven.

Now let us consider the implementation of distributed systems. At present a distributed system may be implemented in one of two ways — as a homogeneous system or as a heterogeneous system.

The *homogeneous* implementation is one where the member systems consist of hardware and software components which are all procured from a single manufacturer. This should guarantee consistency, for a single manufacturer's standards are used throughout. A single manufacturer is responsible for all aspects of the architecture — functional, data exchange and communications — and so it is this approach which is most likely to succeed.

However, this utopian approach has a very real drawback. It is neither politically nor commercially desirable for complex defence systems. It requires expertise in all aspects of the distributed system to be held by a single manufacturer, and defence procurement would be totally dependent upon a single supplier.

The other form of implementation, the *heterogeneous* approach, is one where the individual member systems of the distributed system are procured from different manufacturers with minimum constraints on their choice of hardware and software. This approach is that currently advocated for defence procurement, as it encourages competition and reduces dependence upon any one manufacturer.

But what about the three sub-architectures? The problem associated with heterogeneous systems is determining the degree of freedom to be left to individual manufacturers which is commensurate with achieving a sound design. The individual member systems cannot be allowed to choose in an uncontrolled manner the functions that they are to implement, the format in which they will exchange data, and the communications they will use.

138 Systems design and data management problems

Some form of control must be exercised to produce a viable distributed system, and this should be the imposition of a requirement to conform with a distributed system architecture. So this approach is best described as a controlled heterogeneous approach.

The need for a distributed system architecture and an agency to generate it and administer it will now be considered. The agency's task is to comprehend and interpret the requirement of the distributed system as a whole and to ensure that the overall characteristics of the total system are achieved from its component parts. The agency is concerned that the interaction between the various member systems satisfies the operational requirement, that the required attributes are realised, and that the information within the total system is used to best advantage.

With these responsibilities in mind, the agency can derive a total system architecture consisting of the three sub-architectures. The agency can then generate functional requirements, data exchange standards and communication standards to impose upon the individual member systems which will bring order to the total system. Total system design is concerned with the formulation of a consistent set of requirement specifications for each member system such that the requirement for the total system is met. It is most important that the standards are developed and that the consistent set of requirements is complete prior to any procurement action for individual member systems.

The architecture is not concerned with the detailed physical implementation of each member system, just that each member system performs the functions required of it in the prescribed manner. The total system design does not include the internal design of member systems. Member system designers have to design their systems to meet a requirement specification that has been generated in the total system context.

The agency administering the architecture must be available to answer questions and resolve system-wide problems throughout the design, the development and the testing of individual member systems. The agency must maintain the documentation of the overall system as it progresses from requirement through to implementation. It must be available when the overall system is being set to work.

The agency understands the implications of individual member system problems on the total system, in terms of both its operational performance and its attributes. It is the agency alone that can ensure that decisions are made in the context of, and in the best interests of, the total system. It cannot allow individual member systems to make unilateral decisions that will prejudice the operation of other member systems.

In summary, the role of the architecture and its administering agency is to ensure the coherency and consistency of the distributed system and to ensure that the overall system performs as required. The architecture is the division of the system into a set of component parts and a set of associated standards such that the parts can be procured separately, and then aggregated to provide a total system, with the operational performance and attributes required.

But what if we ignore the need for controls and have no architecture and no

Systems design and data management problems 139

administering agency? Then technical anarchy would prevail. Although, if left to themselves, individual member systems can resolve interfaces, each would be resolved with a very limited perspective. Data would be handled inefficiently and inconsistently: communications would all be point to point, and many opportunities for sharing data would be lost. The resulting total system produced by aggregating the member systems would be somewhat unpredictable and its characteristics would have to be established during setting to work, for it has not been designed as a whole. Consequently, when building a distributed system consisting of complex members, if the total system architecture and the role of a controlling agency are neglected, then the performance of the total system will be a high-risk area. Substantial cost penalties would be incurred if the total system was found to be unsatisfactory and remedial action affecting many member systems was necessary.

Having considered the architectural design requirements associated with distributed systems in general terms, the more specific case of a combat system will be considered.

3.2.5 The combat system design process

Before commencing with a combat system design, it is necessary to have a requirement specification for the total system. This specification would include:

(a) The operational roles to be achieved
(b) The reversionary roles to be achieved in the event of equipment failure or action damage
(c) The availability required
(d) The performance required and the specified scenario in which that performance is to be achieved.

The first step in the design process is the establishment of the functional architecture, that is the assignment of functions to member systems. The designer may be constrained here due to the availability of particular member systems, and may have to accept an already established functional assignment. In the case of a combat system, many member systems are likely to be already available, although currently being used in a centralised architecture. In such cases the total system designer has to achieve a transition from centralised to distributed operation in the most expeditious and cost effective manner.

Following the assignment of functions, the next step is to examine the interfaces between functions. The data flow between member systems that is necessary to meet the normal and reversionary modes of operation is analysed and defined. This analysis ascertains:

(a) The number of messages per second arriving and leaving each member system
(b) The number of bytes per second arriving and leaving such member systems
(c) The total number of messages per second to be conveyed by the communications network linking the member systems.

Consideration of these figures will indicate the feasibility of the total system design, whether member systems can be designed to cope with the communications traffic, and whether the intended local area network can support the traffic requirements.

The next step in the design process is the establishment of the data flow and communications architectures, followed by the specifications for the individual member systems. These specifications are of paramount importance, for it is the aggregation of these member systems that constitutes the combat system itself.

If the member systems are to communicate freely, then they must adopt a common language. There has to be a common communications architecture observed by all the member systems. The international work on the open system interconnection seven-layer reference model for communications (Standard A) has shown how inter-system communication can be best achieved by splitting it into separate layers of protocol. Each layer is independent of the implementation of the layers above and below it, and so the effects of change can be contained.

The use of local area network imposes many aspects of a common architecture upon each member system, but there is a need for the actual data output to the network to observe common standards. Messages from sensors of the same type should use identical formats and use identical units of measurement for data representation. The existence of these common standards will enable a member system to switch from one data source to another with the minimum of effort.

The ability to support evolution will be one of the attributes required of the total system. Indeed, the lifetime of a combat system is considerably longer than its shore-based contemporaries, and a combat system may expect to have changes in its member systems. The accommodation of change requires that the functional decomposition of the total system and the distribution of those functions among the member systems is stable.

Each member system must communicate with others via standard interfaces associated with a generic member type. The generic systems concept is where each type of system has to conform with a general pattern of behaviour characteristic of its class. There should also be internal structure independence. Each system has to be viewed as a black box by other systems so that there is no dependence upon its internal working; internal structures must be hidden behind the interface. Consequently, changes of structure within one system through correcting internal errors or through installing enhancements do not affect other systems. If the standard interface anticipates future developments then the member systems may evolve in performance and capacity with minimum effect upon other member systems. In fact, the architecture divides the distributed system into a set of generic systems performing logical or virtual functions. Each member system then has to comply with its generic identity and perform the associated logical functions. It is also required to interface to the logical functions assigned to other systems.

The more detailed design of the combat system is, therefore, concerned with the generation of a set of standards for data exchange among generic member systems. The result is a set of member system specifications invoking data standards, communication standards and specific data exchange requirements to which the member system has to conform. Within these constraints the designer of a member system is free to design his system.

Systems design and data management problems   141

The decision to fit a local area network as a combat system highway to the type 23 frigate was recognised as an opportunity to introduce fundamental architectural changes from which considerable benefit might be gained. The main architectural objectives were the adoption of the principles of layering, generic systems, and shared data (i.e. no system should be expected to output the same data twice in different formats to suit different users).

The local area network adopted Defence Standard 00-19 (Standard B). It has features specifically intended for combat system use, in particular a broadcast mode with selective filtering by data content. The broadcast mode means that every message output to the highway from any system is received simultaneously by all the other systems. In order to prevent systems being overwhelmed by message processing, Defence Standard 00-19 specifies an intelligent terminal unit with a direct memory access (DMA) interface to a host processor. The host processor is able to load message selection criteria into the terminal unit's filter, and it is the terminal unit that handles the highway messages, only passing to the host the subset of messages selected. Another important feature is extensive error detection and recovery procedures that are also handled within the terminal unit.

The adoption of the broadcast mode of working means that messages on the highway can be received by any member system and data can be shared. It also means that the point-to-point addressing can be discarded, which is a major advantage when reconfiguring the combat system. There is not necessarily a one-to-one relationship between the number of member systems and the number of highway connections (called ports or network nodes). Some member systems have more than one port, normally for reasons of redundancy. The absence of point-to-point addressing means that the actual assignment of the ports of a multiported member system to message reception and transmission is quite transparent to other member systems. In other cases, where it has been impracticable to convert an established system to use the combat system highway, a member system with a port may be required to multiplex its own data with that of the non-connected system. However, in general, there is only one port per member system.

Much work of an innovative nature has been undertaken with regard to the generation of standards, specifications and specification procedures. Practical experience has resulted in three separate activities: a standards activity concerned with generating and supporting Naval Engineering Standards (NESs); a combat system highway project concerned with the provision of the communications service; and a data management task concerned with data exchange between systems.

The standards activity has been responsible for the generation and adoption of a number of standards which are mandatory for users of the combat system highway on the type 23; these will include the command system and the majority of the weapon and sensor systems. This paper will not describe the various standards that were developed and utilised as this has been done elsewhere (see Section 5.1 of this volume; and Hill and Richards, 1985). It is, however, useful to consider an overview and then to illustrate what was found to be required.

142 Systems design and data management problems

The first standard developed was Naval Engineering Standard (NES) 1026 (Standard C) which is a subset of NATO Standardisation Agreement (STANAG) 4222 (Standard D). This standard is concerned with data representation. Its purpose is to define the units to be used, the orientation of co-ordinate systems, and the formats for representing fixed and floating point numbers, discrete quantities and alphanumeric elements. The standard is mandatory for members of the combat system, for their data exchange with other member systems. The representation standard used internally within a member system is entirely the province of the member system designer, although there are obviously advantages to be gained in using the same standard internally as well.

NES 1026 does not define the actual messages. This is the purpose of NES 1028 (Standard E), which consists of three parts:

(a) A message definition language
(b) A message catalogue containing a standard message set for use by individual member systems
(c) The rules and protocols for data exchange.

The message definition language is the formal way in which the individual messages of the catalogue are expressed, and the message formats can be checked by a verification program. The messages in the catalogue are independent of the underlying communications medium. New systems joining the combat system are expected to use the catalogue messages whenever possible; otherwise the generation of a new catalogue message is considered. The rules and protocols for data exchange stipulate the way in which the member systems achieve time synchronisation and the way in which communications status is to be reported.

The third standard produced, NES 1024 (Standard F), has been developed to provide member systems with an embedded communications interface standard, embracing both hardware and software, which is implementation independent. The general principle is that of an intelligent terminal unit and a host computer which communicate via shared memory. The terminal unit is the highway port; it consists of two parts, the host adapter unit (HAU) and the communications unit (CU). The HAU is a simple logic unit interfacing the host DMA bus to the NES 1024 electrical interface. The communications unit contains a microprocessor which handles message assembly, checking and recovery, and the connection to the local area network. For a given network all member systems can use a common CU design, but the less complex HAU has to be designed for each host computer type. This approach means that only a single CU design has to be supported; it also means that the introduction of a new local area network could be effected by redesigning the CU card only and issuing new cards to member systems.

The combat system highway project is concerned with the provision of interface cards to the various member system projects. The card is a communications unit performing the Defence Standard 00-19 protocol and presenting an NES 1024 interface to the host processor HAU. Each individual member system is responsible for the design of its own HAU. Although for individual member systems there may

Systems design and data management problems 143

be mechanical differences in the final assembly of the CU cards, homogeneity has been achieved in terms of electrical/logic design.

The data management task has been concerned with the integrity and coherency of the combat system design and with ensuring that data on the combat system highway is used in the most advantageous manner. It has also been concerned with the generation of combat-system-wide policies where otherwise member systems would perform the same function in different ways (e.g. track numbering, track table initialisation). It has involved the documentation of the combat system data flow and the generation of data exchange specifications (DESs). The DESs specify that the data exchange between the individual member systems in terms of NES 1028 messages. The specification format and procedure have been specifically designed to accommodate multiparty agreements for shared data.

Figure 3.2.5 shows how these various activities, specifications and standards are combined in the total system design process. The combat design is a data management activity which produces the data exchange specifications (DESs). These specifications take into account the relevant messages standards (NES 1026 and NES 1028 part 3).

The DESs are a collection of the multilateral and bilateral data exchange agreements necessary to perform an identifiable function within the combat system. The DES should define all aspects of the data — message structure, message fields, message rates, application protocols, message sources and message sinks. The DES is solely concerned with the data exchange to achieve a function; it is not concerned with the total message flow for a given member system, or with the communications protocols necessary to convey the messages.

An individual member system may be involved in more than one function and will therefore be involved with more than one DES. The DESs may not be totally independent on one another, but they should be organised so that the degree of coupling is loose.

As Fig. 3.2.5 shows, the DESs are used to formulate the communications requirement specificiation for the combat system highway and they are used to contribute to the interface specifications for individual member systems.

The combat system highway design process also provides an input to the member system's interface specification, and the process is associated with the communications standards NES 1024, NES 1028 part 2 and Defence Standard 00-19. This contribution to the interface specification is concerned with electromechanical aspects and communications protocols. It is concerned with interfacing to the highway itself and not with the relationships between member systems, although it is concerned with the total message traffic through the member system's highway port.

With all the preliminary design documentation and standards in place, the member system interface specification can be produced. This is only part of the member system's overall specification, which would also include performance requirements, man—machine interface requirements etc. When the member system specification is complete, its design can commence.

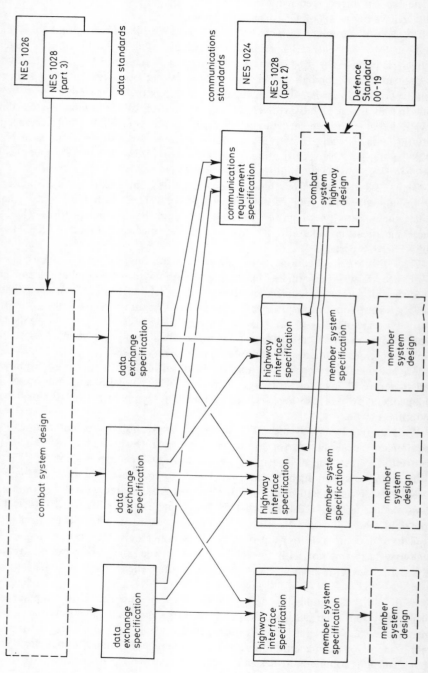

Fig. 3.2.5 The design process

Systems design and data management problems   145

The preceding paragraphs have dwelt upon the design process and its implications for realising the benefits of a local area network. It is worth noting that the adoption of a local area network provides the opportunity for building a portable programmable simulator which can be used to simulate both the highway and other member systems. Such a device can provide a much greater level of confidence in factory acceptance testing, for it allows the policing of standards and the initial testing of data exchange to be undertaken prior to a member system's departure from its factory. The simulator has further application in both a shore-based combat system test facility and in setting the shipborne combat system to work. Although not considered in this section, the actual testing philosophy for the individual member systems and the combat system itself must be developed as an integral part of the design process. The progressive testing from factory through shore development facility to first-of-class ship should result in a considerable reduction in setting to work in the final stage (i.e. aboard ship).

3.2.6 Future development

The first part of this contribution has been concerned with a distributed combat system architecture where the member systems are relatively loosely coupled. In order to communicate, the member systems have to conform with a set of communication standards and a set of data representation standards. The data management task is concerned solely with the messages that are passed between and shared among these member systems. A common language is used between the member systems but, once a message has been received, or before it is generated, it will be in a representation that is under the control of the member system designer. Each member system designer defines his own internal data structures (e.g. a track table within a sensor system). Such are the characteristics of the member systems which comprise today's heterogeneous distributed systems. Currently, combat systems of this type are being implemented for the Royal Navy. It has been possible to introduce a degree of homogeneity by the provision of a common communications unit design to the various contractors supplying the member systems. This has been made possible by the concepts incorporated in NES 1024.

Where the linkage between the member systems is fairly loose and the member systems are largely autonomous, the level of standardisation achieved to date is adequate (i.e. standards for communications protocols and message data content). However, where the coupling is closer and the elements of a distributed system are more integrated, further benefits can be realised by the introduction of integrated database management, and this is the second topic for this contribution. The results of research presented here are a natural development of the distributed combat system architecture now being implemented for the type 23 frigate, and distributed database management could well be implemented in a combat system of the future.

3.2.7 Distributed database management

Already there are some elements of the combat system which are distributed systems in their own right and whose degree of integration is reaching the level where the distributed database management approach is worthy of consideration. It is the command system element of the combat system which is initially being considered for integrated database management. Moreover, as the trend is to a distributed architecture based on a local area network, the need is for distributed real-time database management.

The centralised command systems of the past had one major benefit from the data management point of view: they were either up and working, or down, having failed through either equipment malfunction or action damage. The local area network, on the other hand, although opening up many opportunities has also brought many problems. Parts of the system can be up while other parts are down; although this is a benefit from the system point of view, it makes the integrated management of the database very much more difficult. The network also brings the opportunity to replicate the database for greater survivability, and to ensure local access at each node — but at the cost of further complicating database management.

The *ad hoc* database management solutions of the past were reasonably workable (although extremely resistant to later change) because the database was centralised and also because it was small and simple. However, since those days the users' expectations have moved on. The databases of systems currently being implemented are approaching the order of megabytes and for the future they will be moving into the 10–50 megabytes range.

The following discussion considers a database approach to shared data management over a closely coupled network of distributed processors, such as are found in the command system element of the combat system. The command system, which typically consists of multiple processors and intelligent displays, is a network of processing nodes closely coupled by virtue of the data they process and share. The nodes of such a command system network will comprise, in terms of data passed between these nodes, data producers and data consumers; it is possible to clearly define the data exchange requirements of such a network.

Difficulties arise, however, when such a network is implemented. Each processor which is attached to the network and is consuming network data will have to be aware from which nodes its operational data is available. Each processor will have to either poll each supplier node on a point-to-point basis for information, or assume it will be contacted when significant data is available. In the first case the consumer node, by frequent polling, is creating unnecessary load on itself and the supplier. In the second case, each supplier of data has to be aware of the particular needs of each consumer. The only other alternative is for each supplier to broadcast all updates to the network at some base level of significant change, creating a load on each consumer node that is much higher than necessary.

This dependency on node location and highway standards has another significant drawback. It reduces the evolutionary capacity of the command system, making it difficult to add further nodes without rewriting the message catalogue and reducing overall system performance.

It is considered that a solution to these practical problems of the management of distributed data can be offered by the adoption of logically centralised database

Systems design and data management problems

management facilities at all nodes of the distributed command system, and by placing these facilities in a closely coupled but separate system element attached to each node and to the local area network as indicated in Fig. 3.2.6.

This approach gives the benefits of

(a) Providing a standard interface to all active system processing nodes (i.e. data producers and consumers) on the network
(b) Supplying only the data actually required by the node as a consumer
(c) Accepting all data the node has to give as a supplier.

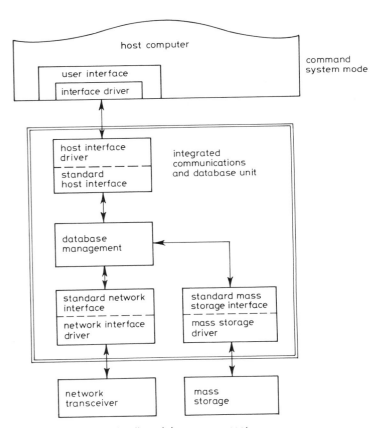

Fig. 3.2.6 *Architecture for distributed data management*

The achievement of these goals will give enormous benefits for current systems, in terms of dramatically reducing the load on current command system processors, and at the same time providing an approach to data networking which leaves the procurement authority completely free of the interfacing peculiarities of individual component systems attached to the network.

148 Systems design and data management problems

3.2.8 Maintaining the integrity of distributed data

In order to provide a database solution to the management of shared data on a distributed system at real-time rates, many major problems must be solved. The main problems for distributed data management arise from the need to replicate data at different nodes of the network for reasons of fast local access and survivability and to keep these replications in a consistent state while coping with updates arriving at real-time rates. Data integrity problems under these conditions can be summarised as:

(a) The need to maintain the consistency and synchronisation of multiple copies of data resident at different nodes during and after the application of updates to the replicated data

(b) The need to maintain the consistency of data having a fixed relationship to other data held on the database

(c) The need to cope with simultaneous update attempts from different nodes to the same data

(d) The need to allow multiple updates to be applied to the database in one indivisible operation, where such a multiple update is the only means of taking the database from one consistent state to the next.

Consider for instance a simple case where three copies of a data item occur on three nodes of the network and updates to this item originate from a fourth node. We need a means to ensure that the update which is transmitted by the fourth is received by the other three and is actioned by them. Now the communications medium may attempt to give us guarantees of delivery but, if we are in a broadcast mode, those guarantees can never be absolute where the network runs asynchronously with regard to the nodes attached to it.

Another example of the problems of multinodal databases is the cases where an update to the database requires changes to be made to more than one element which may reside on more than one node. It is possible that the originator node could fail part way through the update leaving only part of the database changed. In this case we have to be able to roll back the changes that have occurred in order to reinstate a self-consistent database.

Simultaneous updates can be a serious problem in replicated databases if not handled properly. Consider if we have two copies of a data item C_1 and C_2 residing on different nodes and two transactions T_1 and T_2 running on different nodes. If T_1 accesses C_1 and T_2 accesses C_2 and they each update them respectively then, when the updates are broadcast on the network, the second one to be broadcast will overwrite the first.

In order to begin to solve these problems we must start with the intention of having an overall agency responsible for managing all shared data. Notice we say *all* the shared data. Data that is private to only one node does not need to be distributed and replicated, providing that node is both source and consumer.

The problems themselves have been the subject of considerable research over the last few years and much has been published regarding their solution (Report A,

Systems design and data management problems 149

1979; Neuhold and Walter, 1982; and Report B, 1981 all describe partial or total solutions). However, the perspective of these authors is somewhat removed from the requirements of real-time military C^3 systems. They have all been concerned with the more conventional, land-based, geographically distributed, large, disk-based systems. The fact that there are large amounts of disk-based data has usually led to very complex protocols for the temporary staging of updates on other nodes when the ultimate destination node is down. This, and the fact that they are often based on slow telephone networks, has meant that these systems have been large, cumbersome and rather slow.

In the distributed information architecture for ships (DIAS) project we have taken a rather different approach by setting out from the beginning to identify the principal requirements of military real-time applications and by being prepared to compromise on some of the less important issues. Thus the trade-off between availability and reliability has favoured availability. The trade-off between speed and reliability has produced solutions for both, to be used as appropriate. The major lesson of our research has been that 100% reliability all the time is very expensive in terms of computer and communications resources; it lengthens response times and it can cause the closure of the system during recovery periods.

3.2.9 Performance

To realise the full potential of adopting the database approach for current systems in practical terms, two questions of performance need to be addressed:

(a) Does this approach allow the reduction of processor load on network nodes?
(b) Does this approach allow for the simple evolution of current networks to allow further functionality and performance to the limit of network bandwidth?

The answer to both these questions is yes. The removal of the data management load from the host processor may be accomplished by removing data maintenance to a separate processor attached to the host, as discussed in Section 3.2.7. This architecture is described in detail in Moore (1986). The host processor has only to specify for the attached data management system the data that it wishes to receive from the network, and only supply to the data management system the data that it requires to communicate to the network. The data management system can also take over the load associated with filtering out significant updates from all the data arriving at the node by allowing the host to instruct the data management system to pass back only significant data updates as defined by the applications processes running in the host processor.

On the issue of system evolution, the database approach will allow the addition of further nodes to the network without the need to define new messages for the network message catalogue. This catalogue, for the database-controlled network, will consist purely of internal database communications messages controlling the integrity of the networked database and the transmission of data updates between nodes.

Systems design and data management problems

3.2.10 Standards

It is relevant to examine how the data exchange standards for the combat system map on to the proposed database technique. It can be seen that the message catalogue defined in NES 1028 would be replaced by two new standards:

(a) A message catalogue to support the data interchange messages required for database management system internal protocols
(b) A high-level schema definition for the networked database itself.

NES 1026 would remain as it is, as would NES 1024. The importance of NES 1024 is as a standard to enforce uniformity on equipment suppliers at the network interface level. This brings the benefits of independence of network standards. Similar communications interface standards will need to be generated for a communications and database management processor which would be standard for all applications. However, within many current command systems, applications software has no uniformity, typically cannot be exchanged between different manufacturers, and is highly node dependent. The importance of the database approach is that it can provide an approach to networking which allows software reusability and node independence by adopting a database interface standard in terms of a high-level language such as SQL, for which draft ANSI standards are already being formulated (Standard G, 1985).

Now the power of the database solution becomes more apparent. Any applications program for such a system can be developed using a commercial database manager with a SQL interface and with a global schema based on an analogue of the NES 1028 message catalogue.

The database approach will bring the additional advantages of being able to enhance network functionality by a change in a high-level database schema, rather than a lower-level message catalogue. High-level data definition facilities such as those described in Moore (1985) would be available to change the schema, alter data access characteristics as required, reload and reinitialise the network database in one simple operation.

3.2.11 Data definition

Work has already commenced on the definition of a data dictionary for naval command systems. This has proceeded along the lines of a formal entity/attribute/relationship analysis and is published in Tillman *et al.* (1985). This type of analysis provides a powerful unifying feature in the design of an integrated command and combat system, by laying the groundwork for a system-wide data dictionary to support future command and combat system evolution. It is expected that this approach will encompass and supersede the current approach offered by NES 1026 and 1028.

3.2.12 ADDAM (ARE distributed data management system)

The results of our work have so far produced a research prototype which is modest in its size and yet sophisticated and of high performance. It is not the intention in

Systems design and data management problems

this paper to describe the ADDAM system in great detail, that was originally done in Tillman, 1982 and Reilly et al., 1984. However, some features will be described, particularly with regard to the problems already identified in this paper.

ADDAM supports a distributed, partitioned, replicated and dynamically reconfigurable database which is available at real-time access rates via a simple user interface. Data is distributed around the local area network on which ADDAM resides. Any node in the network is able to access any data on the network. ADDAM provides a user interface which isolates the user from awareness of whether a piece of data is held locally or acquired remotely.

ADDAM's partitioning feature allows the whole of the database to be divided up such that each node in the network holds only a subset of the totality of data. Its interface also isolates the user from any concern that a set of accesses may be satisfied from more than one node. The replication features allow several copies of the data to be stored on the network in order to provide backup in the case of node failure and also to provide local access to reduce network overheads.

The dynamic reconfiguration allows the data to migrate around the network as needs arise. This can be to reflect the changing functional distribution caused by the reconfiguration of nodes or recovery actions taken to offset nodal failure.

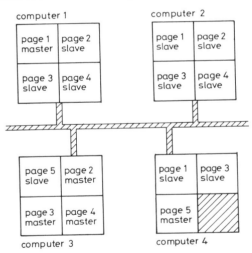

Fig. 3.2.7 *Partitioning and replication in ADDAM*

The database is divided into a number of pages, where each page is a collection of records of the same type. The page is the unit of partition within the database and a particular page is either wholly present or wholly absent with respect to a particular node on the network. The page is also the unit of replication; therefore the same page may occur at several nodes. Figure 3.2.7 illustrates this. Where a page is replicated, one copy exists at any one time as a master copy and is known as the owner, and the other copies are treated as subordinates and known as subscribers. This arrangement is designed to support synchronisation.

152 Systems design and data management problems

The database is maintained internally by a number of protocols known collectively as the control protocols. This purpose is to maintain synchronisation and consistency between the multiple copies of the database stored in the various partitions, to ensure that the number of replications required for each page actually exist, and to ensure that each page has one and only one owner.

The RELIABILITY protocol provides a reliable mechanism for the propagation of database changes to all copies of the relevant pages. It is used to support a number of application facilities including CREATE and DELETE record and the RELIABILITY option of UPDATE. It also provides a concurrent update prevention facility which ensures that two or more processes running on the network are not able to concurrently update the same record occurrence, or create records that have the same key.

This protocol is relatively expensive in terms of response times and message traffic. It has three distinct phases for each update transaction — a distribute phase, a commit phase and a completion phase. It does, however, provide an almost perfect guarantee that a transaction is consistently committed to the database, but at a price.

For those data updates which are independent of the previous value of the data and which are applied at frequent intervals, ADDAM provides a second protocol called the PERFORMANCE protocol. This does not provide the total guarantee of synchronisation of the RELIABILITY protocol, as it relies solely on the highway broadcast facility with no higher-level error checking. The benefits it provides, however, are significantly reduced message traffic and shorter protocol path lengths. It is particularly designed for the sensor update type of application.

In addition to these two protocols there are others which control the propagation of replications, the creation of new owners, the recovery from highway failure and the other various needs of data management.

The research work with ADDAM has shown that the management of real-time data distributed around a network is perfectly feasible. Responses can be achieved in time frames meaningful to the operational system and the cost can be kept to reasonable levels in terms of processor and memory usage. In prototype forms ADDAM has achieved up to 200 updates per second and up to 110 reads per second on a 0·7 MIPS processor. Further study has convinced us, however, that ADDAM's overall performance could be improved by a factor of between 30% and 100% by a combination of tuning and the redesign of some structures to more closely suit particular applications. A detailed analysis of the achieved performance is published in Murrell (1984a) and a more particular comment in a current command system context is contained in Murrell (1984b).

The current laboratory configuration of ADDAM uses a single Ferranti Argus computer to support both applications and database functions at each node of a distributed network, and this prototype ADDAM has proved the feasibility of the database approach for the solution of current problems associated with improving the response and flexibility of current command systems. The development of a physically separate integrated database management and communications system

Systems design and data management problems 153

and the generation of an associated set of interface standards is the logical culmination of the DIAS programme, and is expected to bring significant benefits in the utilisation of local area network architectures for C^2 systems. The adoption of such an approach would introduce a greater degree of homogeneity to the distributed system than has been achieved in the systems being built today.

3.2.13 Conclusions

As was said in the introduction, the fundamental effects of introducing local area networks will affect both the technical and organisational aspects of design.

At the technical level there are problems to be solved. The current work on the type 23 frigate combat system has provided solutions in terms of standards and procedures to achieve data exchange between independently designed heterogeneous member systems. This work has also shown how homogeneity at the communications level can be achieved by the provision of a common communications unit design which is mandatory for all member systems. The next problem to be solved is considered to be that of database management. The time is now right for the introduction of generalised database management packages into military systems. They have been accepted in the commercial marketplace for a number of years. However, investigation has shown that the available commercial products are totally unsuited for military real-time applications, and essentially that is the reason why the ADDAM work was undertaken.

ADDAM of course was aimed not just at generalised database management but at distributed database management. The work has shown that it is perfectly feasible to achieve a compact, efficient, real-time database management system although the development costs of doing it are quite high. To make it cost effective, it is important that the solutions developed are flexible and are designed to be generally applicable in the years to come.

At the organisational level the main conclusion to be drawn is that much more effort will have to go into 'up-front' combat system design. The tendency in the past has been for the navy to decide upon the weapons and sensors and then let a bespoke command system contract to tie it all together. To gain the full benefits of the local area network the combat system must become a design entity in its own right. The combat system design and its associated standards must be well established before the procurement of its member systems commences, and compliance with both the design and the standards must be enforced during that procurement. In effect, either the navy must become the design authority for the combat system or it must contract out that authority; what it cannot do is to leave it to chance during the procurement of the component parts of the combat system.

Consideration of the work in the implementation of a combat system and the research into distributed databases leads to another important conclusion. That is, there might eventually be a conflict between the development of communication standards and protocols and the introduction of distributed database management techniques. This would be because member systems would be interfacing to the database rather than to the communications network. The traffic on the communi-

154 Systems design and data management problems

cations network would be governed by the database design rather than by the data exchange requirements of the group of component member systems. Interfacing to the communications network would be internal to the database design. Although a structured and layered approach would still be necessary (NES 1024 could still be used), it would only be known to, and be the responsibility of, the database designer. All member systems would have to conform with the interface standards and data representation standards of the database. The current data exchange standards such as the NES 1028 message catalogue would be superseded by new database interface standards for those member systems of a combat system adopting the distributed database approach. The first combat system incorporating a distributed database may well be a partial implementation with only a subset of the member systems adopting the new approach. Under such circumstances the NES 1028 standard would be the method of communication between the distributed database and the older systems and between the older systems themselves. As long as there are any member systems who have not adopted the distributed database, there will be a need for a message catalogue, and the NES 1028 standard will coexist with the database standards.

The first part of this contribution dealt with the member systems of the combat system as largely autonomous systems for which communications standards and data exchange standards were adequate to realise the benefits of the local area network. This is firstly a historical perspective, for the individual systems have always previously been considered in this way, and secondly the concept that we can actually implement today. However, the distributed database research indicates that the elements of the combat system are not so autonomous and that they all retain much common data which could be shared.

Furthermore, the nature of the command system needs to be reviewed, for it is fallacious to consider the traditional command system to be a single self-contained member system of the combat system. The traditional command system is nothing more than a collection of combat system functions that had to be grouped together in the centralised computer era. If this view is accepted then the elements of the traditional command system become individual members of the combat system at the same level as the sensors and weapons. This reinforces the conclusion that the combat system must be viewed as a single integrated system with a single integrated, albeit decentralised, database.

Finally, these conclusions highlight two general errors that have been made in military circles during the last decade:

(a) That individual systems which are to form part of a large distributed system can be procured without considering the total system design
(b) That the solutions to data sharing requirements lie in the communications field.

Complex C^3I systems cannot be readily aggregated out of components that were not designed to work together, as is demonstrated by experience in the quest for interoperability. The total system must be conceived as a whole and broken down into its component systems and their communications mechanisms. It must also be

Systems design and data management problems   155

remembered that communications is only a transport mechanism; it is not responsible for the interpretation of data content. Until it is accepted that integration and interoperability require a total system architecture accompanied by database management approach, serious difficulties will continue to be encountered with distributed C^3 systems.

3.2.14 References

HILL, J. S. and RICHARDS, F. A. (1985) 'Standards for naval combat systems', IEE international conference on advances in command control and communication systems theory and applications, April

MOORE, R. B. (1984) 'Study report on task 12 — system layering', DIAS technical report DIAS/TR16, Admiralty Research Establishment, XCC Division, November

MOORE, R. B. (1985) 'High level data definition facility — a route to implementation', DIAS working paper D473T/TN159, Admiralty Research Establishment, XCC Division, March

MURRELL, J. G. (1984a) 'ADDAM performance measurements final report', DIAS technical report DIAS/TR14, Admiralty Research Establishment, XCC Division, October

MURRELL, J. G. (1984b) 'ADDAM and CACS4', DIAS working paper D473T/TN125, Admiralty Research Establishment, XCC Division, May

NEUHOLD, E. J. and WALTER, B. (1982) 'An overview of the architecture of the distributed database system POREL', Second international symposium on distributed databases, Gesselschaft fur Informatik, Fachausschuss 5/7, West Berlin, September

REILLY, M., TILLMAN, P. R., CATT, R. H. L. and MOORE, R. B. (1984) 'Maintaining consistency and accommodating network repair in a self-regenerative distributed database management system', AGARD guidance and control panel 39th symposium, CESME/IZMIR, Turkey, October

REPORT A: Computer Corporation of America (1979) 'A distributed database management system for command and control applications (SDD-1)', semi-annual technical report V, CCA-79-23, CCA, Massachusetts, USA, June

REPORT B: International Business Machines (1981) 'An overview of the architecture', research report RJ3325', IBM Research Laboratories, San Jose, December

STANDARD A: International Standard ISO 7498 'Information processing systems — open system interconnection — basic reference model', International Organisation for Standardization

STANDARD B: Defence Standard 00-19/Issue 1 (1981) 'The ASWE serial highway', UK Ministry of Defence, Directorate of Standardisation

STANDARD C: Naval Engineering Standard (NES) 1026 (1984) 'Requirements for the digital representation of shipboard data parameters', Ministry of Defence (Procurement Executive), Deputy Controller Warship Equipment

STANDARD D: NATO Standardization Agreement (STANAG) 4222 'Standard specification for digital representation of shipboard data parameters', Military Agency for Standardisation

STANDARD E: Naval Engineering Standard (NES) 1028 (1985) 'Part 1: High level language for message construction', 'Part 2: Rules and protocols for inter-communication across digital highways', 'Part 3: Standard messages for inter-communication across digital highways', Ministry of Defence (Procurement Executive), Deputy Controller Warship Equipment

STANDARD F: Naval Engineering Standard (NES) 1024 (1985) 'Part 1: The software interface', 'Part 2: The electrical interface', Ministry of Defence (Procurement Executive), Deputy Controller Warship Equipment

STANDARD G: 'Draft proposed American national standard database language SQL', American National Standards Institute (ANSI), February

TILLMAN, P. R. (1982) 'ADDAM — the ASWE distributed database management system',

Systems design and data management problems

Second international symposium on distributed databases, Gesselschaft fur Informatik, Fachausschuss 5/7, West Berlin, September

TILLMAN, P. R. and CHEESEMAN, A. S. (1985) 'Systems design and data management problems in the utilisation of local area network architectures in C^3 systems', IEE international conference on advances in command, control and communication systems theory and applications, April

TILLMAN, P. R., IRVIN, K. and SENIOR, E. (1985) 'A maritime tactical data analysis', NGCS/TR15, Admiralty Research Establishment, XCC Division, July

4: Communications

Chapter 4.1
Packet radio: a survivable communications system for the forward area

B.H. Davies and T.R. Davies
(Royal Signals and Radar Establishment)

4.1.1 Introduction
A major problem in providing command, control and information systems for the far forward tactical environment is the difficulty of providing secure and survivable communications. This difficulty is due to the fact that the environment is one which, by its hostile and mobile nature, cannot support a communications infrastructure of base stations and unattended repeaters. The difficulty is further exacerbated by the fact that the communications medium is shared with the enemy and flanking forces.

To date, information in this environment has almost exclusively been exchanged by voice over combat net radio (CNR). Much of this traffic is analogue voice using narrowband frequency modulation occupying 25 kHz channels, and there is substantial use of digitised voice (16 kbit/s continuously variable slope delta modulation) occupying 25/50 kHz channels. A major trend in forward area communicatikns is the increasing use of data rather than voice, for two reasons. Firstly, many of the new computer-controlled weapons systems consist of distributed elements (sensors, command and control posts, and weapons platforms) which require accurate and timely data communications to function effectively. Secondly, there is only a limited amount of the radio frequency spectrum (HF and low VHF) that is capable of supporting mobile tactical communications without the use of a complex mobile repeater infrastructure, which would be expensive in terms of men and logistic support. Consequently, this part of the spectrum is grossly over-subscribed. In this context, data systems have a very much higher information throughput per unit of bandwidth than voice systems. Thus there is considerable pressure to transport as much information as possible by data, thereby increasing channel availability for essential voice command and control. A further consequence of the frequency congestion in the forward area is a reduction in the effective range of the radios.

These problems, coupled with the scale of procurement for the forward area, have given rise to the following characteristics being required in the next generation of communications systems for this environment:

1 More efficient use of the RF spectrum
2 Ability to operate in reduced radio ranges
3 Minimum operational overheads
4 Interoperability with other battlefield data systems
5 Affordabillity.

The narrowband packet radio system described in this paper satisfies all five of the above requirements. In summary, because packet radio is a data communications system, it can accommodate greater numbers of users and transport greater amounts of information than an equivalent digitised speech channel. Packet radio copes with reduced radio ranges by delivering packets in a number of transmission hops. It is a self-configuring system which automatically adapts to user mobility and changes in radio connectivity, thus relieving users from having to make frequent radio link engineering checks. It is specifically designed as one element in an integrated electronic battlefield. Finally, the basic combat net radio can be converted to provide packet radio facilities, in addition to those it provides at present, by the addition of a functional module based on signal processing and microcomputer chips. Thus, with advances in VLSI technology the costs of this radical new communications system will be an incremental increase in radio equipment costs.

This paper describes the design and development of such a narrowband packet radio system including early experiences with a small prototype network. Section 4.1.2 describes the basic concepts underlying the application of packet switching techniques to the CNR channel, including the philosophy of the architectural approach. Section 4.1.3 describes how the system appears to the user and the relevant protocols. A detailed description of the channel access, routing and network control algorithms which were developed using a finite state machine simulation is given in Section 4.1.4. Section 4.1.5 summarises typical average performance for various scenarios. Section 4.1.6 describes the signal processing requirements and architecture adopted for the first prototype units, while Section 4.1.7 highlights the main lessons arising from early experiences with these units. Finally, pointers for future work and a summary are given in Section 4.1.8.

4.1.2 Basic concepts

In a CNR packet radio network a number (2 to 50) of microprocessor-controlled radio transceivers, equipped with burst data modems, share a single narrowband channel. The transceivers operate in simplex mode and all transmit and receive on the one channel. The channel itself may or may not be frequency hopped: that is an ECCM issue outside the scope of this paper. The bit rate of the modulation is 16 kbit/s, exactly the same as that employed by digitised speech users, thus making it suitable for use with existing radio transceivers and frequency allocations.

Packet radio: a survivable communications system

All units demodulate and analyse all the transmissions that they receive, and source-to-destination transportation is provided by a store and forward action typical of all packet switched networks. The packet radio units automatically configure themselves so that each is aware of the minimum number of transmission hops to each available destination. This permits automatic and efficient relaying of packets, as illustrated schematically in Fig. 4.1.1.

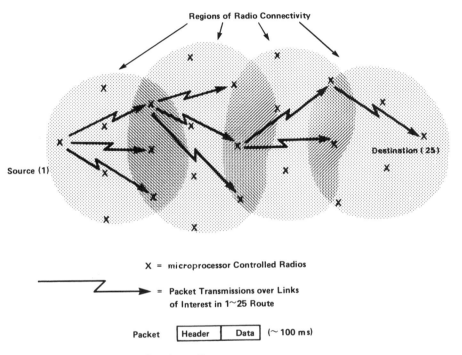

Fig. 4.1.1 *Basic concepts of packet radio*

4.1.2.1 Fully distributed architecture A number of features of the CNR channel, coupled with a primary design aim of survivability in a rapidly changing and hostile electromagnetic environment (EME), has lead to a fully distributed approach to the network architecture in which each unit is functionally identical and there is no central control station. The main task of a control station in a packet radio network would be to obtain connectivity information, calculate routes and distribute this routing information to active users. However, the use of a control station, which for efficiency reasons would have to be centrally sited, could constitute vulnerability. In particular, the control traffic would tend to act as an electromagnetic pointer to the central station, making it the focal point of physical or electromagnetic attack, in addition to chance destruction or failure. Survivability of such a system could be improved by the use of a number of control stations. However, multistation control is inherently complex, resulting in more control protocols which in turn make more demands on the limited channel capacity. Fortunately, the basic parameters of the CNR packet radio system, in which there are a maximum of 50 units and in which no more than 10 packets per second can appear on the channel, lends itself to easy and efficient distributed implementation. A further advantage of the fully distributed approach is that local communications can continue unaffected should the network become bifurcated into two or more collections of units.

4.1.2.2 CNR packet radio in context of battlefield data communications There are a number of different types of radio channel that have different applications on the modern battlefield; these include the high-capacity DARPA packet radio (Kahn et al., 1978), packet satellite (Jacobs, et al., 1978), channels of various capacities and the CNR system described in this paper. These systems are not in any sense competing with each other as they offer differing types and grades of service in differing areas of the battlefield. They all have one particular hallmark, namely that they can interoperate with each other, without modification, given a suitable internet architecture. Although it is conceivable that these networks could adopt a common network access protocol for a given type of service, the fact that the basic communication resource that is being shared in these networks exhibits very different characteristics means that the basic intranet protocols of channel access, routing and network control will all be different. If a channel is to be used efficiently under a variety of conditions then these intranet protocols must be matched to communications channel(s) being used — a fact highlighted by Kleinrock (1978).

The narrowband CNR channel differs in a number of ways from those used by other types of other packet switched networks:

1 The links in a CNR PR network may be subject to very high packet loss rates owing to the inherently high bit error rate of the mobile tactical VHF environment and to transmission collisions between neighbouring units (hidden terminal problem).
2 The semi-broadcast nature of the CNR packet radio environment, in which a number of possible relayers receive a packet simultaneously, is unique to ground-based PR systems.
3 The all-mobile nature of the units, coupled with the possible rapid and dramatic changes in the EME, mean that the network designer has almost no control over the topology of the network connectivity.
4 The usual mathematical tools for modelling networks assume that they are large enough for the laws of large numbers to apply, permitting assumptions such as independence of time of arrival, queue length etc. at each node. This is patently untrue for this very small channel.

The final feature necessitated a heuristic approach to algorithm design, because we did not find it possible to generate a realistic mathematical model which had

Packet radio: a survivable communications system

tractable analysis. In particular, because of the small size of the channel and possible rapid changes in connectivity, we had to investigate the behaviour of the network under non-steady-state conditions, and this was only possible using simulation techniques. The design has evolved through and been tested by extensive computer simulations using a finite state machine representation with the controlling algorithms operating independently for each modelled unit. Despite the heuristic nature of the development, much reference has been made to the literature (Kleinrock and Tobagi, 1975; Tobagi and Kleinrock, 1975, 1977; Leiner, 1980; Boorstyn et al., 1985) on the analysis of similar but simplified systems, and many aspects of the design are practical extensions of well researched techniques.

4.1.3 User services

4.1.3.1 Type of service A packet radio network is capable of providing a number of types and grades of service. We have considered five types of service:

Unacknowledged datagram Only requires hop-by-hop acknowledgment and is suitable for sensor information where timeliness of delivery is more important than reliability. This is the baseline service, inherent in the relaying strategy of the routing algorithm. All other services are build up on this baseline.

Acknowledged datagram Suitable for highly formatted messages which can be contained in one packet. Currently the unacknowledged and acknowledged datagram services are built into the terminal handler, which is accessed via an RS 232 interface.

Multi-address datagram The exact nature of this service is still under discussion. Basically it provides an acknowledged datagram delivery to a primary user, and at least one transmission illuminates those units on the multi-address list. It seeks to emulate the all-informed nature of current CNR operation, should this be required.

Virtual call service Provides a guaranteed delivery of ordered data. Used for database updating, file transfers etc. Currently this is provided by TCP/IP protocols, sitting above the unacknowledged datagram. In order to permit a number of projects to use packet radio without further software development, an X25 access is being developed in which special transnet procedures are employed to minimise overheads. This approach to X25 is very much in the original spirit of the protocol as an access protocol, with the network free to implement transnet procedures best suited to the particular network concerned.

Internetwork service The capability for communicating with hosts on different types of networks (see Section 4.1.3.3).

4.1.3.2 Grade of service The grade of service is a term used to describe the availability, throughput, average delay, delay dispersion and accuracy of data transportation being offered to the user. Unfortunately, it is notoriously difficult to predict the performance of these parameters even for packet switched systems in benign environments. In addition, the grade of service required varies from one type of user and scenario to another. For example, an artillery fire control system may require rapid delivery of data of less than one second and would therefore have to use a lightly loaded network of no more than few hops, whereas a logistic support unit could accept message delivery times of ten seconds or so and could thus use a fully loaded network. The grade of service in a packet radio network is heavily dependent upon the quality of radio links, which may be very variable. The system described below is designed to make the best use of these links.

It is possible to estimate approximate figures for throughput and average delay given the number of active users and the shareable channel capacity. Discussions with potential users have indicated that the majority of data communications requirements in the forward area can be catered for by the use of medium-speed packet radio nets based on 16 kbit/s channels. The users' choice of 16 kbit/s channels employing the same modulation system as is currently employed for digital voice has a number of advantages. Firstly, the 16 kbit/s NBFM modulation system has a proven effectiveness when used on mobile links with low-to-low antennas. Secondly, the users are familiar with the link performance of this system under a wide variety of terrain and electromagnetic conditions, which will greatly help in making optimum use of packet radio. Thirdly, it is possible to convert a current combat net radio transceiver into a packet radio unit by the use of an appliqué unit described below. Finally, it permits coexistence with current users of the band, greatly easing the introduction of packet radio into service.

4.1.3.3 Internetworking An efficient automatic internetworking capability provides two types of service enhancement. Firstly, it permits fast and efficient access to information which may be held on databases of other networks. Secondly, a richly interconnected internet system can be used to enhance the survivability of communications on a single net. This idea of multiple interconnection of the different communications systems in the tactical environment, in such a manner that automatic and rapid reconfiguration makes best use of the available communication resources, is central to the concept of the integrated electronic battlefield. It depends for its realisation on an internetwork architecture for non-homogeneous networks, such as that developed under the auspices of the Defence Advanced Research Projects Agency in the US (Cerf and Kinstein, 1978). The DARPA internet architecture utilises an internet datagram protocol which is implemented by all gateways.

The DARPA internet strategy is based on a layered communications architecture which is very similar to the more familiar ISO open systems interconnection model, but is specifically oriented for interconnecting dissimilar types of network in a highly survivable manner. The DARPA internet architecture is based on the use of a standard gateway. The internet is thus a super-network in which the gateways play the role of switching nodes and the different types of network are the links between these nodes. The survivability characteristics of this architecture stem from replication of basic networking concepts at the internet level. The cost of such a survivable approach is surprisingly low. Any network can be connected without any modifications. All gateways are identical except for the interface modules to the different types of network, and these modules are identical to those realised on all the host computers connected to that network. However, the host computers that require use of the internet services must implement the internet protocol (DARPA, 1980a) for handling internet datagrams and the end-to-end protocols for realising the types of service indicated above. The DoD standard for the virtual call service is called the transmission control protocol (DARPA, 1980b).

4.1.4 Channel access, routing and network control

4.1.4.1 Channel access Carrier sense multiple access (CSMA) is a robust multi-access protocol which can be enhanced to operate in the semi-broadcast environment, with particular benefit when a good capture effect is available as in FM or direct sequence spread spectrum systems. The usual drawbacks of CSMA are the dependence of the throughput on the carrier sensing time and the catastrophic performance under overload conditions (Kleinrock and Tobagi, 1975).

The carrier sensing time is the elapsed time from the decision by a unit to transmit to the detection of the transmission by friendly users. It is composed of the times to protect the unit's receiver circuits and bring its RF carrier up to full power, to modulate that carrier with intelligent unambiguous identification, and for this information to transit the battlefield and be detected by the other users of the network. It is the ratio of this carrier sensing time to the average packet duration, rather than its absolute value, that determines the maximum throughput. Currently, in practical systems with average packet durations of 100 ms this ratio can be kept to about 8%.

The overload problem is solved by limiting the amount of traffic offered to the network so that the offered load is always less than that associated with peak throughput on the CSMA curves. In fact it is generally desirable to limit the offered load at some point below that at which the maximum throughput occurs such that the delay performance can be improved. This is the basis by which a trade-off can be made between throughput and delay performance. The chosen operating threshold will depend on the type of usage required of the network, and is intended to be a configuration parameter. Also, this limiting has to be intelligent if optimum throughput is to be obtained with varying numbers of users with varying traffic. This intelligent limiting is achieved by distributed control procedures which monitor the channel's performance.

4.1.4.2 Distributed channel access control Access to the shared channel is contention based using non-persistent CSMA (Kleinrock and Tobagi, 1975). This exhibits unstable behaviour and degraded performance at high offered loads, and consequently an important design aspect is the distributed control of overall system loading. Each unit generates its own scheduling events and channel access attempts are permitted only at these instants. There are two types of schedule, which we refer to as 'continuous' and 'special'. The continuous scheduler is dominant and runs at a rate which depends on the observed channel loading.

4.1.4.2.1 Continuous scheduler This produces daisy chained schedules which are spaced by uniformly randomised time intervals in the range from zero to a maximum as specified by the system parameter T_s (Fig. 4.1.2). Its effect is to randomise the time distribution of system input irrespective of user characteristic and to limit the input rate of each unit and consequently the system as a whole. The parameter T_s is determined individually by each unit by an adaptive algorithm on the basis of its observation of the channel history. This algorithm has the following outline.

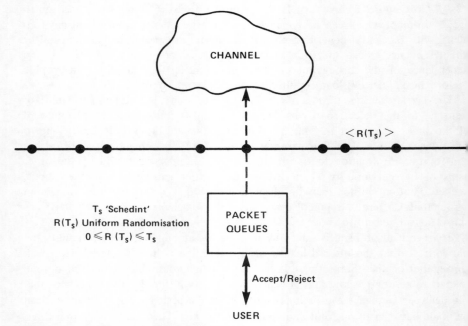

Fig. 4.1.2 *Schematic of continuous scheduler*

Each unit continuously monitors the channel and records the number of cleanly received packets N_r and the number of 'captured' packets N_c (packets resolved from one or more colliding packets by capture switching). The values of N_r and N_c for a suitable integration period are used to calculate the percentage clash ratio:

$$\text{clash ratio} = \frac{N_c}{N_r + N_c} \; 100\%$$

This is compared with a system control constant clash control (found empirically) which corresponds to the desired channel loading. On the basis of the comparison a control action is taken as follows:

1   clash ratio > clash control   action: increase T_s
2   clash ratio < clash control   action: reduce T_s

The value of T_s is confined to fall between preset minimum and maximum limits. The algorithm allows T_s to move towards the required extreme by a fraction of its separation from that extreme (Fig. 4.1.3). This means that the available movement is generally asymmetrical, with greater possible travel towards the further extreme. This feature was found necessary to prevent an alternative but undesirable system state in which a few 'greedy' users took control of the channel with rapid scheduling at the expense of the remaining units with slow scheduling.

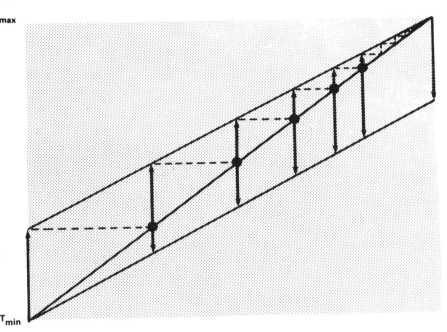

Fig. 4.1.3   *Dependence of adjustment to scheduling interval T on current value*

4.1.4.2.2 Special scheduler This is an event-driven scheduler which is invoked on receipt of packets requiring relay or final hop acknowledgment. Its function is to reduce in-system delays, and consequently the schedule always closely follows the originating reception. If the receipt is at its unique destination then the special schedule is immediate. If it is at one of a number of possible relay units then the schedules are staggered by small offsets assigned in the same order as the ascending unique identifiers of the units concerned. This works particularly well for random traffic patterns, where it has the effect of passing a 'token' between active users. However, for non-random traffic patterns it could lead to hogging between two or more users. (This has not happened in practice because of the significant latency in processing packets.)

168 Packet radio: a survivable communications system

4.1.4.2.3 CSMA distributed control enhancements for multihop networks If a unit has neighbours which are not fully interconnected then it is likely to be required to perform a relay function and handle extra packets. However, because these neighbours are 'hidden terminals' it will also experience an unusually high clash-ratio and the adaptive algorithm will cause its scheduling rate to decrease unnecessarily.

By scanning the routing updates received from its neighbours, a unit can determine the number of possible interconnections that do not exist. It can then characterise their interconnectivity by an integer partition factor, given by

$$\text{partition factor} = \frac{\text{broken links} \times \text{maximum partition factor}}{\text{maximum links}}$$

This ranges from zero for full interconnection to maximum partition factor for no interconnection, and can be used to modify the scheduling rate as a means of correcting for the effect of hidden terminals on the adaptive algorithm. This modification increases the scheduling rate as neighbours become disconnected, as illustrated in Fig. 4.1.4.

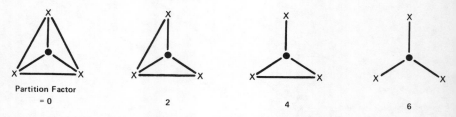

Fig. 4.1.4 *Adaptive scheduling modification for hidden terminal effects*
  N_t total possible neighbour–neighbour links
  N_m number of missing links
  Partition factor $\leftarrow (N_m/N_t)$max. partition factor
  Modification of scheduling rate:
  $T_s \leftarrow T_s/(\text{partition factor} + 1)$

In addition, units at the periphery of a network may not 'see' any collisions because they have only one link into the network. In order to prevent these units from overloading the channel, each unit broadcasts its scheduling interval in the header of every transmission; each unit then uses the greater of its calculated value and the average of the value being used by its neighbours for controlling its main scheduler.

4.1.4.3 Acknowledgment strategy Data packets are assigned unique identifiers (a unique unit identity concatenated with a non-unique local packet sequence number). Acknowledgments are abbreviated and comprise this identifier and the hop count (see Section 4.1.4.7) to the packet's destination from the acknowledging unit. This makes it possible to 'piggy-back' acknowledgments into the first available transmission following the reception of the packets to which they correspond.

4.1.4.4 Retransmission strategy The combination of the special scheduler and the piggy-backing of acknowledgments means that it is possible to support short retransmission time-outs (the first transmission will include the acknowledgment and is likely to closely follow the originating reception).

Apart from imposing a minimum time-out (prevent retransmission before a round trip is possible), retransmission packets are simply queued and their transmissions spaced by the scheduler. The adaption of the scheduler under high load conditions and the additional delays encountered if the retransmission queue builds up both serve as an automatic retransmission control procedure.

4.1.4.5 Transmission packaging To increase the efficiency of channel utilisation, a single transmission may in fact comprise a number of component data packets, a list of abbreviated acknowledgments and possibly a routing control packet, as illustrated in Fig. 4.1.5. Further throughput increases can be obtained by sending acknowledgments more than once, especially when these can be sent instead of padding at the end of an interleaved block.

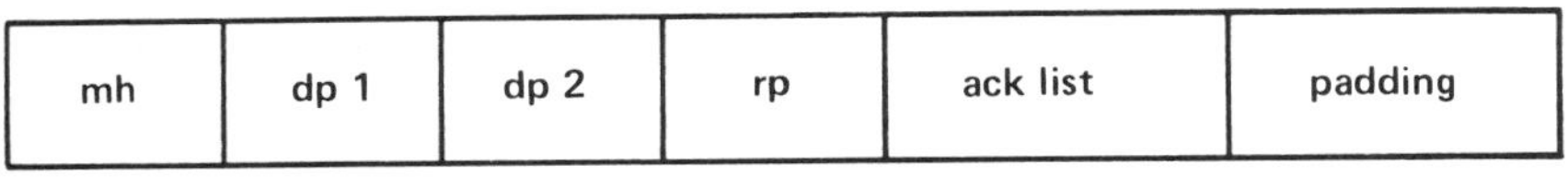

mh	dp 1	dp 2	rp	ack list	padding

mh — main header
dp — data packet
rp — route packet

Fig. 4.1.5 *Transmission packaging*

4.1.4.6 Network control
Congestion control To further safeguard channel stability, it has been found necessary to instigate control actions on the basis of queue lengths. Packets that require handling may either be new user input or be received and require relaying. There are two corresponding controls:

User input control If the total number of packets stored at a unit (including user input that has been accepted and queued but not yet transmitted) exceeds a specified maximum, then further input from the user is rejected.

Relay queue control All units advertise in their transmissions the total number of relay and retransmission packets that they are currently storing. Before transmitting a packet that will require relaying, a unit must examine its record of these numbers for all possible relay units. If they all exceed a specified maximum then the transmission must be postponed; otherwise it may proceed.

4.1.4.7 Routing A basic assumption in our approach of the routing problem is that there are generally a number of units capable of relaying a packet closer to its destination, but, because of the variability of the EME, it is not possible to predict *a priori* which relayer will successfully receive and decode the packet. Thus the routing as well as the channel access is contention based in the manner described below. There are three main parts to the distributed routing algorithm: a traffic forwarding algorithm, a network measurement algorithm and a route calculation and dissemination algorithm.

4.1.4.7.1 Traffic forwarding algorithm The mode of operation of the traffic forwarding algorithm is illustrated in Fig. 4.1.1. In this diagram the crosses represent the packet radio units and the shaded areas define regions of radio connectivity. The arrows indicate the transmission links of particular interest in the delivery of a packet from unit 1 to unit 25. It is assumed that, owing to the operation of the second and third parts of the routing algorithm, each unit has a distance vector indicating the number of hops to each of the remaining units. Two parameters are included in the packet header to aid traffic forwarding: a unique identifier, and the distance from the transmitter to the destination.

Upon receiving a packet not destined for itself, a unit first checks to see if it is a smaller distance from the destination than the last transmitter of the packet. Only if this is the case does the unit put the packet on its relaying queue for subsequent transmission. If a number of potential relayers receive the packet correctly, the first unit to gain access to the channel will cause the other units to remove the packet from their relaying queues (sidestream termination). Each of the relaying stages in Fig. 4.1.1 proceeds in a similar manner until the packet reaches its destination. The transmission at any stage may be used as a hop-by-hop acknowledgment for the previous stage (downstream termination). Some unnecessary relaying may occur with this algorithm if possible relayers are out of range of each other. Any unit detecting unnecessary relaying discards the packet. If substantial amounts of unnecessary relaying occurred in practice, it would be easy to extend the forwarding algorithm to nominate the next 'cluster head relayer', and only those in direct contact with the cluster head would take part in the initial competitive relaying. The cluster head relayer is chosen from the routing tables as being the neighbour with the best quality path to the destination. (This extension to the forwarding algorithm has been implemented in the current prototypes.)

4.1.4.7.2 Network measurement algorithm
The basic network parameter that is measured is connectivity. The successful transmission of packets, in both directions, is used to establish a link. The state of the link is continuously monitored using data and routing update packets.

4.1.4.7.3 Route calculation and dissemination algorithm
From the traffic assignment algorithm it is apparent that each unit has to determine how many hops it is away from all possible destinations. The basic mechanism it uses to determine these distances is that it identifies those units with which it is in direct contact, and from distance information supplied by these neighbours it computes its minimum distances to all possible destinations. This repeated minimisation algorithm is a modified form of the old ARPANET routing algorithm

(McQuillan et al., 1978), and a theoretical analysis including convergence proofs can be found in Abram and Rhodes (1982). The minimum distance vector is broadcast in all routing update packets to the unit's neighbours.

Note that, as for all repeated minimisation algorithms, the only way to avoid oscillation of routing updates when links go down is to employ local initialisation as described in Abram and Rhodes (1982). Unfortunately, this has the effect of declaring the destination to be unavailable for a significant period while the route is reconstructed from the source via other units. Our policy has been to permit some oscillation of the routing updates but to use consistency checking to damp these oscillations. For example, if a unit calculates that its neighbours are offering routes with 1, 2 and 3 hops to A, B and C and 5, 6 and 7 hops to D, E and F, the latter routes are discarded because, with no unit 4 hops away, they are impossible!

4.1.4.7.4 Observations on the behaviour of the basic routing algorithm
The above algorithm works well for slowly changing topologies with good quality links. In developing a scheme to cope under adverse EME conditions, two observations were highlighted by the simulation:

1 If routing updates are event triggered, rapid connectivity changes could cause the whole of the channel capacity to be taken up with routing update transmissions. This can be prevented by placing a maximum frequency on the rate at which any unit can transmit a routing update. However, this implies that the network will have to perform routing using out-of-date information.

2 Simulations have shown that the hop-counter-based routing scheme still works quite well if the units have a pessimistic picture of the network connectivity. This is because the hopcounter information held in the units under these conditions contains signpost information as to the direction in which a packet should be relayed. It can also be shown that under congested conditions packets can still reach their destinations in the minimum number of hops.

The above observations led us to develop a hold-down modification in which the routing tables do not attempt to track fast changes in connectivity, but in which links that have 'flaky' or rapidly varying connectivity are labelled as poor quality links. The route calculation algorithm uses poor-quality links in hop-counter calculations only if there are no good-quality paths. A route quality factor is used to indicate the number of poor-quality links involved.

4.1.4.8 Enhanced routing algorithm
Connectivity measurement The link quality measurement is an attempt to predict the quality of a given link for the next routing update period. In its simplest form this estimate is based on the performance of the link during the last such period:

$$\text{link quality } (i, j) = \frac{(\text{no. of pkts Rx by } j \text{ from } i) \times (100)}{(\text{no. of pkts Tx by } i) \times (100 - \text{clash ratio})}$$

IF link quality $(i, j) > 5/8$ THEN link state $(i, j) = 0$ (good)   ELSE
IF $5/8 >$ link quality $(i, j) > 1/8$ THEN link state $(, j) = 1$ (poor)   ELSE
IF link quality $(i, j) < 1/8$ THEN link state $(i, j) = \inf$ (no link)

The link state can be thought of as a Kalman-filter-like prediction of the state of the link, in that particular direction, during the next update period, and as such can use weighted past values in its calculations.

Action on receiving a routing update packet An update from a neighbour indicates the number of hops to various destinations and the qualities of the routes. It is processed as follows:

(a) The entry for the receiving unit is examined to see if the neighbour received this unit's last update. If not, then the 'transmit update flag' is set.
(b) The unit uses the update to recalculate routes to all other units choosing the minimum hop indications of the highest quality. It notes whether any of the route hops or qualities have changed. If they have changed the unit will broadcast the new update at the earliest opportunity. However, if it has not broadcast an update for T_u (update period) seconds then it sends the new update immediately.

Action when unit is idle If a unit does not have any data packets to send over a routing update period, it forces the sending of a routing update every T_u seconds in order to alert neighbours to its continued existence.

Action when routing information is incorrect If units have a pessimistic measure of the number of hops to a destination, the forwarding algorithm will still deliver the packets, possibly with some unnecessary relaying. However, if a unit has too low a value for the distance to destination, a subsequent relayer may not accept a packet for relaying. This problem is ameliorated by increasing the hop counter in the packet header if no acknowledgment is received after three transmissions. If after seven transmissions no acknowledgment is received, the unit discards the packet.

4.1.4.9 Network control: initialisation and self-configurability The fully distributed channel access and routing algorithms described above mean that no special actions have to be taken during initialisation. After the lifting of radio silence, the units are switched on and the network will self-configure in approximately two minutes depending on the number of areas of radio connectivity and the quality of the links.

This emphasis on self-configuraability rather than optimum network topologies is due to the fact that the geographical distribution of radios in the forward area is primarily determined by the non-communications functions of the forces (soldiers are unlikely to travel along the skyline in the forward area just to improve communications). However, it is possible that existing connectivities will fail to support the data transfer requirements of the users. A privileged command may be entered at any terminal which will result in the display of the connectivity and traffic loadings of the network, thus enabling a signals officer to minimally perturb the locations of some users to obtain the required performance. The aim is that the system should require the minimum of human supervision and, when it does require supervisory input, that it presents system status information in an understandable and easily assimilable form. The current RSRE prototype makes extensive use of the graphics facilities available on standard VT100/200 terminals.

4.1.5 Performance

The RSRE packet radio simulator fully implements all the algorithms described above independently for each unit in the network. A two-dimensional connectivity matrix is used to determine which units are affected by the actions of current transmitters. Packet generation is for random source-to-destination pairs.

Hundreds of hours of simulations have been performed on networks with various loadings, topologies and control parameter settings. A major aspect of the results has been that, although the performance is sensitive to the choice of algorithm, it is not particularly sensitive to the control parameter setting. The main criteria which affect the throughput and delay of such networks will be summarised, followed by particular results for a network subjected to rapid and large changes in connectivity.

The main parameters used in the simulations to which the results refer are:

packet duration	100 units
Tx to Rx switch time	5 units
Rx to Tx switch time	5 units
routing update period	20 000 units
clash control	4 %
maximum partition factor	6

The percentage loadings and throughputs referred to in the next sections are obtained by dividing the total number of delivered bits by the channel bit rate multiplied by the time of the simulation. Thus percentages do not include header or coding overheads. The performance in terms of throughput, delay and delay dispersion depends upon the ratio of the carrier sensing dead time to the average packet duration, the network connectivity and the rate of change of network connectivity. As in all packet switched networks, there is a trade-off between loading and average delay. We have not found the performance to be particularly sensitive to the exact setting of the clash control, maximum patition factor or routing update parameters.

4.1.5.1 Performance with single-hop topology The single-hop or fully connected network is a special topological case which has significantly better throughput/delay characteristics than others. It obtains full benefit from the co-operative properties of CSMA. When traffic is offered randomly by all users, acknowledged throughput rates of about 60% channel capacity are achievable. This figure is

Table 4.1.1 Throughput performance: (carrier sensing dead time)/(average packet duration) = 0·05

State/topology	Delivered information (% capacity)	Messages/hour 20 byte header 80 byte message rate coding
Static, single hop	60%	24 000
Static, multiple hop	25—36%	10 000—15 000
Dynamic, multiple hop	15%	6000

sensitive to the action of the 'special scheduler' which permits any unit which has just received a packet for acknowledgment an immediate channel access opportunity. It has been noted that most transmissions are successful on their first attempt, and therefore the average delivery time from first transmission is of the order of a few hundred milliseconds. However, in a heavily loaded network, packets will suffer significant pre-transmission delays. The throughput figures have been roughly translated into messages per hour (by assuming 80 byte messages, 20 byte packet radio headers and half-rate forward error control overall) in Table 4.1.1.

4.1.5.2 Performance with multiple-hop topologies Although throughput/delay performance for multiple-hop networks depends upon offered traffic patterns and topology, simulations have shown that, for random traffic patterns and topologies with reasonable numbers of relayers, the main factor affecting performance is whether or not the topology is static. This effect is quite easily explained. The dynamic changes have two effects:

1 They cause the network to attempt to route packets when the routing database information is out of date and/or inconsistent. The algorithms to circumvent these problems involve extra transmissions.
2 Routing updates, which take up some of the channel capacity, are only transmitted when changes occur.

Apart from the dynamics of the topology, the maximum throughput and delay are relatively independent of topology. The main reason for this appears to be that, as the connectively becomes more cellular, more units can transmit simultaneously without causing collisions, and thus there is some frequency reuse in non-adjacent areas of connectivity. Throughput rates in the range 25—36% have been achieved in the static multiple-hop topologies.

4.1.5.3 Performance with dynamic topology The results from the simulations in which rapid connectivity changes have been induced have been most promising. Basically the link quality assessment time constants are related to the known time that it takes updates to be disseminated. So unless the connectivity measurement indicates that the link is going to be available for a useful period, it is not included in the minimum distance calculations.

Packet radio: a survivable communications system   175

Thus the routing tables reflect a pessimistic view of the number of hops required, but, when the shorter hop routes are available, the contention-based routing algorithm permits them to be used.

As an illustration of the performance of a network under extremely unfavourable conditions, we have taken the case of a 25-unit network where the connectivity can be varied between a two-hop (good) and a five-hop (bad) scenario, as shown in Fig. 4.1.6. The simulation switched the connectivity between these two states at rates between once every 50 time units (half a packet time) and once every 1 000 000 time units. It can be seen from the time constants indicated in the routing algorithm that, for most of those switching times, routing information

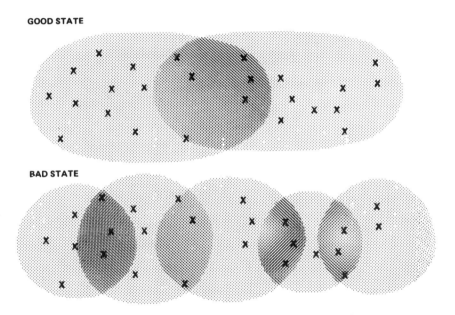

Fig. 4.1.6 Two connectivity states used in dynamic performance evaluation

cannot be distributed from one end of the network to the other! The ratio of the time spent in the good and bad states is the same for all simulation runs so that the results can be compared. The ratio used in the example is 25% in the good state and 75% in the bad state, but the results are maintained for all ratios. Figure 4.1.7 illustrates the throughput, average delay and lost packets for various accepted loads. The lost packets are those which fail to arrive at their destination because somewhere along the route a unit has transmitted them seven times but, because of confusion in the routing tables, has found no 'nearer' unit to accept and relay them. The figures correspond to throughput rates of about 15%. On a 16 kbit/s channel this corresponds to 6000 80-byte packets per hour including half-rate error control coding and header overheads.

4.1.6 Signal processing requirements and architecture

The various functions of the prototype packet radio stations are performed in three subunits as shown in Fig. 4.1.8. The higher-level protocols (routing, internetworking, man-machine interface etc.) are handled by the station processor, at present a PDP 11–23, with about 60 kword of real-time software written in a high-level language, CORAL 66. The critical real-time tasks of the lowest level of protocol, i.e. the signal processing, are carried out by a specially developed burst-data modem. These two together form an appliqué unit for a vehicle VHF radio set, the CLANSMAN 353. The latter has been modified, by means of a single replacement circuit card, to switch from receive to transmit in less than 5 ms, in order to give efficient channel utilisation at high offered load.

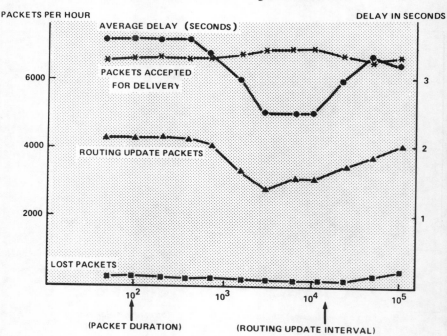

Fig. 4.1.7 Performance of dynamic multiple-hop topology of Fig. 4.1.6

Some of the most important system design constraints arise from the requirement for mobile operation in the highly congested radio spectrum to be found in northern Europe. The combination of the terrain, and the logistic undesirability of the need to rely on extensive well-sighted repeaters, point to the use of the low VHF band (approximately 30 MHz to 100 MHz), and for the same reason most current army voice communications use this band. This in turn has led to the ready availability of radio sets that operate in this band, usually using 25 kHz channelling

Packet radio: a survivable communications system 177

with either narrowband frequency modulation or 16 kbit/s digitised speech, and this 16 kbit/s capability has been used as the bearer for the medium-speed packet radio system.

The modem has to perform the tasks of synchronisation, modulation, demodulation and error control, and also the more specific tasks of carrier sense and capture switching. It is also particularly important to provide the ability to receive correctly consecutive packets with little or no space between them — something that is notably lacking in many realisations of existing packet systems, but very important here in order to maximise the utilisation of the small available channel.

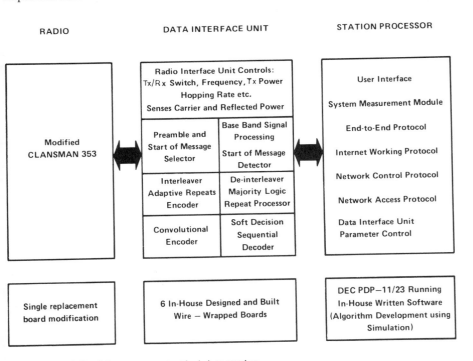

Fig. 4.1.8   *A flexible prototype tactical data station*

In data mode the radio sets provide a DC coupled linear FM channel with a bandwidth of some 10 kHz, and this is modulated directly by the bipolar non-return-to-zero bit stream after filtering by a digital finite impulse response filter. This scheme has the advantage of no error extension or bandwidth expansion, as can happen with partial response, biphase or HDB3 coding, and is also highly tolerant of multipath components up to some $20 \mu s$ spread. The message is preceded by a 64-bit synchronisation preamble, the length having been chosen to give a suitable compromise between performance in noise, false alarm rate and fast signal acquisition. The message itself consists of an integral number of 1024-bit blocks up to a maximum of 16 blocks, each consisting of 512 information bits and

512 parity bits, interleaved to a depth of 32. The first block contains in addition a separately encoded indication of the number of blocks in the transmission. The error control system employed is half-rate convolutional encoding of constraint length 47 and sequential decoding (Forney and Bower, 1971). For a two-block message (1024 information bits) this provides a probability of successful decoding of 0·9 at an error rate of 5%. This forms the base level of error correction; alternative message structures include two or four repeats and majority voting as an outer code to cope with 10% or 18% bit error rates. These alternative message structures are indicated by the using of different synchronisation preambles, the three preambles being chosen to be mutually orthogonal. The incorporation of capture switching requires that the synchronisation circuits are kept running all the time so that, in the event of a stronger signal partially overlapping a weaker existing transmission, the demodulator may switch to the larger signal. This strategy falls down when a false synchronisation event occurs owing to a pattern appearing in a message that is similar to a synchronisation preamble. The problem is compounded by the relatively low synchronisation threshold (to maximise synchronisation probability at high error rates) and also by the tendency for false alarm rates to be higher during random 16 kbit/s traffic than with white noise. Moreover, retransmission of this particular message will cause the same action to occur again. The solution to this problem is, upon an indication of a potential capture, to start a second demodulator running in addition to the current one. The undetected error performance of the error correction decoder can then be exploited to determine whether or not the capture was genuine. It is even possible to completely recover two packets which overlap by not more than 10%, provided that the second one is at least 5 dB stronger than the first.

The signal processing in the modem is all performed digitally; on the receive side the radio receiver output is sampled four times per bit period, i.e. 64 ksample/s or every 16 μs. Many of the operations are bit oriented, which means that a general purpose microprocessor is too slow to perform the required tasks; in addition, at the time of the prototype design, signal processing devices such as the TMS 320 were not available. The prototype was therefore constructed in the form of a number of special purpose hardware units, sharing a common bus under the control of an 8085 microprocessor (Fig. 4.1.9). The bus has a one-bit data path (because so many operations are one-bit operations, and speed is not a problem at these moderate data rates) and a speed of 2·5 MHz. The processing devices are allocated time slots on the bus in a strict time division multiple-access manner, with the sequential decoder being allocated one cycle in two, and the other devices allocated one cycle in sixteen. The pool of buffers has available to it all the clock phases controlling the bus, and the processor can enable any buffer during any clock phase. By this means, the processor connects a buffer to a demodulator and, when it receives an indication that a message has been received, it can connect that buffer to the different processing devices in turn in accordance with the current modem status (self-test, error measurement or normal running) and the level of redundancy of that packet as indicated by the preamble. Thus at any instant

Packet radio: a survivable communications system   179

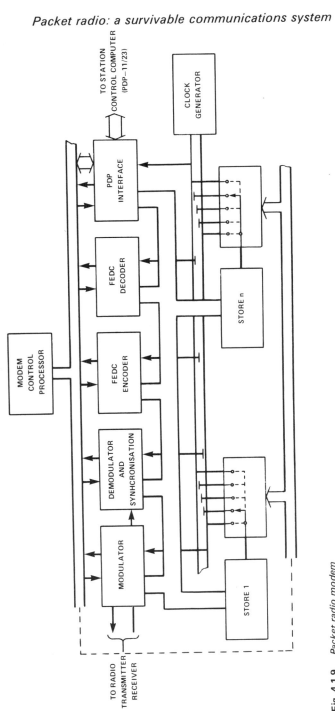

Fig. 4.1.9   Packet radio modem

there may be a number of packets at various stages in their processing, and a snapshot of the queue for the convolutional encoder might show a packet to be transmitted, and one or more received packets awaiting the first stage of error decoding. The existence of a data path, in addition to control and status, between the control processor and the DMA interface permits an extensive self-test facility.

Current work is directed at implementing the modem on two circuit cards, one to carry out modulation, demodulation and synchronisation, and the other for error control. Each card would consist of a processor and a semi-custom support chip; the modulation/demodulation/synchronisation is currently being configured as a TMS 320 plus a semi-custom correlator chip, to communicate with the error control unit via a dual-port memory. The TMS 320 is not ideal for this particular task, so we will investigate the use of small bit slice systems and other devices with a reduced inistruction set, such as the Inmos transputer. A station processor based on a 68000 or similar processor would then complete the appliqué unit, the key aims being to reduce the parts count, cost, power consumption and size compared with the current prototypes. The longer-term aim is to produce a battery powered manpack access unit, which operates a reduced set of protocols (but still needs the full modem functions) and gains access to the main net via the nearest available main packet radio station.

4.1.7 Experiences with prototype equipment

Five prototype units incorporating the hardware and software design described above have been constructed. Initially, basic performance assessment and debugging were carried out in the laboratory with the aid of a matrix switching box which permitted computer control of the connectivity on all of the 20 baseband links. These laboratory experiments demonstrated the effectiveness of the contention-based routing algorithm and validated its ability to handle rapid changes in connectivity as predicted by the simulator. Unfortunately, interconnecting at baseband rather than at RF caused the clash detection to be inoperative and failed to show the beneficial effect of RF capture. We now have the development of an RF switching matrix in hand. The laboratory experiments highlighted one area in which the simulator was deficient in detail. Although the simulator assumed that the whole packet had to be received before it could be processed, it assumed that the processing was performed instantaneously! These early laboratory experiments showed that the combined signal processing and protocol handling produced a significant latency between reception of a packet and the earliest possible transmission of the acknowledgment. This latency was of the order of the shortest packet duration. In particular this has implications for the special scheduler. It will be relatively easy to enhance the simulator to model this latency.

These laboratory assessments have been followed by a very successful series of mobile field trials and demonstrations. For these trials, three of the units have been installed in standard army Landrovers, and two in a transportable flatbed cabin. One of the units is usually demounted from the flatbed cabin to a site with a fixed wire link to a gateway of the DARPA internet. Perhaps the most telling feature of these trials and demonstrations was the fact that no engineering channel has ever been used. In fact the mobiles are equipped with a single packet radio unit only, and it has never been necessary to disconnect the radio from the appliqué and use it in the voice mode. This also bears out the claim made earlier in the paper that packet radio often involves fewer overheads than normal voice operation for the users. These field trials provided confirmation of the mobile communications capability of the network, including its ability to adapt to changes in connectivity generated by units in motion. An unexpected feature of these trials was the ability of the networks to make use of non-symmetric radio links. Thus packets were often delivered in one hop from a source defined to be two hops away (the routing table only accepts bidirectional links), the acknowledgment from the potential relayer terminating the source transmission and that from the destination cancelling the relaying attempt. This had not been observed in the simulation because it had always been run with symmetric connectivity data.

A major feature of the demonstrations to date has been the ability to provide mobile access from remote locations to large computing facilities on the ARPANET, mimicking the ability of field commanders to access powerful C^3I databases from the far forward area. The TCP/IP protocols permit reliable error-free mobile access to hosts via LANs, satellite and long-haul networks. Such long-haul access is performed using local echo and line-at-a-time transmissions. It is a tribute to the flexibility and robustness of TCP/IP that such access, although involving some overheads, was perfectly acceptable to the operator. In fact the main problem with the mobile use of packet radio was entering data correctly via a keyboard. This highlights the desirability of a more sophisticated MMI based upon a speech recognition and synthesis unit.

Interoperability between CNR packet radio and the DARPA packet radio was effectively demonstrated in trials that occurred over a two-week period at the SHAPE technical centre in Holland in March 1985. These trials plus the live demonstration involving half a dozen networks attached to the DARPA internet highlighted the tactical to strategic communication interoperatbility, and the flexibility that can be obtained by designing networks as members of an internet system rather than stand-alone systems.

4.1.8 Conclusions

This paper has described the first iteration of system design and development for a narrowband PR system. The results at all stages have been particularly encouraging. In the second iteration, which is based on a commercial procurement of a 25-station unit of suitable size, weight and power consumption, further refinement and field evaluation of the distributed algorithms will be made.

The algorithms that we have developed have been influenced by the highly mobile and hostile nature of the environment in which systems will be expected to operate. Detailed simulations have been used to evaluate the performance and stability of the algorithms. The channel access and network control strategies as currently implemented are capable of maintaining a useful grade of service over a wide range of operating conditions. In particular, the routing algorithms have been shown to provide a very acceptable level of service when more orthodox algorithms would be generating unsupportable overheads. The strength of the CNR PR, as is the case with the DARPA wideband PR, stems from the synergism of signal processing and communications protocols; the former, using modern VLSI techniques, can match the latter to what at first sight is a very unprepossessing communications channel.

4.1.9 References

ABRAM, J. M. and RHODES, I. B. (1982) 'Some shortest path algorithms with decentralised information and communications requirements', *IEEE Trans.*, **AC-27**, pp. 570-82

BOORSTYN, R. R., KERSHENBAUM, A. and SAHIN, V. (1985) 'Throughput analysis of multihop packet radio networks', Proceedings INFOCOM '85, IEEE

CERF, V. and KIRSTEIN, P. T. (1978) 'Issues in packet network interconnection', *Proc. IEEE*, **66**, pp. 1386-1408

DARPA (1980a) 'DoD standard internet protocol', Defense Advanced Research Projects Agency, IEN-128

DARPA (1980b) 'DoD standard transmission control protocol', Defense Advanced Research Projects Agency, IEN-129

FORNEY, C. D. and BOWER, E. K. (1971) 'A high speed sequential decoder', *IEEE Trans.*, **COM-19**, p. 821

JACOBS, I., BINDER, R. and HOVERSTEIN, E. (1978) 'General purpose packet satellite networks', *Proc. IEEE*, **66**, pp. 1448-67

KAHN, R. E., GRONEMEYER, S. A., BURCHFIEL, J. and KUNZELMAN, R. C. (1978) 'Advances in packet radio technology', *Proc. IEEE*, **66**, pp. 1468-96

KLEINROCK, L. (1978) 'Principles and lessons in packet communications', *Proc. IEEE*, **66**, pp. 1320-9

KLEINROCK, L. and TOBAGI, F. (1978) 'Packet switching in radio channels. Part I: Carrier sense multiple access modes and their throughput and delay characteristics', *IEEE Trans.*, **COM-23**, pp. 1400-16

LEINER, B. M. (1980) 'A simple model for computation of packet radio network performance', *IEEE Trans.*, **COM-28**, pp. 2020-3

McQUILLAN, J. M., FALK, G. and RICHER, I. (1978) 'A review of the development of the ARPANET routing algorithm', *IEEE Trans.*, **COM-26**, pp. 1802-11

TOBAGI, F. and KLEINROCK, L. (1975) 'Packet switching in radio channels. Part II: Hidden terminal problem in carrier sense multiple access and the busy tone solution', *IEEE Trans.*, **COM-23**, pp. 1417-33

TOBAGI, F. and KLEINROCK, L. (1977) 'Packet switching in radio channels. Part IV: Stability considerations and dynamic control in carrier sense multiple access', *IEEE Trans.*, **COM-25**, pp. 1103-19

Chapter 4.2

C^2 communications for the tactical area: the Ptarmigan packet switched network design and development proving

C.S. Warren, S.G. Wells, J.R. Bartlett, B.J. Symons
(Plessey Defence Systems Ltd)

4.2.1 Introduction

The concepts of C^3 in tactical warfare have existed for many centuries and the need for effective communications to achieve responsive command and control is well appreciated.

A considerable change of emphasis has taken place over the last decade. As command and control has moved from manual to automated systems the demands on the service given by the communications system have also changed. For entirely manual C^2 systems the communications system needs to offer an effective voice, telegraph and facsimile capability (to provide for people-to-people contact). For automated C^2 systems, the communications system must offer modern data services such as packet switched access (to provide for computer-to-computer contact).

The current trend towards greater automation of C^2 systems is unlikely to abate, and three significant factors can be seen leading the drive towards increased automation:

1 The mobility and strength of enemy forces in a NATO scenario requires accurate synchronisation of the NATO forces in space (e.g. co-ordination of targets between adjacent forces), time (e.g. co-ordination of a land and air attack) and function (e.g. logistics and operations).

2 Manpower is limited. Staff effort applied to thinking and planning must be maximised, and that applied to the creation of an accurate picture minimised. Most manual C^2 systems have difficulty in keeping up to date and most effort is used in the mechanics of information processing.

3 From a technology-driven viewpoint, the advent (indeed proliferation) of rugged microprocessor systems allows the distribution of intelligent functions. Data processing can be brought to the man who will be situated where he can best carry out his management function.

The technology used in the growth of C^2 systems is also available to the communications systems designer and plays a key role in the provision of effective communications for C^2 systems.

The Ptarmigan packet switched network has been designed to meet the C^2 communications requirement for the tactical area. This contribution examines the design philosophy behind the Ptarmigan packet switched network and the approach to its development proving.

4.2.2 The design philosophy

For reasons of survivability, together with the need to share data over a wide variety of physical locations, the C^2 systems will use distributed architectures. This needs to be supported by a significant communications capability. This could take three forms:

Each C^2 system provides its own communications, i.e. each is a separate C^3 system This allows each communications aspect to be optimised for the particular C^2 system. It does however have significant disadvantages in that it will duplicate design effort, individually characterise each system in terms of spectrum signature/traffic flow and, most importantly, make intercommunication between C^2 systems very difficult.

Each C^2 system uses existing communications but enhances it on an end-to-end basis As an example, the Ptarmigan system offers users direct access to a 16 kbit/s digital circuit switched channel. Each C^2 system would enhance this to provide higher integrity (the error rates for digital voice/telegraph transmission are often not suitable for database transfer of data. Although this improves the traffic flow security it is still likely to involve the duplication of similar work, and does not resolve the problems of intercommunication between the various C^2 systems. Although the circuit switched connection can be set up between any pair of systems, the end-to-end protocols may be different. This approach is also inefficient in terms of use of communication resources.

The communication system provides a range of data services to meet the needs of all C^2 systems The major problem with this approach is to define the complex interface between C^2 systems and the common carrier communications system. The advantages are, however, very significant. It can provide a greater efficiency of use of resources (i.e. less total bandwidth), it avoids duplication of effort and, most importantly, it provides communication within *and* between C^2 systems. This is essential for future growth if the concepts of force multiplication and data fusion are to be realised.

In 1980 Plessey Defence Systems was asked to study these problems in order to determine the impact on the Ptarmigan tactical area communications system (Warren, 1984). After an analysis of circuit switched and integrated data services options it was decided to enhance the Ptarmigan network to provide a range of data communications facilities to support existing and evolving C^2 systems sharing the same tactical area. Fig. 4.2.1 illustrates the approach chosen. The communications system can be considered as a common foundation for each C^2 system and must provide facilities to allow communication between elements of a particular C^2 system and between different C^2 systems. It was further considered that the inherent bit-by-bit switching of totally digital transmission using stored program control would allow Ptarmigan to be easily enhanced to provide a new range of data services. This would also be able to make full use of the redundant area coverage capability of Ptarmigan.

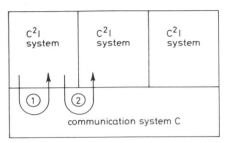

Fig. 4.2.1 *The communication system as a common carrier. (1) Intra C^3I communication, (2) Inter C^3I communication*

At the same time the International Standards Organisation (ISO) published its model of open systems interconnection (commonly called the ISO seven-layer model). This provided a defined framework for the necessary discussions to define the interface between the communications systems and the C^2 systems. The next phase of the work was concerned with the derivation of the detailed requirements (i.e. communications needs) of C^2 systems. Generally, a C^2 system may be considered as a distributed set of processors and terminals as shown in Fig. 4.2.2, where high-integrity data communications is required between:

1 Intelligent VDU terminals and remote C^2 processors for MMI commands and responses
2 One C^2 processor and all others maintaining database alignment using database update messages
3 Two specific C^2 processors, to allow the passage of total databases catering for total losses of information within a C^2 processor.

The requirements for each of these functions can be quite different and, working within practical constraints, there is no solution that fits all needs.

More specifically, typical requirements are shown in Table 4.2.1. For terminal/processor traffic, where the appearance of continuous contact is required but where the traffic rate is low compared with the basic channel transmission rate, virtual circuit access using packet switching is an ideal solution (particularly in view of the fact that the traffic is also 'statistical').

For processor/processor updates a longer response time is usually allowable. However, if a circuit switched connection was required to, say, ten other processors in a serial fashion, then this would not be achieved within even 5 seconds. This requirement is therefore also suitable for packet switching, particularly as most update messages are fewer than 100 bytes.

The last requirement, for processor/processor bulk transfer, contrasts greatly. It is not required on a continuous basis. When it is required the traffic is 'non-statistical', and the data rate needed approaches the basic channel data rate. This is more amenable to a circuit switched solution.

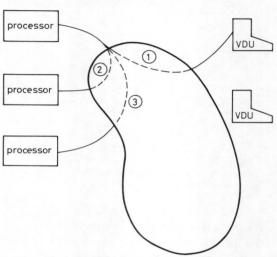

Fig. 4.2.2 *Typical command and control system. (1) Terminal/processor (MMI), (2) processor/processor updates, (3) processor/processor bulk transfer*

The solution, seen from a terminal viewpoint, is shown in Fig. 4.2.3. For terminal processor and processor/processor updates, X25 virtual circuit access is offered. The choice of a civil (CCITT) network protocol was carefully considered and

Table 4.2.1 *Typical C^2 communications requirements*

	Terminal processor	Processor/ processor updates	Processor/ processor bulk transfer
Response time	< 1/2 s (MMI)	5 s	1/4 h
Integrity	10^{-7} BER	10^{-9} BER	10^{-9} BER
Traffic type	Commands and display updates	DB updates. 90% < 100 byte	1 Mbyte continuous
Data rate	50 bit/s typing rate, < 2.4 kbit/s reading rate	≈ 100 bit/s	8 kbit/s for 15 min

chosen because:
(a) Proprietary software and hardware (e.g. HDLC chips) speeds the development process.
(b) It minimises special development of test tools.
(c) It encourages all C^2 systems to use a common approach.

Where enhancements to the civil protocol were deemed necessary (even for the internal-to-network protocols) an attempt was made to ensure that the civil standard was a subset of the military solution.

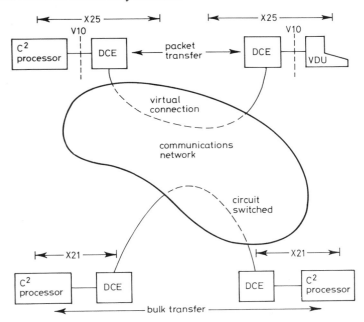

Fig. 4.2.3 *Common carrier solution for C^2 systems*

4.2.3 The tactical environment

In order to design the packet network to be implemented as part of the Ptarmigan system, the speical requirements of a tactical military environment were first identified. These requirements are concerned with the communications needs, the service required and the effects of mobility and damage. These factors lead to significant differences between the tactical military network and a civil network.

Communications need lines exist between the processors and terminals of each C^2 system or other C^2 systems. All these will be referred to as packet terminals. These fall into two categories, namely:

(a) Static terminals having reasonably high usage and justifying a dedicated hardwired access into the packet network

(b) Mobile terminals, or low-usage terminals, which require the facility to demand access to the packet network when required.

The Ptarmigan network provides for both the above categories, but here we will concentrate mainly on the packet network itself rather than the means of obtaining access.

C^2 systems need the 'virtual call' service whereby a number of virtual calls can be established from a subscriber to a number of destinations. These can then exist simultaneously such that data can be passed when required without the need to identify the destination and set up a call each time. Ptarmigan provides each subscriber with the facility for up to 255 virtual calls. Access to the network is provided for static subscribers, operating a protocol based on the CCITT X25 recommendations. This provides the virtual circuit access. Each interface can operate at 16, 32, 48 or 64 kbit/s.

The Ptarmigan system offers the subscriber to the packet network the following:

1 Call establishment and data transfer
2 High availability of a call even under failure, movement and congestion of a link or packet switch
3 Security of data
4 End-to-end flow control for each call
5 Flow control window negotiation
6 Low misrouting probability
7 Low error probability for bit integrity
8 End-to-end delivery confirmation for packet integrity
9 Sequenced delivery of packets for each call for message integrity
10 Multiple simultaneous calls per subscriber
11 Speed changing.

The tactical environment is such that a communications network is in a continual state of movement and liable to damage, necessitating reconfiguration in order to maintain the network. The movement is achieved by a leapfrog principle in that at any given time there is a complete network in operation, but switches and links are continually being added or removed. This movement should not be visible to subscribers who are not themselves involved in any given movement, and so the virtual calls must remain as long as there is a network, although the routes to the destinations may change as a result of the movement or damage. Similarly, subscribers themselves will be removed from one point in the network and be reconnected at a different point. Although this will result in disruption of the virtual calls, requests set up subsequently must automatically find the new location without the need for location information in the call request.

4.2.4 Packet switching overlay in Ptarmigan

Ptarmigan packet switching is an overlay on the Ptarmigan circuit switched network and, as such, utilises the existing transmission, encryption, multiplexing and connection facilities. It therefore offers the advantages of the Ptarmigan network

C^2 communications for the tactical area

design to the subscribers of the packet network. All packet switches are housed in the existing Ptarmigan vehicles.

The addition of a packet switching capability to the Ptarmigan system came about when the Ptarmigan system was already in an advanced stage of development. This, of course, placed a number of constraints upon the design of the packet network, mainly in the areas of equipment interfaces and physical characteristics. Other constraints were the adoption of a civil standard for the access protocol and, of course, the very high engineering standards demanded of tactical military equipment.

There are two versions of packet switch in the Ptarmigan system. These are the primary packet switch (PPS) and the secondary packet switch (SPS). The PPS is contained within the circuit switch installation, and uses the circuit switch network to provide a redundantly connected packet network. It provides through routing and facilities for on-demand access to the packet network from the circuit switched network. The SPS is used mainly to provide access to the network for 16 static hard-wired packet subscribers, but it also has a through routing capability. This permits, besides a connection to a PPS, a second packet trunk to be connected to another SPS (for extension or stacking) or to another PPS (for redundancy). The SPS is located wherever the number of subscribers justifies an SPS. The connectivity is indicated in Fig. 4.2.4.

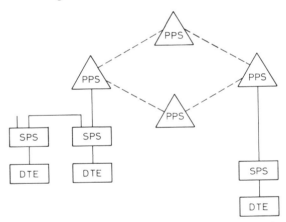

Fig. 4.2.4 *Symbolic representation of Ptarmigan packet switched network*

4.2.5 Network characteristics

In order to operate with the network mobility of the tactical environment, the Ptarmigan packet switching network uses virtual call access for subscribers but, internal to the network, operates a datagram mechanism. This enables packets to be transmitted within the network using datagrams, which may take alternative routes to overcome network changes or congestion. This involves the use of the datagram protocol to route to the destination switch (with no knowledge of the virtual call or subscribers) and a trans-network protocol to operate between the originating and destination switches. This trans-network protocol links the virtual call access protocols at each end of the virtual call and makes it appear to the subscribers that there is a constant virtual call across the network.

Datagrams are routed through the network using the delegated routing principle, whereby each switch transmits a datagram on the link which it considers to be the best link for the given destination (considering local connectivity information and congestion). The means by which the switches obtain routing information is described in the section on network management.

Another principle of the network design is that, in order to achieve fast trans-network delivery times, the packet queues in each switch are kept small. This means that there are situations in which packets are discarded and recovery is achieved by retransmission of the packet from the source.

4.2.6 The protocols

The packet switching network protocols can be described in terms of a layered hierarchy. The hierarchy is such that each level of protocol utilises a header transmitted before the information field. This combination of header and information is itself the information field of the lower-level protocol.

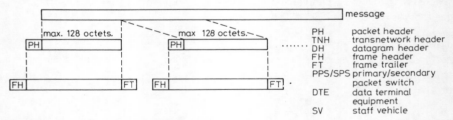

Fig. 4.2.5 *Hierarchy of headers for access protocol*

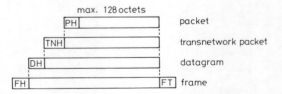

Fig. 4.2.6 *Hierarchy of headers for trunk protocols. For key see Fig. 4.2.5*

The hierarchy of headers for the access and trunk protocols is shown in Figs. 4.2.5 and 4.2.6. The relationship between the protocols and the Ptarmigan packet switching equipments is shown in Fig. 4.2.7.

The access protocols consists of the internationally agreed CCITT X75 frame protocol (level 2). The level 3 protocol is an amended form of X25, taking cognisance of the user requirements and the particular nature of the tactical military environment.

The trunk protocols at level 2 consist of CCITT X75 frame protocol and a

BCH/ARQ protocol. Both level 2 protocols provide error protection for the higher-level trunk protocols.

The low error probability is provided by use of the CCITT X75 frame protocol on every link, both access and trunk, in the network. This frame protocol provides error detection and recovery from errors by retransmission. Additionally, on trunk links with potentially severe error rates (e.g. radio paths), a further level of error protection, the BCH/ARQ protocol, is provided. The BCH/ARQ includes a powerful error detection mechanism, and again recovery from the errors is provided by retransmission. However, the number of bits involved in a retransmission is small compared with that for the frame protocol and therefore, at a given high error rate, the addition of the BCH/ARQ protocol maintains the integrity of the data while achieving a higher throughput.

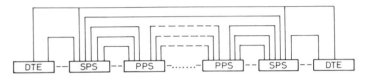

Fig. 4.2.7 *Hierarchy of protocols and equipments. For key see Fig. 4.2.5*

There are two level 3 trunk protocols, namely datagram and trans-network. The datagram protocol routes the packet to its destination packet switch. The trans-network protocol joins together the two access protocols involved in a call across the trunk network, making it appear as a direct connection between the two access protocols.

Calls are established by the subscriber transmitting a call request packet. The packet network automatically locates the called subscriber, if attached to the network, and the call is set up according to the procedures of X25. Data may now be transferred between the two subscribers. Movement of packet switches within the network will not cause the call to clear as long as a route exists between the two subscribers.

The packet access protocol provides flow control and delivery confirmation. The trans-network protocol therefore passes this information between the switches at the ends of a virtual call in order to link the two access protocols such that these facilities are provided end to end.

The CCITT X25 packet protocol virtual circuit service involves a sequenced delivery of packets for each call. Thus the subscriber does not have to reorder the received packets to re-form the message. The Ptarmigan packet network provides independent routing of each packet in order to provide a survivable network despite failures, movement and congestion. Packets for a call may therefore arrive at a destination switch in a different order than that in which the packets were originally sent from the originating switch. These packets are reordered into the original sequence by the trans-network protocol, and are then delivered to the subscriber in the original sequence.

The low misrouting probability (i.e. data arriving at the incorrect subscriber for the call) is provided by robust trunk protocols. All routing information is protected from errors by the same error protection method (i.e. level 2) as applied for subscriber data, therefore giving a low probability of errors in the routing. If, however, an occasional error is not detected by the level 2 protocols, the trans-network protocol is still subject to an extensive check before passing data to the subscriber access protocol. This involves checking the validity of the particular packet with respect to the information held for the call. This results in a very low misrouting probability.

4.2.7 Network management

In order to enable the packet switches to locate the wanted subscribers and to route through the network, network management is provided. This is completely automatic and involves only information exchange autonomously between switches. This should not be confused with system management, which is concerned with facilities provided for manual control and monitoring. All the network management functions must be sufficiently responsive to cope with a dynamic network where subscribers, switches, links and congestion conditions are constantly changing.

The datagram routing protocol requires information at each switch concerning the preferred output link to be used for a given destination switch. This required information is provided by a procedure whereby each switch, at intervals of five seconds, transmits to its neighbours the value of the length of the shortest route to all other switches of the network based on the similar information which that switch itself received. It can be seen that, after several iterations, every switch will know the length of the shortest route to any other, and will also know which output link provides that shortest route. This information is used by the datagram protocol to provide delegated routing functions which can overcome changes of network topology.

Each switch holds a list of subscriber numbers which are being served by that switch. This involves information being input by means of the control panel of the switch and also the status of the link to the subscriber terminal.

When a subscriber terminal transmits a call request to a switch, that switch, by means of a flood technique, searches the network in order to obtain a response from the switch to which the wanted subscriber is affiliated. This enables the call set up to proceed using the identity of the destination switch. Further, so as to reduce the number of times this flood facility is used, a list is kept of the destination switches of the most frequently called subscribers.

Owing to the dynamic nature of both the network and the offered traffic, congestion in parts of the network may occur. This is controlled and its effect minimised by a congestion control procedure, whereby congestion information is passed around the network using procedures similar to these described for connectivity information. Depending on the level and extent of this congestion, the control mechanisms cause one or more of the following:

(a) Modification to the routes of datagrams to avoid the area of congestion

(b) Exercise of the flow control procedures at packet level, to indicate to a transmitting terminal that the route to a given destination is congested and traffic on that particular virtual call should wait

(c) Exercise of flow control procedures at frame level, to indicate that all traffic from a terminal should wait (this level is extreme and shortlived).

These procedures result in the ability to carry high traffic levels with only graceful degradation when subjected to peak overload conditions. This is achieved without incurring the penalty of long delays caused by queues within the network.

4.2.8 The need to prove

4.2.8.1 General: Today's data communications products are inherently extremely complex. It is essential that these products (be they switching nodes, terminal equipments or complete networks) are proved comprehensively prior to being released to a customer. The reasons for this are manifold, but the most important are:

(a) To maximise operational effectiveness by minimising limitations discovered in the field

(b) To ensure that the collections of specifications and drawings which were used to construct the development model of the equipment are suitable for future production of the equipment.

The Ptarmigan packet network proving activity covers three distinct areas:

1 The proving of requirements placed on both the equipments and the network by the customer from the Ministry of Defence (Procurement Executive) (MoD).

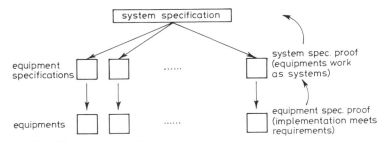

Fig. 4.2.8 *Specification and proving structure*

This is a two-stage process. The network requirements are contained within the system-specification placed on the prime contractor by the MoD. These requirements have to be shown to be met directly to the MoD. The network requirements are translated into requirements for equipments and installations by the prime contractor and then placed on the subcontractors as equipment or installation specifications. These requirements have to be seen to be met by the subcontractor to the prime contractor. This structure of specification and proving is shown in Fig. 4.2.8.

2 Confidence testing of the design of the network during development. This will consist of the testing of the various hardware and software modules prior to their being incorporated into an equipment, as well as the testing of the various equipments together to prove that they will interoperate satisfactorily. In terms of the Ptarmigan packet switching network the protocols are tested as separate entities, i.e. the level 2, level 3 and network protocols, by the subcontractors responsible for the relevant equipments. These equipments are then brought together by the prime contractor and tested for true interoperability. The latter process is necessary because although each contractor will have proved that their interpretation of the requirements has been implemented correctly by them, it is the responsibility of the prime contractor to ensure that the equipments interoperate and that the overall protocol design is satisfactory.

3 Characterisation of the behaviour of the network. This is done in order to determine how the network will perform under certain conditions and stimuli. A typical example of this characterisation is to set up a sample network and progressively load it with test traffic to investigate response times, delay times, probabilities of lost packets etc. This traffic loading would be variable and would vary from extreme under-utilisation of the network (possibly simulating just a couple of dozen virtual calls with no more than one virtual call per user) up to extreme overloading of the network (simulating the maximum number of users on the network with each user employing several virtual calls). For logistic reasons this characterisation would be conducted on a fairly small network (in comparison with the anticipated fielded network) and a theoretical exercise, using a computer model, would extrapolate the results into those expected for a full network.

4.2.8.2 The proving of requirements: Prior to the conduct of any proving activity it is necessary to examine the specifications, to identify the requirements and to determine how each requirement will be proved.

The Ptarmigan proving activity splits the system level requirements into two categories for proving purposes:

Requirement to be proved by testing This is an evaluation of the actual equipment.
Requirement to be proved by validation This is a theoretical exercise, albeit with actual infeed data obtained from testing.

Testing should, in all cases, be the preferred method of proving since it will demonstrate the actual operation of the equipment rather than the theoretical proving which validation offers. However, there will always be a close interaction between the two. This can be seen specifically by way of the infeed data that the testing activity will provide for the computer simulations of the network (switch response times etc.).

Both testing and validation can be split into two further areas of proving namely:

Functional testing/validation This deals with the operation of the network and the elements of the procedure as contained in the protocol specifications.
Performance testing/validation This deals with the performance of the network in terms of throughput, error rates, delays etc.

Generally, the latter activity deals with quantitative requirements whereas the former activity deals with qualitative requirements.

Performance testing will deal with requirements such as:

The packet switch shall be able to support a queue of 2 kbytes of packets on any one output port.

The mean delay through a packet switch, when subject to a constant throughput of 100 packet/s, shall not exceed 10 ms.

These requirements are testable but require a test tool able to generate the required traffic. The performance measurements are conducted on fully mature equipments with the completed software builds since this will yield true performance results. This means that the performance testing will be one of the last proving activities to be conducted, although every confidence will have been obtained using validation techniques.

Functional testing deals with requirements such as:

It shall be possible to pass non-voice-mode traffic at 16, 32 or 64 kbit/s through the network over packet channels between subscribers' equipment.

The packet switch shall be capable of storing, in a local archive, data concerning equipment and protocol failures.

The functional testing activity will also embrace protocol testing. Functional testing of requirements will, generally, be conducted on fully mature equipments. However, in the case of protocol testing this might not be possible since it is often necessary to isolate one level of protocol from the others in order to test specific operations.

Using the terminology defined in the CCITT X25 standard, initial testing will be centred around the level 2 protocol. This will not be conducted on mature hard-

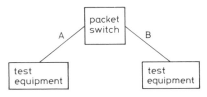

Fig. 4.2.9 *Initial testing configuration*

ware since it is necessary to isolate the higher levels of protocol from the level 2 frame and transmit random data in the information field. The tests are configured as detailed in Fig. 4.2.9, the proprietary test equipment being connected to any two packet loops on the packet switch. The proprietary test equipment will then simulate the operation of a DTE by generating the correct sequence of frames. This will include the exchange of frames simulating set-up and close-down of links, information transfer and acknowledgement (checking for correct $N(R)$ and $N(S)$ values), busy conditions, and resetting of links (using REJ). It is necessary to create a loop-round within the packet switch to connect the test equipment on port A with the test equipment on port B. This could be either a hardware or a software connection.

The above tests will re-create the operation of level 2 protocol under normal operating conditions. However, it is also necessary to check that error conditions can be dealt with by the protocol. Examples of protocol sequences that are generated to test the error handling capability of the packet switch are:

1 Set up link, test equipment sends information frames, then UA or DM to check for receipt of FRMR.
2 Test equipment sends I-frame with information field greater than the maximum permissible.
3 Test equipment sends frames outside the maximum permissible window size.
4 Test equipment sends I frames out of sequence.
5 Test equipment uses invalid addresses.

In fact it should be possible, within the test, for the test equipment to send any frame at any point in the test and for the packet switch to recover satisfactorily (in many cases this recovery strategy will simply consist of the packet switch discarding the invalid frame).

The next stage in the protocol testing activity is to test the level 3 access protocols. This will be conducted using the same configuration as was used to test the level 2 protocols. The reason for this is that, at this stage, it is not necessary to involve the network level trunk protocols, the test equipments acting as local subscribers to the packet switch.

Since the level 3 access protocols are more complex than the level 2 protocols it is necessary to build up to the full test, of which the following is an example:

1 Set up a single virtual call between subscribers. Check allocations of logical channel numbers. Close down the virtual call.
2 Set up a single virtual call between subscribers; pass data in one direction only. Check that the packet switch permits data to be sent up to the window size and acts correctly upon the information received from the test equipments. Close down the virtual call.
3 Set up a single virtual call between subscribers; pass data in both directions; create certain invalid packets to check that the packet switch reacts correctly (e.g. out-of-sequence packets, packets too long etc.). Close down the virtual call.
4 Set up two virtual calls between subscribers; pass data in both directions. Perform the same tests as in 3 above. Close down the virtual calls.
5 Set up multiple virtual calls between subscribers; pass data in both directions. Create many deviations from the protocol to ensure that the packet switch provides the correct response. Close down the virtual call.

In addition to the above tests, the error conditions must also be checked. The easiest way of doing this is to test the X25 state tables, i.e. forcing the interface to the packet switch to enter a specific state and then sending the packet switch various packets, thus generating error conditions. In order for this activity to be totally successful it is necessary to obtain an output from the switch detailing the states that have been entered at any particular time. This again means that it is not possible to use the final software loads since they will not output the appropriate information.

Functional testing will also include the engineering testing which is necessary for a military network. This testing includes:

Environmental tests These will cover vibration tests, high and low temperature tests, dry heat tests and damp heat tests. The equipment is subjected to certain environmental conditions and tested for correct operation before, during and after the test conditions have been applied.

Electromagnetic compatibility tests for the equipment.

Reliability tests

Nuclear survivability tests where the components and modules are subjected to nuclear radiation at controlled levels and then tested for correct operation.

Performance validation is the proving activity which is used to prove the performance requirements that cannot be proved by testing. This could be for many reasons, but the most likely use for performance validation is in the proving of the operation of large networks. Sample data is obtained from the performance testing activity; this is then used in a computer model of the packet switching network to theoretically prove certain system requirements that cannot be proved practically.

Typical requirements that can be proved with performance validation simulatins are:

The time from the first bit of a call request packet leaving the calling terminal to the arrival of that bit at the called terminal plus the time from the first bit to the call accepted packet leaving the called terminal to arrival of that bit at the calling terminal shall not exceed five seconds through eight switching points.

This requirement cannot be proved by testing as the available number of development models causes difficulty in configuring a large network with eight switching points.

The probability of the loss of a data packet after retransmission shall be assumed to be less than one packet in 1 000 000.

This is a very difficult requirement to prove by testing owing to the enormous quantities of data that would have to be transmitted over a network.

Functional validation is designed to prove the functional requirements on the system that cannot be proved by the testing activity. This is a theoretical exercise and uses techniques of logical argument and references various specifications to prove the requirements. A typical requirement which is proved by functional validation is:

> The access protocols used in the Ptarmigan packet switching network shall be based on the CCITT X25 protocols.

This cannot be proved by testing and requires the examination of both the Ptarmigan protocols and the CCITT protocols to determine the differences (if any).

4.2.8.3 Integration testing In addition to the requirement proving of the individual equipments it is necessary for the prime contractor/system integrator to bring the equipments together and test them in limited networks to prove that they interoperate satisfactorily.

In terms of the packet switching equipments it is one of the prime contractor's/system integrator's functions to ensure that the protocols that have been implemented are not only correct in their implementation but also compatible with each other. This activity can highlight two major problems:

1 The difference in the implementation of the network protocols necessitating work to rectify the fault.
2 A basic design error in the network protocols. Both the implementors could have implemented the specified protocols correctly but the protocols themselves do not perform the function for which they were designed.

Initially this integration testing is conducted on the first development models to ensure that the initial protocol design is satisfactory. This initial testing will show that the level 2 protocols are satisfactory. This testing is undertaken on equipments which are fitted with the actual level 2 software and dummy level 3 software since it is not necessary to set up and close down virtual calls.

This testing will progress in defined states both in terms of the protocols being tested (increasing from the level 2 to the level 3 (access) to the network protocols) and in terms of the equipment configurations being used (increasing from simple configurations of one packet switch and one DTE to more complex configurations of several packet switches and several DTEs).

For efficiency and cost effectiveness the integration testing requires that the implementators have already proved that their protocols work correctly within their own developed equipments. The function of integration testing is simply to test the correct interoperation of the equipments and not exhaustively to test the protocols within the equipments.

4.2.8.4 Network characterisation In order to provide a potential user of the Ptarmigan packet switching network with a statement of expected service, the network is characterised. This is done in the presence of background traffic and measured in terms of network delay and the probability of the terminal needing to retransmit a packet.

The network delay is measured from the reception of the last bit of a good

C^2 communications for the tactical area

packet at a source packet switch to the consequential transmission of the first bit of the packet at the sink packet switch. The network delay is measured for all packets and their responses.

The probability of a terminal having to retransmit a packet due to a time-out expiring is measured.

Both the measurements above are calculated with background traffic present. This background traffic is set to various levels of loading from zero to overload.

This characterisation is conducted using both the packet switching network model for the large networks which require theoretical examination, and the traffic generator for smaller networks which can actually be tested.

4.2.9 Test tools

Both testing and validation require special tools of various types in order to conduct the activities. These tools can be anything from a simple attenuator to a minicomputer with specially developed software.

The tools utilised in the testing activity include the following:

The developed equipment under test   This could be a production model or, in the early stages, a development model which is of a sufficiently advanced stage of development to be tested.

Protocol testers   Often proprietary equipment which is able to test the X25 protocol and is able to be adapted in order to test the Ptarmigan protocols (which are based on X25). The proprietary equipment simulates a DTE. It is attached to the Ptarmigan packet switching network via the appropriate modem and attempts to communicate with another similar piece of equipment across the network. The equipment used was only able to cope with seven virtual calls; thus it is limited in the amount of testing that it can undertake. However, it is suitable for testing certain basic attributes of the network such as the access protocols, the ability of the switches to cope with more than one virtual call per packet loop, and the ability of the switches to route packets across the network. The protocol tester is also useful in analysing the protocol and identifying any problems and shortcomings. It is the proprietary test equipment that is used to test the protocol error conditions. The protocol tester is not expected to be able to generate a large amount of traffic since it is only used to test protocols and to prove that the network is operational. A traffic generator able to produce a large output is also required.

Pseudo-traffic generating equipment   This is a minicomputer which contains software to generate and receive traffic, as well as operating the correct Ptarmigan access protocols enabling the traffic to be passed across the packet switching network. This equipment provides up to 16 packet loops which are able to access the packet switches using the correct protocols, and is able to generate up to 200 packets per second in order to stress the network. The test engineer is able to create a test traffic file which will specify the configuration data, the number of packet loops to be used, the identities of the host on each packet loop, the speed of the packet loops (up to 64 kbit/s) and the traffic data for each loop.

The traffic data consists of statistical values such as the percentage of transient calls, the percentage of continuous calls, the mean duration of the calls, the mean inter-arrival time for the calls, the packet rate per call and the pattern of data packets in a message (all long packets (128 octets), all short packets (25 octets) or a mixture of the two). The computer will then generate a traffic profile for each loop which will detail exactly when the calls will be initiated and to whom, and the quantity of data that is to be passed. The computer will access the network by way of the packet switches using a suitable Ptarmigan cable interface. The computer is able to attach itself to as many as 16 packet switches (one switch per loop) but generally it is only practicable, for space reasons, to connect the computer through two or three packet switches. However, each loop will have a different address. The computer is able to set calls up to itself across the Ptarmigan packet switching network. The test will progress using a sample network and the test traffic file generated by the computer. During the test operation phase the computer will store various statistics enabling it to provide certain parameters during the test analysis stage. The test analysis stage will provide data on the number of calls set up (both transient and continuous), the number of unsuccessful call set up attempts, the number of packets transmitted per loop, the number of resets transmitted and received, the number of restarts transmitted and received, the number of retransmissions etc.

User equipment This consists of packet terminals and processors from user projects which will be connected to the Ptarmigan packet switching network in the field. These equipments will be used to demonstrate that the two projects will interoperate and also to provide a degree of confidence that the protocols are operating satisfactorily.

Miscellaneous hardware The most common miscellaneous items used during the testing activity are attenuators, which are used to degrade the performance of the radio links in order to test the throughput requirements and the congestion aspects of the network.

Software to run the relevent test equipment This software will be for the proprietary test equipment (enabling it to run certain tests), the traffic generator (the test traffic file), and possibly certain test software within the switches to activate loop-rounds etc.

A typical testing configuration is detailed in Fig. 4.2.10. This shows the sample network of primary packet switches (the switches used to route packets around the network) interconnected by trunk radio links (two of which are able to be degraded using attenuators). Also connected to these primary packet switches are the secondary packet switches (the access points for subscribers). The protocol testers, the traffic generator and the user equipment are connected to the network via the secondary packet switches as shown. During the test operation phase the protocol testers will generate virtual calls between themselves, creating protocol inconsistencies to examine how these will affect the network. The traffic generator will provide a background traffic for the network which can be varied from under-utilisation of the network to overloading of the network. The user equipment will provide actual traffic for confidence purposes.

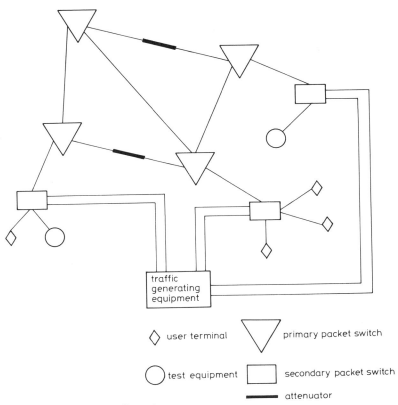

Fig. 4.2.10 *Final testing configuration*

4.2.10 Ptarmigan packet switched network model

The model is an event-based simulation embodying the majority of the characteristics of the real Ptarmigan packet switching system. Thus, for a relatively small cost, a model of any proposed packet switched network based on the Ptarmigan protocols can be built and tested. This can demonstrate the performance of the proposed system under various traffic loads and enable the examination of possible alternative configurations. It provides greater visibility to customers and management at all stages of network development and gives confidence that the network will perform as required.

The model is written in standard Fortran to run on a VAX machine. It has two facets, namely data capture and simulation.

The data used by the model falls into three categories:

Table 4.2.2 *PSN model parameters*

Traffic compression factor		30
Input buffer size	(1–10)	10
Link buffer size	(1–20)	8
Link buffer congestion point	(1–20)	5
Switch processing time (time slots)		2
Number of time slots per second	(1–n)	50
Congestion control frequency (time slots)		250
Length of offered traffic queue	(1–39)	39
Packet retransmission interval (s)		0
SPS–SPS delay (s)		0
Link speed multiplication factor		4000
Average no. channels into each node		2
Packet length (octets) (0 = random)		128

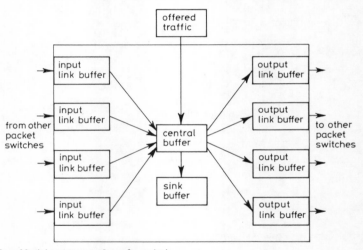

Fig. 4.2.11 *Model representation of a switch*

(a) Network connectivity data
(b) A matrix of need lines defining the traffic between nodes
(c) Packet switched network characteristics.

The network connectivity data is used to generate link weighting tables. They define the shortest path length to every other node in the network for each output link. The traffic data, consisting of a matrix of need lines, is used to characterise the customer's traffic profile. These need lines will be used by the model to generate representative traffic patterns.

The packet switched network characteristics are specified by the parameters given in Table 4.2.2. These parameters can be adjusted to enable the model to be used to carry out sensitivity analyses.

The model simulates the passage of packets between nodes of the network by transferring blocks of data representing packets from one array to another, each array representing a node or a link between nodes (see Fig. 4.2.11).

Once the generated traffic has been placed in the offered traffic files the simulation time is started. The first item in each offered traffic file is searched until one is found with a sourced time less than or equal to the simulation time. This traffic item is transferred into an input buffer representing the node. As the packet information is transferred from the offered traffic file to the input buffer the time it would take to transmit a packet between a DTE and a switch is calculated and added to an elapsed time variable. associated with the packet.

As the simulation time continues the first item of traffic in each of the input buffers is searched for an item that has an elapsed time less than or equal to the simulation time. If such an item is found a decision is made, by reference to routing tables, as to which directly connected node the packet should be passed to. Once the routing decision has been made the packet information is entered in a link buffer representing the transmission link between the two nodes. As the packet information is transferred the elapsed time variable is increased to the simulation time plus a figure representative of the switch processing time.

The link buffers are periodically searched and items of traffic are transferred into the input buffers representing the destination nodes of the links. As the packet information is transferred from a link buffer to an input buffer the elapsed time variable of the packet is increased to the simulation time plus the calculated transmission time over the link.

Once an item of traffic has reached the input buffer of its destination node, the switch processing time is added to the packet's elapsed time and the packet information is transferred into a sink buffer. The transmission time of the packet and the number of links over which the packet travelled are noted. An acknowledgment packet is then generated to be sent back to the source node and the original packet is discarded.

Each packet has associated with it a link count, which is set at the source node to the minimum number of links between the source and sink nodes plus six. Each time the packet is transmitted over a link the link count is decremented by one. If at any stage the link count of a packet is reduced to zero or to less than the minimum number of links remaining in the packet's journey to the destination, the packet is discarded.

Each link buffer in the network has a limiting capacity. Once a link buffer contains a predetermined number of packets the buffer is considered to be congested. If the congestion point is reached the only packets that can be entered into the buffer are those with a destination node at the end of the link. It is possible for a number of packets to be entered into a congested link buffer such that the buffer capacity is reached. If this happens and another packet is posted to the full buffer, it is discarded.

If a packet is lost because it has been discarded due to a full link buffer or link count expiry, and the packet has not been retransmitted three times, the

Table 4.2.3 Traffic profile

Needline		Packets per hour
From	To	
S1	S4	500
S1	S6	500
S1	S8	500
S1	S10	500
S2	S3	1000
S2	S7	250
S4	S5	500
S4	S10	750
S5	S7	250
S5	S10	250
S6	S6	250
S6	S10	750
S7	S1	1000
S7	S3	250
S9	S2	500
S9	S6	300

packet information is replaced in the offered traffic file of the source node as a retransmitted packet. A variable associated with the packet is used to record the number of times a packet has been retransmitted. Once the packet has been retransmitted three times and is lost again, the count of the number of packets lost after possible retransmission is incremented by one.

At the end of each simulated second the model outputs the number of packets offered, sourced, sunk, retransmitted, lost, and lost for a third time. At the end of the simulated period, summary statistics of the above are printed.

An example of a model run using a network of ten primary packet switches and ten secondary packet switches is now given. The traffic profile (see table 4.2.3) and the packet switched network characteristics (Fig. 4.2.12) are also supplied to complete the picture. The network characteristics are non-realistic and have been chosen so that all of the model's facilities are exercised.

Some points of interest which appear in the output from the model are as follows (see Table 4.2.13–15):

(a) The details of the packet which encountered the longest observed delivery time shows that it did not travel directly from its source node to its sink node but went via a route one link longer than the shortest route.

(b) The average end-to-end delivery time is far less than the maximum observed time, which suggests that the delivery times are skewed towards the origin. This is borne out by the histogram of delivery times.

C^2 communications for the tactical area 205

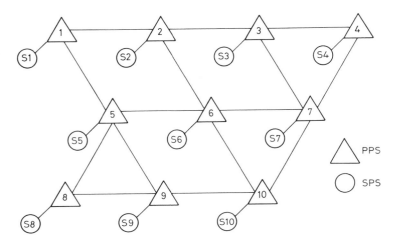

Fig. 4.2.12 Ten-node packet switch network

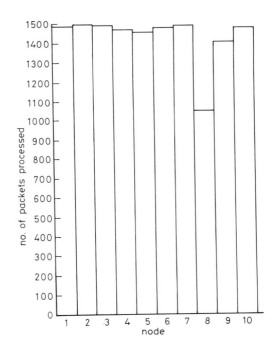

Fig. 4.2.13 Packets processed by each node

206   *C² communications for the tactical area*

(c) The minimum end-to-end delay is for a local packet which did not have to queue to be processed.
(d) The links that overflowed can be compared with the traffic profile and comparisons drawn.
(e) The cause of packet losses can be attributed to high utilisation of the packet switches, because most switches process very near their maximum of 1500 packets per minute.

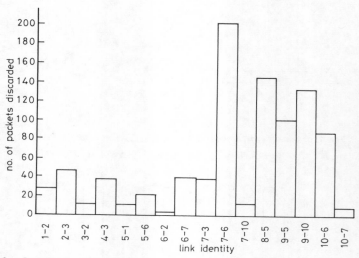

Fig. 4.2.14   *Packets discarded*

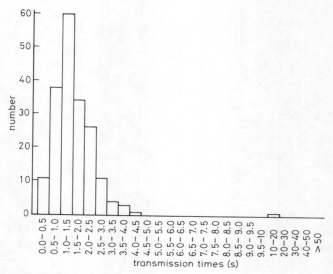

Fig. 4.2.15   *Transmission times*

(f) The histogram of delivery times clearly shows a very skewed distribution with 90% of packets being delivered in under 3 seconds.

4.2.11 Status

Development of the Ptarmigan packet network took place between 1982 and 1985. This was concurrent with the production activity for the circuit switched and message switched system with which the packet network had to be integrated. A design target clearly stated at the beginning of the development was that the new features should be quickly and cheaply retrofittable in production or at least before delivery to the field). By reorganisation of units within cabinets before production started, a full shelf was made available for later retrofit with a primary packet switch in the Ptarmigan circuit switch. For the vehicles which were to house the secondary packet switch, and were therefore to terminate an increased number of subscribers' loops, a new patch panel and termination panel was designed prior to production. Space was of crucial concern (particularly in the armoured installations) and, to gain confidence that the retrofit would be possible, a power consuming, weighted space model was produced early in the development process.

Following a successful period of development and production the retrofit was completed satisfactorily and the first deliveries of the Ptarmigan system to BAOR included the full complement of packet switches.

Table 4.2.4 *Simulation output*

Total packets sourced during simulation	3009 (max. 2518 + 1432)
Total packets retransmitted during simulation	954
Total packets lost during simulation	938
Total packets sunk during simulation	2118
Average no. of packets remaining in system	185
Maximum transmission time	46.640
Link count remaining was	5 links
Packet was sourced at	5.8900 s
Packet was sunk at	52.5200 s
Distance travelled was	4 links
Average transmission time	1.7208 s
Minimum transmission time	0.160 s
Linkcount remained was	6 links
Packet was sourced at	0.1600 s
Packet was sunk at	0.3200 s
Distance travelled was	0 links

The first C^2 system to use the Ptarmigan packet network was Wavell and following extensive interoperability trials, the decision was taken that both systems should be fielded at the same time. Thus the Ptarmigan packet network became the world's first tactical packet switched system to enter service, and since it has been deployed in networks consisting of many nodes and a large number of Wavell subscribers (both processors and terminals).

Since that time extra features planned include full multiple access from a single subscriber loop (i.e. voice, telegraph, fascimile *and* packet), virtual circuit access from a radio subscriber (single channel radio access) and a bulk transfer capability utilising further circuit switched connnections.

This ensures that the Ptarmigan packet switching network continues to provide the services required by all of the evolving C^2 systems and that their smooth introduction into service will be enhanced by the provision of the advanced communication facilities necessary.

4.2.12 References

WARREN, C. S. (1984) 'The Ptarmigan system' *International Defence Review,* Special electronics issue 1, 'Millitary communications'

5: Standards

Chapter 5.1
International standards in military communications

D. F. Bird
(Lynwood Scientific Developments)

5.1.1 Introduction

5.1.1.1 Background: With the introduction of computer-based systems in the field of command, control and communications, a vastly increased flow of information is observed within and between the services, both nationally and internationally. This stems from the need to co-ordinate activities between different role cells, command centres, countries and so on. The result is an urgent requirement for the exchange of data between systems each of which operates under the control of a different organisation. Such organisations function according to diverse procedures. Divisions in responsibility between organisations include:

Strategic and tactical deployment
The split between the three services
UK and NATO interests.

Within the C^3 sphere applications are typically concerned with the manipulation of large databases which are updated by information collected from a variety of sources. Furthermore the databases must in general be duplicated, for reasons of survivability, at widely separated locations. The systems for central command and control, information gathering systems and interconnection tend to be designed independently. In addition, a policy of competitive procurement, pursued in the interests of obtaining the best system at the least cost, ensures that there will be a wide range of vendors involved in the supply and support of the components involved. The components include transmission systems, switching systems and supporting management systems.

It is clearly important that the interfaces between the individual systems or subsystems can be specified unambiguously within an internationally recognised framework. This contribution reviews some of the issues involved in specifying these interfaces. It includes a brief survey of the range of communications techniques that are in use and an indication of the potential difficulties involved in interconnection. Civil standards and the open systems interconnection programme are discussed. The impact of specific military requirements are then considered, and finally some of the initiatives originating within the UK, the EEC and NATO are described.

5.1.1.2 Evolution of communications: The concept of widespread and flexible communications between computers has become a reality only in the last ten to fifteen years. Improving technology (in particular with regard to integrated circuit design and construction) has made available digital switching and transmission systems of increasing performance at decreasing cost.

Early computer communications made use of point-to-point links using the voice grade circuits provided by the national telephone network. A substantial number of systems still do! The infrastructure of the telephone network in the UK is being upgraded to make use of digital transmission and switching, but the local subscriber loops remain analogue. Additional digital services are, however, becoming directly available to the user. Most telephone administrations are aiming towards the provision of an integrated services digital network (ISDN), giving digital access to switchable transparent connections which may carry either voice or data.

A more efficient use of the bandwidth available over individual circuits (which may perhaps only support a modest data rate) is obtained by using packet switching techniques. Packet switching involves separating a data stream into discrete packets each of which is sent from source to destination over a logical channel. The data associated with a number of logical channels can then be statistically multiplexed on to one physical link. Networks are constructed by interconnecting packet switching exchanges; this is well suited to the transmission of 'bursty' traffic, where the mean data rate between subscribers is low.

Within a small geographical location, use may be made of a local area network (LAN). In general LANs provide shared access to a high-speed transmission medium, using packet based protocols. National and international command and control subsystems, rather than use dedicated interconnection links, often employ a network that can provide more flexibility with regard to logical connectivity, as well as offer more reliable communications. The remainder of this contribution assumes the use of this sort of configuration.

5.1.1.3 Problems of interoperability: To achieve communications between computers, some of the important aspects to be considered include:

(a) The electrical and physical characteristics of the connection to the transmission medium or network
(b) The signalling used to ensure the reliable transmission and reception of data (given that no transmission medium is error free)
(c) The means of effecting flow control in order to align the rate of data exchange with the processing capabilities of the computers.

International standards in military communications 213

Table 5.1.1 *Some principal standards organisations*

CCITT	International Consultative Committee for Telegraphy and Telephony
ISO	International Organisation for Standardisation
ECMA	European Computer Manufacturers Association
BSI	British Standards Institute
CEPT	European Conference of Postal and Telecommunications Administrations
CEN	European Committee for Standardisation
CENELEC	European Committee of Electrotechnical Standards

Needless to say, each network provider or computer manufacturer tends to have his own answer to each of these. In order to provide some means of standardising these items, a number of national and international bodies are actively promoting the standardisation of some of the aspects of intercommunication between computers.

5.1.2 Open systems interconnection and the ISO reference model

Open systems interconnection (OSI) is a programme aimed at solving the problems of interoperability (including those highlighted above) and thereby enabling easy and efficient exchange of information between computer systems. Table 5.1.1 provides advance warning of the abbreviations of the names of the organisations involved. A brief description of these and some others is to be found in Section 5.1.7.

5.1.2.1 The ISO reference model: Underpinning nearly all work for communications standards is the basic reference model for open systems interconnection (ISO, 1982). The model, which was developed by technical committee 97 of ISO in the late 1970s, is illustrated in Fig. 5.1.1. The diagram shows the important principle of 'layering', whereby communicating entities in one layer, using the service offered by the layer below, offer an enhanced or expanded service to the layer above. This is achieved by operating suitable protocols between the entities. The functions of each of the layers are summarised in Section 5.1.8. As can be seen, there is a division of layer 3 (the network service layer) between protocols which are operated end to end and those which operate between intermediate network elements. A tutorial on the upper layers of OSI is given by desJardins (NATO, 1985a), and many other aspects of OSI for military application in NATO (1985b).

The concept of layering was not invented by ISO. It is in fact a fairly obvious approach to building upon simple transmission facilities in a structured way, in order to offer a more sophisticated service (including error detection and correction, for example) to an application. Unfortunately, until the arrival of the reference model, each manufacturer tended to create his own layered model with variations in the nature of the service offered at each layer, resulting in an inevitable incompatibility between systems.

Following the publication of the initial draft of reference model, ISO was joined by CCITT in the promotion of OSI. CCITT has produced a set of recommendations which parallel the equivalent ISO documents; some of these are listed in Table 5.1.2. Because of the different way the two organisations work there are occasional problems in achieving alignment of standards. There can also be a reluctance to update a standard by one party for fear of temporarily losing the alignment.

To support the reference model (which is simply the underlying architecture of OSI) it is necessary for ISO to produce standards which define the services precisely and specify the appropriate protocols to achieve such services. Attention was originally concentrated on connection-oriented communications (i.e. those in which there are three distinct phases: call set-up, data transfer and call clear-down).

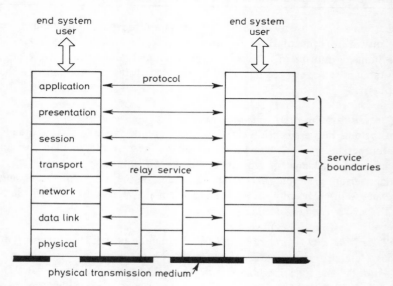

Fig. 5.1.1 *The OSI reference model*

Subsequently addenda have been produced to cover connectionless data transfer, in which units of data are exchanged in isolation with no call set-up and clearing. Lately there has been consideration of another variant, multipoint communication, where three or more systems are in communication, each using a single connection.

Input to the supporting standards comes from far and wide. For example ECMA was responsible for early specifications of transport service protocols. Although

Table 5.1.2 *Some of the ISO/CCITT communications standards*

Layer	ISO	CCITT	
1–7	IS 7498	X200	Reference model
5	IS 8326/8327	X215/225	Service definition/protocol
4	IS 8072/8073	X214/224	specification
3	IS 8348	X213	Service definition

International standards in military communications 215

the network service was defined at an early stage there was some difficulty in determining the appropriate protocols. This was because of the need to encompass a wide range of existing networks which already offered various lower-level services and which in real systems would be used to interconnect end systems. The problem has been resolved by defining 'real subnetworks' and by specifying the means by which the ISO Network Service is operated over one or more of them in tandem. The theory is described in ISO (1985).

5.1.2.2 Recommendation X25: A commonly used protocol which, in its latest version, supports the ISO network service is the CCITT recommendation X25. This defines an interface to a packet switched data network, which has been widely implemented. Although it first appeared in draft form in 1974, thus predating the ISO reference model, it is structured in levels which correspond in principle to the lower three layers of the ISO model.

The packet level of X25 may be used an an end-to-end protocol (rather than as a network access protocol) so that X25 can be used over any subnetwork, whether it be an actual packet switched service or just a local area network. This makes the job of the designer and supplier of end systems very much easier.

A potential problem with X25 is compatibility, which arises because the recommendation has been revised a number of times. The most widely implemented version is that published in the recommendations of the 1980 CCITT plenary assembly (CCITT, 1980; known as the *Yellow Books*). Even this can be found in many variants. This situation arises because manufacturers in a particular country will aim for compliance with the public packet switched service provided by the local national telecommunications agency (commonly known as the PTT) – and each agency has a slightly different preferred implementation.

Continuing co-operation between CCITT and ISO has ensured that the latest version of X25 (following the 1984 CCITT plenary assembly, and published in the *Red Books*; CCITT, 1984) is capable of providing the Network Service. The two versions are usually referred to as X25 (1980) and X25 (1984), respectively.

Support for X25 (1984) is not yet widespread, because the PTTs have to accommodate the large number of existing X25 (1980) users. End systems using X25 (1980) may interoperate over a 1984 subnetwork but the reverse is not true.

Version X25 (1984) is increasingly being specified for private networks, where interoperability is not an immediate problem but where the designers are keen not to be tied to an out-of-date standard when considering future expansion.

5.1.3 Military requirements

Civilian standards are formulated in response to civilian requirements. However, these often overlap with military requirements, but with deficiencies in certain areas. A set of such military requirements has been identified by NATO for the reference model (the background to this activity is described in Section 5.1.5.3). The list of requirements is as follows (NATO, 1983):

Precedence and pre-emption   In order to counteract congestion, particularly in a damaged network where resources are at a premium, it is desirable to be able to allocate resources on the basis of the priority levels assigned to the connections being routed through the congested area. A facility is therefore required to associate a priority level with a connection when it is established.

Security   Protection measures are required to prevent unauthorised access to information and to mitigate against denial of service.

Multihomed and mobile host systems   Multihoming is a mechanism for attaching an end system to two or more network access points without the need for a system setting up a call to it to be aware of the extra connectivity. As well as enhancing survivability, this facility may be used to support mobile hosts such as aircraft and ships.

Multi-endpoint connections   In order to transmit data to a number of recipients (a common occurrence in signal handling) it is usually necessary to establish several connections and send the data on one at a time. A more efficient use of the communications resources (and a better performance in terms of minimising delay) is obtained if the sender has to transmit the data only once and the network then takes care of routing it to the various destinations.

Internetworking   Mechanisms are required to facilitate the interconnection of various NATO systems at the boundary points between networks.

Network/system management functions   Management functions are required that are of greater sophistication than is considered satisfactory for civilian networks. A fast response to changes in network topology (caused by damage to the switching or transmission components) is essential to maintain important connections.

Robustness and quality of service   The need to maximise the survivability of the network is related to the management requirements. The aim is to maintain an adequate quality of service to users (or at least to the users operating at or above a certain priority level) in the face of a damaged or partitioned network.

Real-time and tactical communications   Certain applications (generally tactical in nature) are prepared to sacrifice such aspects of quality of service as sequencing and guaranteed delivery, for the minimum possible transit delay. An example is the collection of radar data, where the information for each scan replaces that for the previous one after each transmission.

Not all of the above requirements are evident in every system, and the mechanisms to implement them do not necessarily impinge upon the user's interface to the network. Certain of the requirements may be fairly easily realised as an extension to the network layer protocol. For example, a priority indication for an X25 virtual call (to enable precedence and pre-emption to be operated within the network) may be specified by a parameter in a user facility field in the call request packet. The attraction of this approach is that the protocol still conforms to the published standard. An alternative approach is to define a special purpose (and hence nonstandard) version of the protocol, tailored to a particular set of requirements. As a short-term measure this approach permits optimum performance to be achieved, but in the long term interoperability may be made problematical or impossible.

5.1.4 Gateway issues

Even with the existence of standards for services and protocols, problems are still likely to arise in the interconnection of two networks. Figure 5.1.2 shows a general scenario in which two computers are separated by more than one network. It is

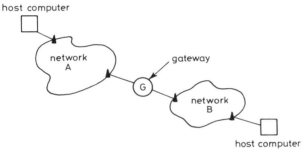

Fig. 5.1.2 *Use of a gateway between dissimilar networks*

unlikely to be possible to effect a direct connection between the two networks (thus creating one large network) for various reasons, mainly concerned with management procedures — routing, accounting etc. (Such a connection can be effected between two public X25 networks using the procedures of CCITT recommendation X75, but this is of limited use in the private domain.) The gateway shown in the figure is therefore interposed between the networks. This immediately raises the question of ownership or administration of the gateway: is it considered part of network A or of network B?

Where gateways have been introduced, their design has tended to be tailored to a particular environment. The discussion below indicates the reasons for such specialised designs. It is assumed that the gateway is operating in the approved OSI manner — that is, interconnecting two subnetworks as a relay at the network layer.

The responsibilities of a gateway include:

Mapping of services The complexity of this task depends on how disparate are the services provided by individual networks. The fact that both apparently offer the same service may not eliminate the need for this task. For example, in the case of X25, one network may support the fast select option while the other may not. Each end system may have to sacrifice some of the aspects of the desired service offered by its own network in order to achieve a connection.

Address conversion Each network may use its own addressing scheme, and *ad hoc* mechanisms for passing an external address to the gateway may be demanded.

This problem is tackled by ISO (1986), which defines the syntax and semantics for a global network layer addressing scheme.

Flow control and congestion control   The gateway must make use of the appropriate facilities provided by each network to control the flow of data between the networks.

Enforcing security policy   In the defence environment it is likely that each network will have its own security policy. This will be enforced by whatever facilities an individual network provides for access control, authentication etc. A strategy must be formulated for the gateway, which has then to be trusted (by the administrators of each network) to carry it out. This raises issues of computer security such as validation and verification.

Management   The management functions of the gateway permeate all the tasks listed above and lead back to the original question of who determines how decisions are made or how the supporting information (such as routing tables) are to be obtained. Work within ISO on management issues is still at an early stage, with no more than a basic architecture as yet agreed.

The job of the gateway is rendered even more onerous if one or both networks has been enhanced to cater for some combination of the military requirements listed earlier. For example, a mobile host wishing to move across network boundaries poses problems of ensuring that a connection is maintained at all, let alone at the original quality of service.

The gateway approach to internetworking has been pursued quite successfully within the DARPA internet system in the USA. User data transfer is provided by two protocols: internet protocol (Dod, 1981a) and transmission control protocol (Dod, 1981b). The internet protocol is supported by all gateways within the system. A demonstration of this sytem operating between the USA and Holland was given in 1985 (NATO, 1985b).

5.1.5 Initiatives

Many organisations have an interest in expediting OSI implementations now, while some of the standards are still in the process of formulation.

5.1.5.1 UK Department of Trade and Industry: Following an investigation of information technology, the UK Department of Trade and Industry (DTI) established the Information Technology Standards Unit (ITSU) in 1982 with the intention of supporting the development and establishment of OSI standards and their adoption within British Industry.

The activities of ITSU include:

Formulation of an intercept strategy
Publication of technical guides
Organisation of implementation working groups.

International standards in military communications 219

The intercept strategy (DTI, 1985) lists those ISO and CCITT working documents which, though they are not yet accorded formally issued status (i.e. in the case of ISO, they have not yet reached the IS stage), are considered sufficiently stable to be used as input to the design of communications systems.

Intercept recommendations are then published within technical guides, each of which considers a particular layer or service within the reference model. Most of the guides published so far deal with the lower three layers of the model, including the use of X25 in various configurations (such as within a local area network).

In co-operation with the National Computing Centre (NCC), ITSU has set up a number of OSI implementors groups aimed at bringing together representatives from industry. The following areas are currently covered:

Transport layer
Local area networks
File transfer
Common application service elements
Message handling systems.

Of the other activities, one of considerable importance is concerned with the subject of conformance testing. What is claimed to be the world's first commercial OSI testing service is being run by the NCC with the support of the DTI (Davidson, 1985).

Potential implementors wishing to obtain the relevant ISO documents can make use of drafts for development (DD) published by the BSI. These are simply copies of the ISO documents, to be replaced by British Standards when they are finalised by ISO.

5.1.5.2 The European programme: Within the European community there is a programme to harmonise the use of IT standards. This programme is jointly organised by CEN, CENELEC and CEPT. It is focusing on the development of a range of application-related functional standards. It is the intention of the DTI to align its intercept strategy with the European programme.

5.1.5.3 NATO: Within NATO there are many organisations concerned with ensuring that effective use of standards is made in the design of communications systems. The guiding document is the NATO interoperability management plan (NATO, 1982), issue 2 of which is to be published at unclassified level.

There is an additional NATO document (NATO, 1984) which details a transition strategy for the implementation of standards. It provides recommendations which mirror the ITSU UK intercept strategy. It includes a set of tables for each of the seven layers of the reference model which are updated on a regular basis.

That document is an output from study group 9 (SG9) of the NATO Tri-Service Group for Communications and Electronic Engineering (TSGCEE). This study group is responsible for formulation of standards for data processing and distribution. Of the other study groups, SG8 deals with satellite communications, SG2 with radio communications and SG1 with tactical ground area communications. The brief of SG9 includes consideration of both strategic and tactical systems. Input to SG9 is provided from the UK by the Ministry of Defence.

One of the aims of SG9 is to influence the production of civilian standards with regard to the incorporation of military features. This is achieved by member nations approaching their respective national standards bodies (who are in turn represented within ISO) with proposals for amending or enhancing standards under development.

A number of STANAGs are in production to provide NATO implementors with recommendations for services and protocols within each of the layers of the reference model. STANAG 4250 is the NATO equivalent of the reference model; it introduces the supporting STANAGs and directly references IS 7498. The STANAGs for each layer are intended to follow the developing civilian standard, defining the NATO interpretation and any enhancements as necessary.

5.1.6 Summary

The formulation and use of international standards for communications is a difficult enough matter in the civilian environment, let alone when military requirements are taken into account.

There is a conflict in development and procurement. On the one hand, those responsible for military equipment adhere to a policy of obtaining off-the-shelf commercially available equipment, with the benefits that this brings in terms of price, maintenance, enhancibility and flexibility. On the other, military considerations seem to demand special developments to satisfy stringent security and survivability requirements. This conflict may be alleviated if the military requirements (which often prove to be of interest to civilian designers) are incorporated within the international standards with which most manufacturers are now conforming.

5.1.7 Appendix A: Standards bodies

Kearsey and Jones (1985) provide a detailed examination of the principal organisations involved with the formulation of standards. The following list gives a brief outline of some of them.

International Organisation for Standardisation (ISO)   ISO was set up in 1946 as a specialised agency of the United Nations. It is responsible for all aspects of standardisation except those regarding electrical and electronics engineering, which are the work of the International Electrotechnical Commission (IEC). Its 89 members are the national standards bodies of the member countries (for example the British Standards Institute in the UK). It functions through the work of a number of technical committees. Technical committee 97 deals with computers and information processing.

International Telecommunications Union (ITU)   The ITU was founded in 1865. It was recognised as a specialised agency of the United Nations in 1947 and currently

International standards in military communications 221

has 160 members. Its principal aim is to maintain and extend international co-operation in the field of telecommunications.

International Consultative Committee for Telegraphy and Telephony (CCITT) The CCITT is one of the arms of the ITU. It was formed by a merger of the individual telegraphy and telephony committees in 1956. Its members are national telecommunications agencies. It produces standards for the telecommunications industry in the form of recommendations, a complete set of which is published every four years at the conclusion of each plenary session. The recommendations are produced by the efforts of various study groups. Recommendation X25 was produced by study group VII.

British Standards Institute (BSI) The BSI is the national standards body for the UK. It is an active contributor to ISO on the subject of OSI. The Office Information Systems (OIS) committee comprises a set of working groups which correspond to the subcommittees of ISO TC97.

Institute of Electrical and Electronics Engineers (IEEE) The IEEE is one of the many bodies responsible for setting standards in the United States. Its impact on OSI stems from work done in project 802 to define standards for emerging local area network technologies. The output from this project has since been incorporated by ISO into the 8802 series of standards.

European Conference of Postal and Telecommunication Administrations (CEPT) CEPT was founded in 1959. Its members are 26 European PTTs. It is one of the parties to the European programme (together with CEN and CENELEC) and it has provided input to CCITT on ISDN.

European Committee for Standardisation (CEN) CEN was formed in 1961 by representatives of the national standards bodies of the EEC and the European Free Trade Association.

European Committee of Electrotechnical Standards (CENELEC) CENELEC is a regional version of the IEC.

European Computer Manufacturers Association (ECMA) ECMA was founded in 1960. Its membership is drawn from European industry but includes US interests by way of their European outposts.

5.1.8 Appendix B: The seven layers of the ISO reference model
1 The *physical layer* deals with the transmission of a bit stream across a physical medium.
2 The *data link layer* is concerned with the reliable transfer of data across a physical link. This involves the provision of error detection and correction etc.
3 The *network layer* provides for transmission of data between end systems possibly separated by any number of relay systems and data links. The functions of the network layer include addressing and routing. The service provided to the

end system is independent of the particular technologies used in the intermediate elements.
4 The *transport layer* provides an end-to-end service which makes optimum use of the supporting network services to provide the user with the required quality of service.
5 The *session layer* handles the organisation of the dialogue between users. Synchronisation and recovery facilities are provided.
6 The *presentation layer* resolves the differences between the representations of data used by the end systems.
7 The *application layer* supports the application processes, for example by providing a file transfer service.

5.1.9 References

DAVIDSON, I. (1985) 'Testing conformance to OSI standards', *Computer Communications*, 8(4), p. 170
DOD (1981a) 'Dod standard internet protocol', RFC 791, Information Sciences Institute, USC
DOD (1981b) 'Dod standard transmission protocol', RFC 791, Information Sciences Institute, USC
DTI (1985) 'Department of Trade and Industry (UK), Information Technology Standards Unit — Intercept Strategy'
CCITT (1980) "Yellow book, Fascile VIII.2 Data Communication Networks, Services and Facilities, Terminal Equipment and Interfaces", Recommendations. X.1—X.29
ISO (1982) 'ISO/IS — 7498 information processing systems interconnection — basic reference model'
CCITT (1984) "Red, book, Fascile. VIII.3. Data Communication Networks, Interfaces", Recommendations X.20—X.32
ISO (1985) "The internal organisation of the network layer", DIS 8648
ISO (1986) 'Network service definition — addendum 2', DIS/8348/DAD-2
KEARSEY, B. N. and JONES, W. T. (1985) 'International standardisation in telecommunications and information processing', *Electronics and Power*, September, pp. 643—51
NATO (1982) 'NATO interoperability management plan (NIMP)', Allied Data Standards Interoperability Agency, D/1
NATO (1983) 'Military requirements for open systems', NATO TSGCEE subgroup 9, note AC/302/(SG9)/D19
NATO (1984) 'Standards transition strategy', NATO TSGCEE subgroup 9, note AC/302/(SG9)/D30
NATO (1985a), desJardins, R. 'Tutorial on the upper layers of OSI', NATO symposium on the interoperability of ADP systems, SHAPE Technical Centre, the Hague, Netherlands, March
NATO (1985b) Proceedings of the second international symposium on the interoperability of ADP systems, SHAPE Technical Centre, the Hague, Netherlands, March, session 3

Chapter 5.2
C³I and the upper layers of the OSI

H.J. Pearson
(Smith Associates Ltd)

5.2.1 Introduction

Considerable effort has been expended over the past three years or so within the ISO community in progressing standards within the upper layers (session, presentation and application) of the OSI reference model. As a result an initial, although by no means complete, set of protocols in each of these layers is now available, either as published international standards or as technically stable drafts. These protocols are therefore candidates for use in military C³I systems.

The previous contribution has provided background to:

(a) The evolution of communications and the problems of interoperability
(b) OSI and the ISO reference model
(c) Military use of standards and their special requirements
(d) Gateway issues
(e) National, European and NATO initiatives to expedite OSI implementations.

This paper develops some of these themes in the context of the upper-layer standards. It also discusses the structure of the standards making up these layers of the architecture and their relevance to military C³I systems.

5.2.2 The military use of standards

There is a clear case for the use, in specifying military C³I systems, of common standards that are agreed, well tried and tested, well structured and comprehensive, in order to facilitate:

(a) The specification and development of multivendor systems
(b) Interoperation between different systems
(c) Maximum commonality between different systems, even if they are not required to interoperate
(d) Compatibility with future replacements for, or extensions of, any particular system.

The arguments for making use of the emerging civilian OSI standards are primarily that:

1. They will meet the basic criteria listed above.
2. Off-the-shelf equipment (both hardware and software) will be available, hence potentially reducing development and production costs and technical risk.
3. Potential suppliers should be familiar already with the protocols and interfaces used, hence minimising system integration times and costs.

However, before it is concluded that the emerging civilian OSI standards will provide a universal panacea for many of the problems associated with procuring military C^3I systems, the following issues, some of a fundamental and some of a practical nature, must be addressed.

Suitability of civil standards Existing OSI standards may not meet specific military requirements, and need modification or augmentation in order to do so. This issue is discussed in detail in Section 5.2.3.

Compatibility with existing systems National and NATO military authorities have an enormous investment in existing communications systems which will, for the foreseeable future, remain in use and have to interwork with newer, OSI compatible systems. Some of these current systems use protocols which are extremely difficult or impossible to map on to a layered architecture and provide data transfer facilities (services) very different to those of the OSI standards. Integrating these systems into an overall, OSI-based communications architecture will pose a major problem.

Choice of new standards It may not be immediately apparent which of the variety of emerging application standards would be the most appropriate to meet a particular requirement. For example, in some cases it may not be clear whether a message handling system (X400 or MOTIS) or a file transfer standard (FTAM) (see Section 5.2.7) would be the best to specify.

Completeness of new standards At present the OSI standards are very new, particularly those in the upper layers, and so the general user community has little experience of their use. This inexperience may lead to the incomplete or misleading specification of systems based on OSI protocols. For instance, an incomplete set of protocols may be specified initially. Alternatively, where a standard provides one or more optional areas of functionality, insufficient information may be given in the system specification as to which, if any, of the optional features must be provided.

Of the above four issues, the first two are of a fundamental nature whereas the final two are of a (hopefully) short-term practical nature.

Specific features that have been identified as being required by military systems, and the attempts that have been and are being made to include these requirements in the OSI architecture and standards, are discussed in the next section. The impact that these military requirements are having on refining the OSI architecture and standards is also discussed in the next section.

Progress may seem to be slow to those who must make decisions about project specification details according to externally imposed procurement timetables, and who cannot necessarily wait for the final deliberations of standardisation bodies. However, the military impact on the OSI architecture and the functionality of some of its component standards should not be underestimated.

Major examples of existing communication systems that are likely to prove difficult to integrate directly into an OSI environment are those based on some of the NATO digital data link protocols (for example Link 11, Link 16). These systems not only represent large investments in terms of both development effort and hardware, but also have been optimised for the particular operating conditions under which they are used. This optimisation has tended to lead to highly non-layered protocols in which the functions of all layers of the OSI reference model are intermingled.

Issues in the choice of and completeness of new standards are closely related and, although they should only be associated with the initial period of introduction of OSI-based standards, they may well cause unnecessary confusion and uncertainty during the initial stages of the procurement of new systems. They can only be overcome by education, experience and the careful drafting of the relevant NATO STANAGs.

5.2.3 Military requirements

The previous contribution outlined the list of eight military features introduced in 1983 by subgroup (SG) 9 of the NATO Tri-Service Group for Communications and Electronic Engineering (TSGCEE) (see Section 5.1.3). These were felt to be problem areas which the then emerging OSI standards did not adequately cover. The requirements are

Multihomed and mobile host systems
Multi-endpoint connections
Internetworking
Network/system management functions
Security
Robustness and quality of service
Precedence and pre-emption
Real-time and tactical communications.

These were originally identified in the context of the network and transport service standards. Further work within SG9 has, and is currently, examining the military requirements at the other layers. This work is not yet complete, particularly for the upper levels but no new requirements have been found that do not fit within the existing categories.

The principle of operation of SG9 in its approach to the use of OSI standards has been to attempt to get military requirements included as modifications or extensions of the basic ISO standards whenever possible, rather than to develop a separate military reference model and protocols. It could be argued that, to date, the efforts of SG9 have borne little fruit in terms of the inclusion of text reflecting military requirements in published international standards. However, the efforts of the NATO community, co-ordinated by TSGCEE SG9, have had considerable impact within ISO in helping to expose deficiencies in the reference model and specific layer standards.

These efforts have been instrumental in setting up new work items within ISO for the progression of addenda to the reference model to cover:

Connectionless operations
Security aspects
Multi-peer (multi-endpoint) communications.

Technical support within ISO has and is being provided in each of these areas by individuals and organisations connected with TSGCEE SG9.

The requirement to support mobile host systems is now receiving some attention within the ISO community, and is an area where further military input would be timely.

Protocols for connectionless data transfer must support systems in which units of data are exchanged as individual items, rather than as parts of a structured communication between the parties involving the setting up of a specific entity — the connection — to embrace the exchange of a stream of data items between the parties (connection-oriented data transfer). The OSI reference model was originally developed on the basis of connection-oriented communications; however, there are clear military requirements for connectionless operation of protocols at some or all layers of the model. These requirements arise for a number of reasons, including:

(a) Applications such as the transmission of sensor data, telemetry data or packet voice where minimum transmission delay is at a premium and the occasional loss of data, bit errors and out-of-sequence delivery of packets are more acceptable than the delays introduced by connection-oriented protocols to detect and recover from such errors
(b) Applications operating over transmission media (for example, combat radio nets or local area networks) which, because of their broadcast nature, naturally provide a connectionless service
(c) Applications which, owing to special circumstances such as radio silence, require strictly one-way transmissions.

The connectionless addendum to the reference model is now in place, and addenda to standards in layers 2–6 to provide connectionless services are now being progressed. The author is not aware of any inherently connectionless application layer standards being worked on within ISO at the moment. However, it seems likely that the use of distributed directories of names and addresses at both network and application levels will involve connectionless operation.

The security of data transfers is of paramount importance to many military systems and is, to say the least, a complex issue. Work, with considerable military input, has led to the drafting of a security architecture which can now form a framework for the necessary amendments to individual layer standards. Security implications for the presentation layer standards are discussed briefly in Section 5.2.6, and a short review of security in military OSI networks is given in Section 5.3. The provision of adequate security features in application standards is an area which requires urgent work.

Military requirements for robust communications systems may be seen in terms of quality of service (QOS) and management. Precedence and pre-emption issues — encompassed in the term 'priority' in ISO terminology — may also be considered aspects of QOS. In these two areas of QOS and management, ISO standards are either deficient or still at a relatively early stage of development.

It has become clear that QOS is currently a problem area for the potential military user of OSI standards because insufficient architectural consideration has been given to this issue, which is inherently multilayer in its scope. TSGCEE SG9 is attempting to tackle the problem in two ways:

(a) By providing specific input to ISO for amendments to the network service definition to allow implementation of required QOS features
(b) Through initiation of a more general review, initially within NATO, of QOS requirements.

Communication management has long been recognised, in ISO and related bodies, as a crucial area for standardisation. However, for a variety of reasons, although work started on the topic as long ago as 1980, concrete progress has been slow. Current work is progressing on three fronts:

The OSI management framework
Management information services
Directory services.

Documents in these areas have not yet (May 1986) reached draft proposal status. The implementation of C^3I systems therefore cannot yet make use of any off-the-shelf standards in this area. However, military input to these standards is important and will be effective if made available in the near future.

It may be seen that affecting the content of civilian OSI standards is a long-term process requiring careful co-ordination and expenditure of considerable amounts of technical effort.

5.2.4 Structure of the upper layers
Before turning to the features provided by specific standards in the upper layers of the reference model, it is useful to consider briefly the architecture of these upper layers. The relationship between the layers and their internal structure is showed schematically in Fig. 5.2.1.

The structure of the session and presentation layers is straightforward and, as with the lower layers, is strictly hierarchical. There is precisely one protocol standard sitting above and using the services of the immediately lower layer. This is not the case in the application layer. Many different standards are being produced, each capable of using the presentation service. These standards may, in general, be interrelated such that one standard builds on and uses the services of another. This does not necessarily imply a strictly hierarchical relation. In some applications, standard A may use standard B. In some applications the reverse could hold. In some applications, A might use B which uses C, whereas in other applications A could use C directly.

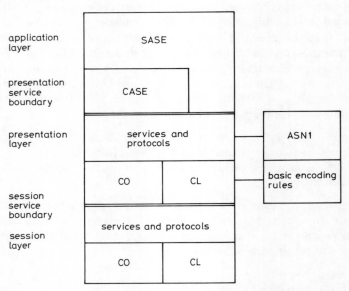

Fig. 5.2.1 *Structure of the upper layers. CO: connection-oriented, CL: connection*

These features of the application layer standards are a manifestation of the richness of functionality available from them, but must be borne in mind by anyone specifying or implementing a system based on them.

It was recognised by ISO that some of the operations of different standards are similar, and could usefully be standardised separately as a set of elements forming a tool-kit to be referenced by other standards. This led to the concept of common application service elements (CASE). CASE protocols are discussed in Section 5.2.7. Other application standards are referred to as specific application service elements (SASE).

Precise details of the architecture of the application layer have been, and are still, a fertile area of debate within ISO, and the final picture is not yet entirely clear. However, sufficient progress has been made for it to be possible to build a system using protocols from all seven layers of the OSI architecture.

Returning to Fig. 5.2.1, the provision of both connection-oriented and connectionless services by the session and presentation layers has been indicated. As was noted in Section 5.2.3, no connectionless SASE standards have yet to be identified by ISO. This has led to some question as to whether connectionless services are required in the upper layers. However, as a military requirement appears to exist, any attempt to drop these protocols should be resisted strongly by those seeking military use of these standards.

The standards for abstract syntax notation one (ASN1) and its basic encoding rules are shown as being closely associated with the presentation layer, as the two sets of standards are closely bound. The use of ASN1 is discussed in Section 5.2.6.

It is worth noting that it has always been envisaged that other abstract syntax notations may be needed for applications with very different primitive elements and construction elements from those naturally represented by ASN1 (which is oriented towards text-like or numerical data structures). For example, the primitives required to represent graphic information may be points, lines, colours, shadings, line markings (dotted, dashed, full etc.), point markings (triangle, cross, square etc.), and the constructional mechanisms may be polygonal forms, splines, infills, ellipses and so on, together with overlaying, windowing and notation. Other requirements for differing abstract syntax notations occur in videotex communications and digitial voice work.

It is also envisaged that different encoding rules may be required in some cases to provide for example:

(a) Encryption
(b) Data compression
(c) Verbose but simple and readable encodings.

These are all areas of direct relevance to military C^3I systems, but as yet nothing is available as a standard. However, the overall structure is in place for these extensions to be handled easily.

5.2.5 Session layer

The session layer consists simply of two standards, for the services provided to the presentation layer and the protocols that implement these services using the transport service. The session documents have been published as full international standards, but currently addenda to them are being progressed to cover:

(a) The provision of connectionless services
(b) Amendments arising from the requirements of particular application standards.

All OSI application standards use the session (and presentation) standards and so will inevitably be required by military C^3I systems.

The function of the session layer is to provide dialogue control, which involves:

1   Connection establishment
2   Management of data transfer, including synchronisation and recovery following transport connection failures
3   Orderly release of connections.

Both two-way simultaneous (full duplex) and two-way alternate (half-duplex) data transfer can be supported.

The session service provides a very wide range of primitives, grouped into what are called functional units. All these primitives are made available to the application layer by the presentation layer as pass-through services. The precise set of functional units required on a connection depends on the application layer standard involved. Implementations of the session service need only implement the functional units needed by the applications they support. In practice this may lead to considerable reductions in the quantity of protocols to be implemented in comparison with a full session service implementation. Consequent size and cost savings may be useful.

5.2.6 Presentation layer and abstract syntax notations

Application layer standards must specify the transfer of some form of data structure on a connection. This data structure may be anything from a relatively small piece of application protocol (consisting of perhaps one or more optional parameters) to a relatively large data structure consisting of a lengthy report or part of a file or database. The definition of the meaning to be transferred (semantics) is entirely an application layer matter.

However, the application standards are not involved in determining the details of the representation (precise bit pattern) used to transfer the information between communicating systems. This process of converting the data to be transferred from whatever internal representation is used within the end system into the form used for communication (the transfer syntax) is carried out by the presentation layer.

The definition using a well defined notation of the data structures passed by the application layer to the presentation layer constitutes what is called an abstract syntax. The well defined notation for the definition of abstract syntaxes (an abstract syntax notation) could, in theory, correspond to any programming language with a sufficiently rich data structuring notation (for example Pascal or Ada) or a variant of Bachus-Naur form (BNF).

For the purpose of the presentation layer, any abstract syntax notation needs one or more accompanying sets of encoding rules that determine algorithmically for any set of data structures defined using the notation, the transfer syntax to be used. Programming language (and BNF) notations for data structure definitions are not provided with such encoding rules.

CCITT and ISO have defined abstract syntax notation one (ASN1) and a corresponding set of basic encoding rules. ASN1 is used extensively in upper-layer standards for specifying both the protocols themselves and the standard data types to be transferred.

The question of whether NATO standard message formats could and should be respecified using ASN1 is an area currently under study. There is no requirement to employ ASN1 associated encoding rules for defining abstract and transfer syntaxes when using the presentation layer protocols. However, as indicated in Section 5.2.4, there are possible advantages in making the distinction between abstract and transfer syntaxes explicit, in that a variety of encoding rules (for example to provide encryption or compression) can then be employed with the same abstract syntax notation.

As with the session service, the presentation service must be made available in any C^3I system using application layer standards, at least notionally. For compatibility with existing CCITT standards such as teletex and X400, in which the transfer syntax is effectively specified at the application layer, the presentation service may be used in such a manner that the presentation layer is effectively null, with no additional protocol control information added and no encoding of data taking place.

Military requirements for security features and multipeer connections (see Section 5.2.3) will impact on the presentation layer, and future revisions to the standard will be required.

5.2.7 Application layer standards

5.2.7.1 Overview: This section surveys the range of application layer standards currently being progressed by ISO which are candidates for use by military C^3I systems.

None of these standards (except the graphical kernel system) has yet (May 1986) been published as a ratified international standard (IS). However, concerted efforts are being made within ISO to process a package of standards consisting of CASE (kernel and CCR); file transfer, access and management (FTAM); and presentation. The aim is to make them available by early 1987 so that complete implementations of ISO file transfer will be possible.

Except in the case of message handling systems (see Section 5.2.7.8) the author is aware of little work on enhancements to application standards needed to meet military requirements. The major area for these enhancements is almost certain to be the provision of additional security features. Security enhancements may, eventually, be required in almost all of the standards.

5.2.7.2 CASE: The current ISO CASE standards for service and protocol each have three parts:

1 A very short introduction to the CASE standards
2 The association control CASE standard, commonly known as kernel CASE
3 Commitment, concurrency and recovery (CCR) standard.

Kernel CASE is concerned with overall control of the use of a presentation connection by one or more SASE standards and associated CASE standards. As such, although required as a set of protocols for any OSI information transfer, the details of its operations are only of technical interest.

However, commitment, concurrency and recovery CASE provides a much richer functionality. The CCR protocol is used to enable successful completion of activities distributed across several open systems, taking into account possible communications network failures and end computer system crashes. CCR standardisation is well advanced and full IS status should be achieved by mid 1986.

CCR protocols are designed to be used in close conjunction with many of the

SASE standards. This use of CCR may be optional (as in the case of FTAM) or mandatory (as in the case of job transfer and manipulation (JTM)).

CCR protocols are concerned with resolving two types of problem which may occur with the following simple protocol: an initiator requests an action, and a responder either performs the action and returns an acknowledgment or refuses the action and returns a diagnostic. The two types of problem arise when one or other system crashes, or the initiator needs to work with two or more responders simultaneously (for example to debit one bank account and credit another).

In the case of system crashes with a simple protocol, the initiator does not know whether

(a) The action was performed but the crash lost the acknowledgment; or
(b) The action was refused and the diagnostic was lost; or
(c) The request for the action was lost.

CCR ensures that following a crash the initiator can determine what has happened and can recover the situation. This recovery protocol would normally be achieved by storing status and other data on disk at critical points in the transaction so that they survive a crash. This ensures that actions are neither lost nor duplicated on restart after a crash. Related to this recovery aspect, CCR protocols also ensure that other users are prevented from accessing data that may be in an inconsistent state in different parts of a distributed system because of a crash. (This is a concurrency aspect of CCR.)

The second problem area with simple protocols is when, for consistent operation of a distributed system, changes requested by an initiator (master) of two or more other systems (subordinates) must either occur simultaneously or not at all. In this context 'simultaneously' means before any other user accesses the data being changed (another concurrency aspect). In the simple protocol case the master commits himself to the action possibly taking place as soon as he makes the request to any of the subordinates. An inconsistent state may then be reached if some of the subordinates accept the action and some refuse it (as an example, think of the money transfer case).

CCR gets round this commitment problem by using a two-phase commitment process. The master system initiates the action and receives a commitment (or refusal) to perform the action if ordered by each of the subordinates.

If the master receives an offer to commit from each subordinate, he can then order commitment. Once a subordinate has offered commitment, concurrency controls are imposed (for example, file locking) to ensure that no other action can remove the ability of the subordinate system to perform the action to which it is originally committed.

If one or more subordinates refuse the action the master then issues a rollback command which restores all resources to their original state and releases all concurrency controls.

It seems likely that the CCR protocols could potentially be an important component of many future distributed C^3I systems, as the features they provide —

correct recovery from system crashes or disruptions, and the maintenance of database consistency — are certainly required. Indeed CCR (in conjunction with FTAM) has already been specified for use in the UKAIR CCIS project.

5.2.7.3 FTAM: The file transfer, access and management (FTAM) standard is concerned with the manipulation of identifiable bodies of information called files. It provides sufficient facilities to support file transfer, and establishes a framework for file access and management. It is expected that future revisions of the standard may lead to enhancement in the means of file access and management.

FTAM does not specify the interfaces to a file transfer or access facility within a real local system. The architecture of FTAM involves the concept of the virtual file store — really a data structure definition, with large parts of the structure, the files, left undefined. The overall concept is, at one end, of a user of FTAM (the initiator) issuing service primitives, and, at the other end, of corresponding service primitives being issued whose semantics are defined in terms of operations on the data structure (the virtual file store).

The virtual file store could represent a wide range of real-world devices, including:

1. Real discs
2. Magnetic tape decks
3. Line printers
4. Plotters
5. Card readers
6. User programs maintaining or generating documents (files) in whatever way they wish.

However, it is likely that most FTAM implementations will map the virtual file store on to real disks or other mass storage devices.

The virtual file store consists of a number of structured files, each of which has associated with it one or more attributes. File attributes which may or may not be supported include:

File name (mandatory)
Access type (random or sequential)
Access control
Account (for charging purposes)
Data and time of creation
Data and time of last modification.

Each file is composed of a structure of file access data units (FADUs). Each FADU may or may not contain a single data unit, and may or may not contain other FADUs. Hierarchical trees of FADUs or flat sequences of FADUs may therefore be built up. Each leaf FADU must contain a data unit.

A data unit is a piece of data of any form whatever, wholly unconstrained by FTAM.

FTAM allows the following operations on a file:

Read any FADU
Insert new FADUs
Replace any FADUs
Delete any FADU
Add data to a data unit
Read file attributes
Change attributes (some attributes only)
Delete whole file.

The access control attribute can restrict a particular type of file access operation to:

Specific user identifiers
Specific password holders
Specific calling addresses.

However, it seems likely that additional security features will be required to ensure that FTAM meets specific military requirements.

As indicated above, there is already some military interest in the use of FTAM, and this seems likely to grow. It is not yet clear to what extent FTAM (possibly extended) could be a tool for supporting the operation of distributed databases within co-operating C^3I systems.

5.2.7.4 JTM: Job transfer and manipulation (JTM) is a complex standard that provides a set of communication-related services which may be used to perform work in a network of interconnected systems. This work can include the running of traditional jobs. The JTM protocol not only covers the movement of job-related data (input and output) between open systems, but also provides for the movement of data concerned with monitoring job-related activity, and controlling and manipulating the progress of this activity.

The author is unaware of any military C^3I applications for JTM yet identified.

5.2.7.5 Virtual terminal protocols: CCITT work on protocols for supporting remote terminal access to computers has been in place for some years, and is in widespread use over X25 networks. These are the so-called triple-X recommendations — X3, X28 and X29. Triple-X is closely tied to traditional scroll-mode ASCII terminals operating over asynchronous lines.

ISO has taken a rather different line to CCITT and has adopted the virtual terminal (VT) approach. In the VT model the program and the terminal intelligence share a common data structure, broadly modelling the screen. Both are able to read it and, subject to access controls of the token variety, both can write to it. How the information is presented on a real display is left up to the terminal.

A number of classes of VT standard are at various stages of development within ISO. These are:

Basic class

Forms class
Graphics class
Image class.

Only basic class VTP has reached draft proposal stage, and the details of the others are still far from clear.

The display object (data structure corresponding roughly to the terminal screen) in basic class VTP is a one-, two- or three-dimensional array of elements. Each element is either null or has a value for a primary and some secondary attributes. The primary attribute is a character from some specific character set and font. The secondary attributes specify:

Level of emphasis
Foreground colour
Background colour.

Images in this context include text, line graphics, videotext graphics, facsimile and even voice (sound).

There has been little military input to these emerging protocols as yet, although C^3I systems would seem to be potentially major users of such standards.

5.2.7.6 Graphics: A major piece of graphics standardisation was completed in 1985 with the final approval of the IS text for the graphical kernel system (GKS). GKS language bindings for several languages (Fortran, Pascal and Ada) are also at an advanced state. GKS is a relatively well known standard, and a wide variety of implementations are already available.

There is now interest within ISO in another graphics standard, PHIGS (programmer's hierarchical interactive graphics system). This has considerable functional overlap with GKS but some significant functional extensions. PHIGS is designed to support applications needing a highly dynamic, highly interactive operator interface, and expects rapid screen update of complex images to be performed by the display system. There is already some military interest in using PHIGS for C^3I systems.

5.2.7.7 Database work: The support of distributed databases is a key area for military C^3I systems. However, work within ISO on relevant standards is still at a relatively early state. Recently, work has started in the following areas:

1   An architectural study of the protocol support needed for distributed databases
2   A protocol for access to a remote database.

Work in the second of these areas has begun with a substantial document by ECMA (see Section 5.1), which has rewritten and extended FTAM, CCR and ASN1 to provide an independent standard in this area. ISO work is likely to attempt to align this work to use the actual FTAM and CCR standards.

Although no standards are yet available from ISO in this area, it is one in which the military community already has considerable expertise of advanced requirements and so should be able, through technical inputs to ISO, to make a mutually beneficial impact on the future development of standards.

5.2.7.8 Message handling systems: Message handling systems (also referred to as text interchange or electronic mail systems) are being standardised by both CCITT, as the X400 series of recommendations, and by ISO, as the MOTIS (message-oriented text interchange system) standards. Series X400 is concerned with providing message handling over public data networks, whereas MOTIS, which is based strongly on X400, allows for message transfer over both private and public networks.

There is considerable military experience at both a national and a NATO level in the use, interoperation and standardisation of message handling systems. Recently a number of projects which involve new message handling systems in various NATO nations (UK, Norway and Germany) have led to an initiative coordinated by TSGCEE SG9 to develop a standard for a military message handling system (MMHS) based on X400/MOTIS. The method of working has been:

(a) To identify extra military requirements not covered by the current civilian standards
(b) To develop enhancements to the civilian standards to meet the military requirements
(c) To input these proposed changes to the civilian standards to CCITT/ISO via national representatives.

The overall military requirement was to provide compatibility with existing (ACP 127) messaging systems. This involved enhancements to the civilian standards in the areas of:

1 Priority parameters
2 Security features, including support for various levels of encryption and routing control
3 Increased end system functionality to support the signal office concept.

Considerable technical effort has been put into developing these ideas over the past year, and text has been input to ISO defining the enhancements required. However, although there has been support of at least some of the proposals from the civilian community, because of the current MOTIS standardisation timetables it is not clear whether any changes to the standards can be effected in the short term. In this case, unfortunately, the military systems currently being procured will not be able to make use of emerging ISO standards.

Related work within CCITT which may be of future relevance to C^3I systems concerns the possible development of standards for distributed conferencing protocols and file transfer protocols based on X400 series protocols.

5.2.8 Conclusions

A broad range of civilian OSI application standards is emerging which are potential candidates for use in military C^3I systems. In general these will not meet certain specific military requirements, particularly in the area of security, at least for initial versions of the standards. The process of getting military requirements incorporated in civilian OSI standards requires considerable technical effort and cannot usually be achieved in a short time scale. However, much of the groundwork, in architectural terms, for such enhancements of the standards has been or is being laid. Furthermore, the long-term benefits of being able to use civilian standards in military C^3I systems are such as to make it sensible to continue the process.

The emerging standards offer rich functionality and many general features that will enhance the performance of future C^3I systems. However, continuing education of system specifiers and implementors to improve their knowledge of the scope and interrelation of the standards is required to obtain maximum benefit from their use.

Chapter 5.3

Security in military OSI networks

T. Knowles
(CAP Scientific)

Many military computer systems hold classified information and must be protected from attack. When connected to communications networks these computers will be exposed to potential threats from a very wide population, especially when the networks are interconnected to public switched networks. The technology to communicate with and gain access to these computers is now available in high street shops, and there is a growing band of expert 'hackers' who possess and share expert knowledge of computer systems in order to break in to computer systems for fun. Although these hackers are not necessarily a threat in themselves, their expertise could easily be misused by those with more serious intentions.

In the non-military world, computers can also hold information which is sensitive. As a result, ISO has for some time now been examining the question of securing systems. Part of this work has been an extension of the OSI reference model to incorporate security features as a precursor to enhancing OSI protocols. The ultimate aim is to ensure that open systems can be made secure without deviating from published ISO standards.

The question has to be asked as to whether it is sensible to standardise security mechanisms. Traditionally the military have felt that by keeping their operating procedures and in particular their security precautions confidential they are improving their security, since fewer people know how to attempt to breach that security. This leads to the view that security enhancements to the OSI architecture and protocols should not be standardised.

However, it is the case that the most determined attacks will come from those who will, be able to find out what these operating procedures are, and thus good security will be obtained only if the security measures can withstand determined attack from a knowledgable intruder.

Thus I see no problem with the concept of standardising security procedures, although clearly in a military environment crypto algorithms would remain secret.

Security in military OSI networks 239

It is a fact that the addition of non-standard security procedures to an otherwise OSI system *makes the whole system non-standard* and negates the advantages of of using standard OSI systems.

This contribution not only reviews the work of ISO but also tries to put it into a military context. First there follows a quick overview of the various ISO activities in the area of security.

One activity has been to expand the OSI reference model to define the types, location and means of achieving protection of communications. This work is being done in an *adhoc* group within ISO/TC 97/SC 21/WG 1, whose output is reported in the security addendum ISO 7478 Part 2.

Another activity is concerned with the production of standard encipherment algorithms, with priority attached to standardising DES, the USA federal data encryption standard (ISO DIS8227), and RSA, the so-called public key algorithm devised by Rivest, Shamir and Adleman (ISO DP9307). Also this committee, ISO/TC 97/SC 20, is responsible for the enhancement of protocols to incorporate the use of encryption.

5.3.2 Types of security protection

5.3.2.1 Protection needed The aim of the OSI security work has already been stated to be to ensure that a secure system can be constructed using OSI communications protocols such that the external communications are no less secure than the hosts or endsystems.

This section will list the types of security protection or services that can be required by a pair of communicating applications.

It is worth observing that all of the official security policites of UK and NATO relate just to the type of protection described below as confidentiality.

It is also worth noting that one major issue which is often of concern to military system builders is not mentioned below at all. This is denial of service. It has been omitted from the OSI security work because it is not felt to be relevant. In the very early versions of the security addendum it was included as a type of protection but subsequently it has been determined that there were no protection measures that were visible within an instance of communication between two end systems. Protection comes from measures such as having alternate processing sites and diverse alternate routes within a network.

The types of protection visible in an OSI environment are as follows:

Confidentiality The protection of transmitted data from accidental or deliberate disclosure to unauthorised persons.

Confidentiality is subdivided into a number of categories, as follows:

(a) Confidentiality of selected fields in an SDU
(b) Confidentiality of a single connectionless SDU
(c) Confidentiality of all SDUs on a connection
(d) Traffic flow protection.

Traffic flow protection ensures that an observer cannot gain any useful information from observation of the number, path or frequency of data transmitted, as opposed to the data itself.

Integrity   The assurance that data received is exactly as sent by an authorised entity, i.e. contains no duplications, insertions, modifications or replays.

As with confidentiality, integrity is subdivided as follows:

(a) Integrity of selected fields in an SDU
(b) Integrity of a single connectionless SDU
(c) Integrity of all SDUs on a connection.

Peer entity authentication   The identification of a remote entity (this requires that there be a means to detect a simple replay of a previous authentication sequence).

This function does not cover the machine identification of humans, e.g. by finger printing. It only covers identification of a remote entity that is trying to communicate via OSI protocols. When a human is using a terminal to access a remove system, the identification of the human to the local virtual terminal application is a matter which is outside the scope of OSI.

Access control   The ability to limit and control the access to host systems and applications via communications links. To achieve this, each entity trying to gain access must first be identified, or authenticated, so that access rights can be tailored to the individual.

Alternatively the intending user may possess credentials, the presentation of which is sufficient to grant requests for access to resources.

Non-repudiation   Giving protection against the recipient of data later denying that it was received or the sender later denying that it was sent.

Data origin authentication   The provision of assurance that the source of data received is the one claimed. This can be important, for example when receiving commands and instructions.

5.3.2.2 Protection mechanisms   To many people, security protection is synonymous with encipherment. However, although it will play a large part in any high-security system, it does not offer complete protection on its own. For example, encipherment plays no direct part in access controls (although it can be used for the entity authentication that precedes it).

Incidentally, the word 'encryption' does not translate well into French, one of the three official languages of ISO, and therefore the word 'encipherment' is used exclusively in ISO.

This section will present the protection mechanisms relevant to OSI and which are expected to be incorporated into ISO protocols.

Clearly many means can be used to protect systems, only some of which will be visible in the communications procedures between open systems. It is only the latter that are discussed.

Security in military OSI networks 241

One of the most powerful techniques is encipherment, and this can be used not only to give confidentiality but also to play a part in providing a number of other protection services. This section is not restricted to consideration of encipherment, however.

Passwords can clearly play a part in authentication. In order that they cannot be monitored, great care should be taken to disguise them, possibly by using encipherment. Their advantage over encipherment-only mechansism is that they do not cause great system overheads in their use. Thus a system with no confidential data can protect itself against misuse by insisting on entry of a password before access is granted. Because passwords are not entered frequently they can be software enciphered before transmission without too great a system overhead.

Another advanrage is that passwords often need changing less frequently and therefore the management support is less burdensome.

Passwords have some specific weaknesses which are exploited by hackers, however. These relate to the weaknesses of the humans who use them. People find it difficult to remember passwords, and get round this problem in two ways. The first is to choose a password that is easy to remember, and these very frequently are things like the names or birth dates of close family. Where the user does not choose his password and is given a meaningless string of letters or numbers, he frequently writes it down!

Traffic padding and spurious message generation can be used to disguise traffic flows and possibly to remove some covert channels.

Manipulation detection codes are used to detect deliberate modification, insertion, deletion or replay of data from a data stream. Again, to be really effective against a knowledgable attacker the MDC should be protected by encipherment, or better still cryptographically generated and in addition supported by a recovery protocol.

Even when protected by encipherment, special measures must be taken to detect simple replay of a previous good message. This sometimes involves incorporation of sequence numbers and time stamps into MDCs.

Physical security measures such as locks and secure rooms and cabinets will always be necessary as part of a completely secure system. Physical security is costly to provide so the objective of security architectures is to minimise the need for it.

Security audits can be used to analyse breaches of security in order to identify the perpetrator and method of attack. This can have a strong deterrent effect.

Security alarms, which alert a security centre whenever suspected attacks occur such as a detected manipulation or repeated entry of invalid passwords, can be used to catch attackers, and again have a deterrent effect.

Digital signatures, which are a cryptographically derived digest of a piece of data, can be used to protect against repudiation (i.e. later denial that the data was sent) by the sender of data. Asymmetric (public key) ciphers are used to create digital signatures so that the key used to create the digital signature is held only by the sender but the key used to verify the signature can be widely distributed.

Notaries are trusted third parties who can be used to give assurance of the origin of data and its contents.

Using this technique, all communications between two entities are conducted via the notary, who records the transactions. When data is received via the notary, the recipient will accept its stated origin and contents.

The notary's record can be used later to give proof of the transactions should one or both parties get into a dispute.

Trusted functionality is the name given to hardware and/or software which a user has assured himself can be trusted to carry out its tasks without jeopardising system security.

All the preceding mechanisms implicitly assume that their implementation can be trusted. For example, it is no use having an encipherment device which occasionally, owing to software error, transmits data in clear. There are many techniques that can be used to establish trust in hardware and software. These range from thorough testing to the use of formal methematical proofs of correctness. OSI is not concerned with the method used to establish trust. It does, however, aim to minimise the amount of trusted software needed to implement a secure system.

5.3.3 Architecture – placement of security services

The OSI security addendum to the reference model (ISO 7498) identifies:

(a) Which of the security services and mechanisms are relevant to OSI standardisation
(b) Placement of the relevant services in the layers in such a way as to enable secure communications to be achieved cost effectively using OSI, and giving guidance to the various working groups as to what addenda are needed to layer services and protocols in order to achieve this objective.

The following matrix shows the current allocation of security services to the OSI layers. Where a service is shown to be present in a layer, this indicates that the particular layer is directly responsible for the achievement of the service. Clearly in these cases the higher layers must also offer the service to their users, but they will achieve that protection service by passing the request down to the lower layers.

Service	Layer						
	1	2	3	4	5	6	7
Peer entity authentication	N	N	Y	Y	N	Y	N
Access control	N	N	Y	Y	N	Y	Y
Confidentiality	Y	Y	Y	Y	N	Y	N
Traffic flow security	Y	N	Y	N	N	N	Y

Integrity   N N Y Y N Y N
Data origin authentication   N N Y Y N Y N
Non-repudiation   N N N N N Y N

Key:
Y Yes, the service will be included in the layer as an optional service.
N No, the service will not be provided within the layer.

5.3.4 Non-OSI aspects

A number of important security measures are currently deemed to be outside the scope of OSI standardisation, as has been indicated earlier. This does not necessarily imply that no international standards will be produced to cover these aspects. It means only that they have no visibility in the OSI environment and therefore have no effect on the protocols used between two end systems.

One of the prime examples of this is encipherment algorithms. These are being standardised by ISO independently of OSI. However, the OSI architecture should be such that any suitable encipherment algorithm can be used at the defined places in the architecture, and it therefore does not specify the algorithms to be used. In particular, for military applications secret algorithms can be used but slotted into the standard architecture and protocol.

The following security measures are outside the scope of OSI:

(a) Physical security, e.g. locks and guards
(b) Personnel identification methods, e.g. finger printing or badge readers
(c) Operating system security, i.e. ensuring rigorous separation between software modules
(d) Security audit analysis programs
(e) Suppression of electromagnetic radiations.

5.3.5 System security

When planning a secure distributed system, a number of steps must be taken before the most appropriate security measures can be determined. OSI security services and protocols needed can only be determined as part of this overall security risk analysis.

The risk analysis will:

(a) Determine the cost to the application of each possible type of risk. Risks include disclosure of information, modification, unauthorised access to assets, denial of service and destruction.

Assets are all the items of value to the user including the system hardware, supporting services such as air conditioning, and data.

As an example, consider a computer system providing a directory service for a distributed network. As all network users are allowed to use the directory, access and disclosure are not problems. On the other hand, denial of service and destruction could seriously impact the effective use of the whole network and are therefore of great importance.

As a further example, a command and control system receiving sensor data and controlling the aiming and firing of weapons will be most concerned about modification and destruction of data and masquerading by unauthorised users.

(b) Determine the value to be derived from breaking security by a possible attacker. The values assessed will depend on the type of attacker.

For example, if the system contains commercial secrets then the value to a rival organisation may be considerable and thus sophisticated and determined attacks may be mounted. If the system contains examination results then the attacks may be determined but less sophisticated.

This exericise will determine the level of threat, which in turn may be used to assess the likely frequency and strength of attack.

(c) Consider the system weak points. These could include physical access to terminals and other equipment, radiation, system software loopholes, use of broadcast technology (satellites or local area networks etc.), subversion of staff, wire tapping (active or passive), mutual distrust of communicating parties (where contracts are involved) etc.

For example, if the system contains commercial secrets but resides entirely in a secure building then the only form of external threat is from radiation. If the intersystem communications are considered to be weak points (e.g. attacks via other equipment on the same network or attacks via line tapping or equivalent) then OSI security measures must be considered.

Thus it can be seen that OSI security has to be considered as just one part of an overall system security policy. Factors such as security classification of objects, management of access rights of subjects, and permissable actions of subjects on objects have not been treated since they will be of concern only to the applications within end systems and are outside the scope of this book.

There are a number of publications on the subject of risk analysis which cover the subject in much more detail, e.g. FIPS (1979).

5.3.6 Protocol enhancements

The work of ISO aimed at achieving an architecture for secure OSI communications has been discussed above. This is only a start for the real work of defining service and then protocol enhancements.

At the time of writing, the work of protocol enhancements has not made great progress. The physical layer standard is at DIS stage, but enhacnements for the other layers have not yet got as far as establishing a base document within ISO. In the USA ANSI have fairly advanced drafts for enhancements to the transport and presentation layers, however.

5.3.6.1 Physical layer encipherment This document, currently ISO DIS 9160, specifies the interoperability requirements for encipherment at the physical layer.

Security in military OSI networks 245

The main body specifies requirements in general terms, not specific to any encipherment algorithm. An annex specifies the particular requirements for using the DEA-1 (ISO DIS 8227) algorithm.

The general requirements are specified for a number of physical layer standards (V24/X21bis, X21, synchronous, asynchronous etc.). For each of these, the interface conditions are defined under which an initialisation variable is sent and upon which encipherment is commenced. For full duplex operation, two initialisation variables are sent, one for each direction of transfer; for two-way alternate operation, one initialisation variable is sent each time the line becomes ready for sending data.

The additional procedures for use of DEA-1 specify a number of options for how long the initialisation variable should be. The mandatory requirement is for an initialisation variable of 48 bits. Use of 64 bits is an additional option.

5.3.7 Securing military systems

It has already been observed that military and government security policy is at present exclusively concerned with 'confidentiality'. The policies were formulated before the concept of a layered architecture was standardised in the civilian area, and do not therefore make any recommendations as to where protection mechanisms should be placed within such a layered architecture.

The OSI addendum does contain an annex giving advice on the choice of placement of encipherment, since the addendum potentially concedes its use at every layer except 5! It is probable that many military applications will require the use of encipherment at more than one layer. Physical layer encipherment (equivalent to what is often called link-by-link encipherment) will be needed to protect links against traffic analysis, but does not give fine enough granularity to separate need-to-know groups; this may require encipherment at the presentation layer.

There is also a terminology difference between the military security world and that used in ISO. To a great extent this is deliberate. The ISO standardisers were at pains to ensure that general protection concepts were not expressed in terms that directly connote specific and existing techniques, civilian or military. As an example, the terms 'security level' and 'classification' are avoided because they are too closely associated with military security policies.

In fact the OSI security addendum allows for both types, and also conceives of a further type where the application can request a required level of protection for each instance of communication. In a military environment this would of course only be permitted where the application was a trusted multilevel implementation.

OSI also recognises the alternative approach where an application requesting communication has a level of protection imposed by external, management, dictate.

The OSI work also goes beyond current military security policies in that it addresses issues of protection against active attacks. A military system designer is therefore free to use whatever mechanisms are appropriate from the ISO repertoire for this purpose.

5.3.8 Conclusions

It has been shown that the ISO OSI security addendum to the reference model addresses all aspects of current military security policies where they impact on computer communications. It also addresses issues of concern which are not currently covered by national or NATO security policies.

I believe that the security addendum therefore provides a sound basis for the achievement of military security.

Although not much work has been done in ISO on protocol enhancements, it is to be hoped that they too will be suitable for military use, leaving only the encipherment algorithms unique to military applications. ISO protocol enhancements will in any case have to allow for the use of many encipherment algorithms, since each will only have a finite life and there will always be communities of users who wish to use private algorithms.

5.3.9 References

FEDERAL INFORMATION PUBLICATION SOCIETY, (FIPS) (1979) 'Risk analysis of ADP systems:, National Bureau of Standards.

ISO 7498 Part 2 'Information processing systems – open systems interconnection – basic reference model, Security addendum'.

ISO DIS8227 'Information processing systems – data cryptographic techniques – specification of algorithm DEA 1'.

ISO DP9307 'Information processing systems – data cryptographic techniques – specification of DEA 2'.

ISO DIS9160 'Information processing systems – data crytpgraphic techniques – physical layer interoperability requirements'.

Chapter 5.4

Standards for naval systems

J.S. Hill and F.A. Richards
(Admiralty Research Establishment)

5.4.1 Introduction

A warship, reduced to its essentials, is a mobile platform bearing an assortment of sensors and weapons and a means for using them effectively to carry out the ship's current task and at the same time to defend itself against any threat that may arise. These systems (sensor, weapon and command) together comprise the combat system of the ship.

5.4.2 Background

The leisurely days in which the main sensor was the mark one eyeball, the main weapon was the cannon and the command system was the ship's commander and his staff have long since gone. Weapon and sensor systems of great complexity and sophistication must now interact, and be managed, in time scales which will enable the ship to counter missiles which may be capable of approaching at several times the speed of sound.

Until quite recently the interaction between the components of the combat system was effected by point-to-point dedicated data connections, nearly all of which were between the weapon or sensor and the action-information organisation (AIO). This situation is illustrated in Fig. 5.4.1. The AIO therefore was the hub of a star-connected network which was created piecemeal, with different standards applied to each link. At the time these systems were designed, computing power was both expensive and bulky and tended to be concentrated in the AIO, which did much of the data processing required by the weapons and sensors in addition to its proper job of managing the combat system as a whole.

Advances in technology have changed the picture in two very significant ways. Firstly, local area networks have been developed to a point at which they can be taken seriously in a real-time process control environment, of which the combat system and the world about it are an example.

248 Standards for naval systems

Secondly, computing power has become cheaper and smaller and consequently is being incorporated in the design of sensors and weapons to perform data processing close to the point of application.

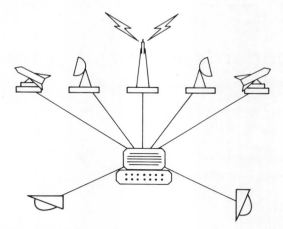

Fig. 5.4.1 *The combat system as a star network*

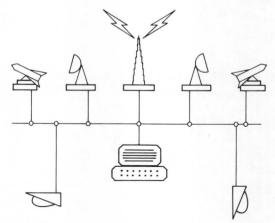

Fig. 5.4.2 *The combat system as a total system*

5.4.2.1 Introducing the local area network: These changes have made it possible to introduce the concept of a combat system highway as a local area network to replace the point-to-point links, and to regard the combat system as an aggregation of *member systems* interacting with and supporting each other to form a *total system,* as illustrated in Fig. 5.4.2. While the command system retains its role as the management element of the combat system, the introduction of a local area network makes it possible for all the member systems to exchange data on an equal basis. Shellard (1985) describes some of the current thinking in the design of naval command systems.

5.4.3 Constructing a system

Two approaches to the construction of a total system are possible. One is mainly applicable to the design of systems at the hierarchical level of a member of the combat system, i.e. to an individual sensor or to the command system. Systems of this type are generally designed as a unit, and the designer is able to treat their components (many of which will be computers) as a closed system. This opens the door to the possibility of creating a unified database within the system, which is logically centralised (data is accessible equally to all components) but which is in practice distributed and replicated for efficiency of use and for reliability through redundancy. Such a distributed database is described by Cheeseman and Catt in Section 3.2.

The second approach is more relevant to the design of the combat system itself. It accepts that the member systems are separately (and largely independently) procured, and that such control that the combat system designer has over them is limited to the data and control interfaces that they present to one another. (It is often an uphill task to exercise any control even over these.) There is therefore in effect an open system interconnection problem to be solved. The need to produce practical solutions in the time scales of the design for a new frigate (type 23) led to the setting up of the combat system data communication study, in which staff from the Admiralty Research Establishment, the Chief Naval Weapon System Engineer's team, Ferranti Computer Systems Ltd and Software Sciences PLC co-operated to define a set of standards and guidelines for data exchange over local area networks. These are being applied initially to the type 23 combat system, but have been developed with a view to extending them to similar naval systems and perhaps to process control systems in general. The output of the study forms the subject of the remainder of this paper, in which frequent reference will be made to the ISO reference model of open system interconnection (ISO, 1982).

5.4.3.1 Hierarchical design: Consideration of the two approaches outlined above suggests the possibilty of a hierarchical extension to the concept of the total system.

A member system may well have its own internal data communications system, but this would be hidden from the other members of the combat system, as would all other details of its internal structure. Such a member system must be treated as a black box. Its functional properties will be known to other systems only through the data interface to the combat system highway. (This data interface must cover all levels of data exchange from the physical to the applications.)

This principle of 'hiding' is of considerable importance, because it is a fact of life in naval procurement that systems are usually designed in a closed environment. Indeed, current policy is to encourage purchase of weapon and sensor systems off the shelf, without regard to their internal structure. There is thus little encouragement to design combat system data flows in the knowledge of the roles of

250 Standards for naval systems

the components of member systems, despite the fact that one of the main objectives of open system interconnection is to facilitate direct communication between such components.

It is therefore not only possible but desirable to regard each member system as a total system in its own right, as far as its internal communications are concerned. At the combat system level, to take an example, it is of no concern how the designer of each member system assembles the components of his own total system — the computers, consoles, trackers or launchers and so on. It is of interest to consider extension of the concept upwards as well as downwards; in this case the combat system as a whole becomes one of several member systems of some larger total system which represents, perhaps, the ship. Fig. 5.4.3 illustrates this concept.

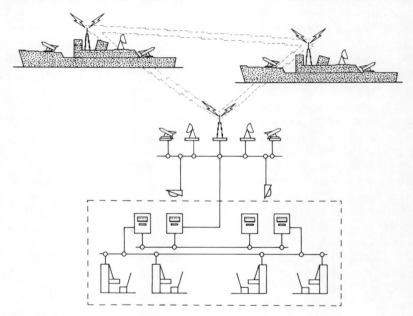

Fig. 5.4.3 *Hierarchical design of ship systems*

Thus there are several layers in a hierarchy:

(a) Within the command system, taken as a representative of a member of the combat system, the peer systems are the displays and the main computers, linked by some sort of data transmission system; in the type 23 this will be a UKRN local area network known as the ASWE serial highway (UK Defence Standard 00-19) and in the single-role mine hunter it will be Defence Standard 00-18 part 2 (MIL-STD 1553B). Similar ideas apply to weapons and sensors.

(b) At the combat system level, the peer systems are the individual weapons and sensors and the command system as a whole.

(c) At the ship level, the peer systems are the combat system, the machinery control system, the message handling system etc.

Standards for naval systems 251

(d) At the force level, the peer systems are the ships. Remember that ships of various navies will be interacting via, say, Link 11 or JTIDS (Link 16) and there is no way in which the design of data flow in an RN ship can recognise the internal structure of Dutch, Danish or US vessels.

It must be accepted that direct communication in the OSI sense between, say, one of the command system displays and the control computer of a missile system is not practical; communication must be from the command system to the weapon system as entities in their own right. This topic has been discussed at some length because it has coloured the whole approach to the selection and specification of data exchange standards for the type 23. A strict adherence to the layering principles is necessary if the quality of service (performance) of the network is to be guaranteed during its working life.

5.4.4 Framework for standards

5.4.4.1 Open system interconnection: Very little experience or material to draw upon was found when the search for suitable standards for a combat system highway was commenced. It was decided to try to use the OSI model as a guide in deciding how to organise communications between members of the combat system, how to incorporate such standards as were available, and what standards (in the form of Naval Engineering Standards (NES)) needed to be produced.

The OSI layers (Fig. 5.4.4) can be divided into two parts. Layers 1 to 3 are concerned with a single data network and the protocols for its operation; layers 4 to 7 define the end-to-end connection (possibly across several networks), how the communications system is to be used and the messages and data that are to be transferred between systems. The transport layer 4 has little to do in a hierarchical

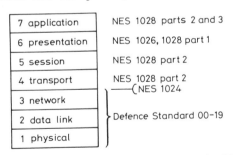

Fig. 5.4.4 *Location of the Naval Engunineering Standards within the OSI framework*

system, since there is normally only one network. Since there is a natural partition between layers 3 and 4 it was convenient to consider an interface standard at this point so that layers 4 to 7 could become independent of the particular local area network that was chosen for layers 1 to 3.

5.4.4.2 The standards produced: Layers 5 to 7 are concerned with the data to be communicated. NATO STANAG 4222 (NATO, 1985), at the time in draft form, defined standards for digital data parameter representation. This covered the aspects of data representation required for layer 6 (presentation) and it was decided to adopt as much of it as possible. The RN implementation of this was to become NES 1026 (MoD, 1984).

Message formats were also regarded as capable of standardisation, again within layer 6. Part 1 of NES 1028 (MoD, 1985a) contains a formal language for message specification, providing a set of rules for the construction of messages conforming with NES 1026. These two standards constituted the definitions required for the presentation layer 6 in the combat system highway.

A catalogue of standard messages was defined in NES 1028 part 3 for use in the combat system, and as such resides in layer 7 (application).

The need for rules and protocols for communication between applications processes at layer 7 (application) was foreseen, and these are placed in part 2 of NES 1028. Sundry protocols at layers 4 (transport) and 5 (session) are also incorporated in part 2, largely as a result of difficulty in determining where they belonged in the OSI model at the time the NES numbers were allocated.

The host-communications interface is specified in NES 1024 (MoD, 1985b). It was decided after some debate that this interface is located between layer 3 (network) and layer 4 (transport), though as mentioned previously the hierarchical approach to system construction leaves the transport layer with little work to do.

The new standards which had to be produced for the combat system highway have now been introduced; their overall relationship with the reference model for OSI is shown in Fig. 5.4.4. They will be described in more detail later. Before this the choice of local area network to support the lower three layers of the ISO model must be discussed.

5.4.5 Choosing a local area network

A number of data transmission systems for shipboard use are available, but the two which are most appropriate for use as a local area network in the combat system are those specified in Defence Standard 00-18 part 2 and Defence Standard 00-19. The first of these is more widely known as MIL-STD-1553B, and is primarily aimed at systems in which a central computer is controlling a set of relatively unintelligent peripherals; its use is not confined to this, however, and it is being deployed in several naval applications including a new class of submarine. The second, the ASWE serial highway, was developed specifically to provide a high-integrity data network connecting computer-based, asynchronously operating systems. Since this is the situation obtaining in the combat system, and since 00-19 has three times the bandwidth of 00-18 part 2 and is already deployed in the type 22 batch 2 frigates in the computer-assisted command system (CACS 1) (Shellard, 1985), its choice for the type 23 combat system was natural. It will also be used in the type 23 command system.

Defence standard 00-19 subsumes the facilities and services of layers 1 (physical), 2 (data link) and 3 (network) of the OSI model. The highway operates at 3 megabits per second, with a message length of 32 sixteen-bit words (three of which are reserved for addressing). Both broadcast and point-to-point addressing are available; broadcast messages are selected by the recipient on the basis of a message type number which is included in the three-word header. The standard provides for dual redundant highway controllers and triple redundant highway cables, and includes comprehensive error detection and recovery. These capabilities allow a highly comprehensive communications service which is transparent to the users to be provided.

5.4.6 Network/transport layer interface

Defence standard 00-19 is unique among local area network standards in including, in its part C, the specification of a software interface to the highway. This has been so successful in practice that it was deemed desirable to produce a general specification of an interface between a host computer and a communications medium. This specification not only would replace 00-19 part C, but would be applicable to a whole range of LANs including 00-19, CD-CSMA nets, nets based on frequency division multiplexing, packet switching and so on.

NES 1024 is designed to enable a given local area network to be interfaced to any of a range of host computer types and vice versa. It enables the type of computer or network to be changed with a minimal effect on the data and protocols passing over the network. The interface is defined at the message level (as distinct from the word/byte or even bit level). It is divided into two parts: the specification of software interface data areas and protocols; and the specification of an electrical interface which can be made a contractual boundary.

5.4.6.1 Software interface: The software interface assumes the concept of a terminal unit (Fig. 5.4.5), which may be implemented in hardware or software. The terminal unit has direct memory access to at least part of the main memory of the host computer, and implements layers 1 to 3 of the OSI reference model. Two facilities are provided:

(a) A host may place messages in the interface to be sent to the intended recipient(s) without prior notice (at this level). Messages are entered in a single output queue and removed one by one by the terminal unit as it finds itself able to access the local area network and dispatch them. A simple handshake mechanism, using a field within the message, allows the host and the terminal unit to enter and handle messages asynchronously in the circular buffer holding the queue. Exactly the same mechanism operates in the input queue, filled by the terminal unit as messages arrive and served by the host in its own time. Since the size of the buffers containing the queues is necessarily bounded in any practical system, messages are limited to 64 bytes maximum length; a few bytes within this limit are reserved for interface control information (destination address, source address, message length and message type number). The locations of the output and input buffers in the host memory are contained in the control and status table, which (as shown in Fig. 5.4.6) contains or points to all the other interface control information.

254 Standards for naval systems

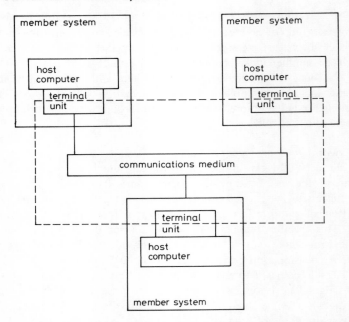

Fig. 5.4.5 *The terminal unit as an interface between a system and the communications medium*

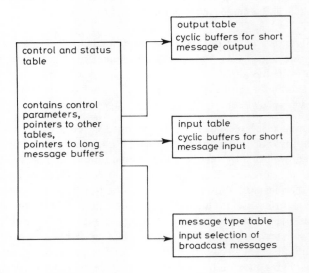

Fig. 5.4.6 *The structure of the software interface of NES 1024*

Standards for naval systems 255

(b) A host may send and receive one message at a time by prior arrangement with the recipient(s) or sender. Data space must be set aside before the transmission commences, and the message length can therefore be much greater; a limit of 65535 bytes applies in this mode. Once the communicating hosts have arranged the transfer (and the protocols for this are not included in NES 1024), the message handover between host and terminal unit is again effected by an asynchronous handshake. The flag for this, and the location and size of the data area to be used and the source and destination addresses, are contained in the control and status table.

A particular feature of the software interface is the use of the message type number. Each short message bears this number; the host computer is able to set a field in the interface control information data space which instructs the terminal unit which type numbers of broadcast messages it is to accept. Incoming broadcast messages are filtered by the terminal unit. Point-to-point addressed messages are always delivered to the intended recipient only.

5.4.6.2 Hardware interface: The hardware interface assumes that the terminal unit is divided into two hardware entities, the host adaptor unit and the communications unit (Fig. 5.4.7.) The electrical connections between the two are specified, together with the signalling protocols to be used to pass data over them.

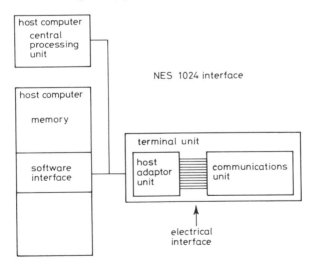

Fig. 5.4.7 *The structure of the hardware interface of NES 1024*

The communications unit handles all the protocol and signalling requirements of the local area network, converting the messages between the form required by the software interface specification and the form required by the local area network. The host adaptor unit need do no more than convert the electrical signals defined by the hardware interface specification into a form appropriate to the host's backplane bus, ported store or other DMA facility; the interface is designed to minimise the complexity of the circuitry required for this.

The interface provides the ability to transfer addresses up to 32 bits wide from the communications unit to the host, data streams in a sequence of 16-bit (2-byte) transfers in either direction, and interrupts in either direction. All transfers are fully handshaken, the intention being to provide a technology-independent interface which can be driven at a rate at least as great as that required by the real-time constraints of any local area network. In addition, the interface enables the communications unit to ascertain, and to respond to, the number of bytes per computer word in the host. This data, and all the other information required by the communications unit to construct addresses etc., is contained in the control and status table specified in part 1.

The specification leaves it open for the designer of the communications unit to provide message buffering within it, to lessen the real-time constraints on the host's direct memory access or central processors, and for the designer of the host adaptor unit to provide memory areas within the unit.

5.4.6.3 Grade of service assumptions: NES 1024 makes certain assumptions about the services provided by layer 3 and below of the seven-layer model:

(a) It assumes that once a message has been accepted by the communications unit, that message will be delivered uncorrupted to its recipient(s) if this is at all possible. NES 1024 contains no provision for error detection and correction.
(b) It assumes that any segmentation required by the local area network will be effected in or below layer 3. NES 1024 does not provide any segmentation or concatenation facilities. This applies mainly to the long-message facility, since the short-message maximum length is chosen to be not greater than the natural message length of most currently available LANs. Similarly, if the application wishes to pass longer messages via the short-message facility, they must be broken up and reconstructed at the transport level (layer 4).
(c) It assumes that any message marked as broadcast is potentially available to every other subscriber to the local area network, with a guarantee of delivery equal to that for point-to-point addressed messages. NES 1024 does not offer a group addressing facility.

It is now appropriate to review the Naval Engineering Standards, outlined earlier, which were required to support layers 4 to 6 of the OSI model.

5.4.7 Presentation layer
The presentation layer describes how data is to be formatted and constructed into messages which pass over the system.

5.4.7.1 Data representation: NES 1026 is a fully compatible subset of draft STANAG 4222. This was made possible at least in part by inputs to the development of the STANAG through its draft stages by ARE and Ferranti. It differs from the STANAG only in that it omits the optional data representation key of the

Standards for naval systems 257

latter, and does so because the message specification language of NES 1028 (of which more later) renders it unnecessary. Since it deals with data representation in intersystem communications, NES 1026 is clearly applicable to layer 6 (presentation) of the seven-layer model. It relies heavily on international civil standards which are already widely accepted.

NES 1026 covers three aspects of data representation. These are:

(a) The units to be used in data exchange, which are SI (metric) units as defined in ISO 31. Two qualifications are noteworthy; these are the unit of angle, which is defined as the full circle (1 FC = 360 degrees = 400 grads = 2π radians), and the ordinal date and time of day, which are as defined in ISO 2711 and ISO 3307.

(b) Measurement conventions, which are based on right-handed Cartesian coordinates and are derived from ISO 1503. Rules are given for geometrical orientation, linear and angular measure from a reference point or plane, and geographical co-ordinates. Polar co-ordinates are not covered and a separate clarification of these has been produced as shown in Fig. 5.4.8.

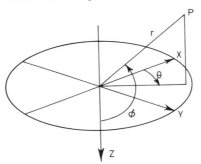

Fig. 5.4.8 *The relationship between Cartesian and spherical co-ordinates. P is the target position, X is north/ship's head, Y is east/starboard*

(c) Representation of fixed and floating point natural numbers, binary-coded decimals, alphanumeric strings and discrete codings. Fixed point numbers are binary two's-complement with the length to be an integral number of 8-bit bytes and the binary point to be on a byte boundary (not necessarily within the bytes actually used). Floating point numbers are defined in accordance with IEC 559 single- and double-precision forms (the extended forms are excluded). Binary-coded decimal digits are as specified in ISO R963, and alphanumeric string characters are as specified in the NATO 7-bit code (STANAG 5036), which is based on ISO 646. Finally, rules are given for the construction of discrete codings (in which abstractions such as 'RED' or 'J-band' are given numerical values).

NES 1206 is designed to give a reasonable compromise between efficiency of encoding, efficiency of data transmission and standardisation of data interchange. It applies only to data transmission; for example, the fact that speeds are communicated between systems in metres per second does not imply that they must be displayed to operators in that form.

5.4.7.2 Message construction: NES 1028 part 1 is the high-level language for message construction (HILMEC). It assumes the data representation definitions of NES 1026, and enables its user to specify a message's name followed by a declaration of its components in the form of named parameters, each parameter being specified as one of the types allowed by NES 1026. The sizes of the fields (e.g. number of bytes and position of the binary point in fixed type) are specified where this is appropriate, and units may be included in the declaration. The language is formally defined syntactically and is machine readable; software implementing this has made it possible to check every message declaration for correctness and completeness, and automatically to generate a corresponding field map. Sample messages, with actual values inserted, can be produced. It is possible in principle, though not yet undertaken in practice, to create message construction and analysis software directly from a message declaration in a manner analogous to the use of a compiler.

An optional feature of HILMEC is the ability to specify control information in standard form. The fields available comprise the source and destination addresses, the message type number referred to in NES 1024, a message catalogue number as used in NES 1028 part 3, a service field and, optionally, a time of validity field. Use of this facility enables a standard header to be defined for a given total system, which can then be applied to every message.

5.4.8 Application (and other) layers

5.4.8.1 The message catalogue: NES 1028 part 3 contains the catalogue of standard messages for intercommunication across digital data highways. As such it is expected to be a 'living' document, from which formats will be discarded when they are obsolete and to which new formats will be added as required. It is likely that different parts of the catalogue, and different standard headers, will be appropriate to different total systems; it is unlikely, for example, that many of the standard messages in a combat system will be of use in a machinery control and surveillance system. It is therefore difficult to do more than highlight the existence of the document and note that systems will be required to use message formats from it, contributing a new format if it can genuinely be shown that no existing format is appropriate.

5.4.8.2 Message rules and protocols: The rules and protocols for communicating messages across the network are specified in NES 1028 part 2. They do not fit neatly into a single layer of the OSI model, some aspects being relevant to layer 4 (transport), some to layer 5 (session) and the rest to layer 7 (application).

NES 1028 part 2 describes rules and protocols for intercommunication across digital data highways. In particular, it covers:

Standards for naval systems 259

The use of message catalogue numbers   Each message in the catalogue (NES 1028 part 3) carries its catalogue number in a defined location. This is the key whereby any member system receiving a message at its short-message input is able to determine the format of the message and therefore what it is about and how to decode it (layer 7).

The use and assignment of message type numbers   The message type number facility is defined in NES 1024, but the way in which it is to be used on a total system basis is not. Rules are therefore provided which, if followed, ensure that changes in assignment philosophy are not encountered if, for example, a weapon system is deployed on more than one class of ship (layer 7).

The establishment and use of system time   Data will be subject to delay both within and in transmission between member systems. This is taken into account by labelling time-sensitive data with a time of validity, which may be predictive. Rules and guidance are given for the maintenance of a total system time reference, and for the accuracy and precision required (layer 7).

Communication status messages   Point-to-point connected systems have tended to use *ad hoc* test messages over each link to establish whether the link was functional and the systems capable of exchanging data. Use of a local area network means that a total system facility is required, and this is provided by the regular broadcast of a communications status message by each member system. Any system requiring the information can determine from these whether another sysem is capable of communicating and, if it is, whether it is receiving the first system's communication status messages. This mechanism replaces conventional test messages.

Message precedence facilities   NES 1024 provides only one output queue for short messages. If any need exists for queue jumping (though this is expected to be exceptional) it must be implemented according to the rules given (layers 4 and 7).

Datagram transfer protocols   These cover the use of the short-message facilities of NES 1024, and in particular the segmenting and concatenation of datagrams which are longer than that allowed by NES 1024 (layer 4).

Datablock transfer protocols   These will give the protocols at layers 5 and 7 to be used by member systems to negotiate long-message transfers. Although the long-message facilities of NES 1024 exist and can be used these protocols are not yet included in the standard, as this is an area of continuing development.

5.4.9 Other areas of standardistion

The Naval Engineering Standards described briefly above are the more formal standards resulting from this activity. There is a further area of standardisation which is of considerable importance; this is the use of a standard format for member systems' data interface specifications.

When communications between systems were by dedicated point-to-point link it was only to be expected that the data interface specification was negotiated privately by the two parties concerned. They could use any data formats, data representation standards etc. that they chose, and the format of the specification was a matter of private agreement. Usually one of the two was chosen to be custodian of the interface in a contractual sense.

The introduction of local area networking has made it necessary to move somewhat from this position, since data is at least in principle accessible by other systems and the messages, formats and data representation standards are now global in effect. Member systems' interfaces are now constrained to take a common form, though the content will of course vary with the nature of the system. This common form comprises two parts:

(a) The combat system highway interface specification, providing a common aspects document and a member system specific interface specification.
(b) The data exchange specification for the member system.

5.4.9.1 Common aspects: This document contains detailed information which is common to all users of the combat system highway. This includes, for example, reference to Defence Standard 00-19 and to all the NESs described above, and defines certain member system responsibilities in observing these. It also defines vaious options which are open either to the combat system highway authority (such as the actual ranges of message type number that may be used, and their meaning) or to member systems in support of their particular requirements (such as use of time synchronisation, or the means of setting terminal addresses). Preferred options are stated.

5.4.9.2 Member system specific aspects: Each member system has, in standard form, a document specifying the options actually taken up and the data communication parameters of the system. This includes, for example, whether or not the data block transfer protocol will be implemented, and a summary of the input and output data flows.

5.4.9.3 Data exchange specification: Finally, data exchange specifications defining the data flow between the originator of the data and its main consumers are drawn up. The concept of custodianship of this part of the interface definition has been retained, but each member system interface includes a copy of all the relevant data exchange specifications.

From these documents it is possible to establish both the nature and the rate of the data flow in and out of each system, and the total load on the combat system highway. The latter is of particular importance, in that it is essential to monitor the build-up of traffic during development of the combat system to ensure that the local area network does not become overloaded, especially during periods of peak activity.

5.4.10 Status and conclusions
All three Naval Engineering Standards discussed here have been formally issued. The use of the OSI model as a discipline for constructing the standards has been an

Standards for naval systems 261

interesting experience. The OSI definitions have been lacking in some areas, especially in the handling of recipient-selected broadcast messages. Nor do they offer any assistance in supporting time reference between member systems. Some difficulty was also encountered in maintaining a clean distinction between the layers, as will be appreciated by anyone reading the standards. Nevertheless, the use of the reference model was an invaluable aid in the work of the study group.

5.4.11 References

ISO (1983) 'ISO 7489 — reference model for open system interconnection'.
MoD (1985b) Naval Engineering Standard 1024 'Data transmission — direct memory interface standard issue 1', DTS(WE) Sec TE112b
MoD (1985) Naval Engineering Standard 1026 'Requirements for the digital representation of shipboard data parameters', DTS(WE) Sec TE112b
MoD (1985a) Naval Engineering Standard 1028 'Standard for intersystem communication protocols', DTS(WE) Sec TE112b
NATO (1985) 'STANAG 4222 — standard specification for digital representation of shipboard data parameters', AC/141(1EG/5)/D132.
SHELLARD, D. J. (1985) 'C^3 within a naval ship', Proceedings of IEE conference on advances in command, central and communications systems, Bournemouth, England.

6: The man–machine interface

Chapter 6.1
Man–machine aspects of command and control

W. T. Singleton
(University of Aston)

6.1.1 Introduction
Historically, military systems have often been idealised as pyramidal hierarchies with one commander at the apex. This commander is supposed to have an intellectual grasp of the total situation and to deploy his available resources in some optimum fashion to defeat the enemy. Up to the end of the nineteenth century the general sat on his horse at the top of a convenient hill and the admiral stood on the bridge of his flagship. They directly observed the manoeuvres of their forces and those of the enemy. On the basis of this data they ordered adaptive responses to match the continuously changing situation. For most of this century there has been extensive reliance on telecommunication of verbal data used to update central mapping facilities. These maps and their associated movable symbols formed the basic information display used by the commander whose orders were also transmitted by telecommunication.

Technologically, the engineers have recently developed more and more elaborate ways of transmitting, storing, manipulating and presenting information. Information technology has become a separate profession and, inevitably, has developed its own objectives and standards. A system which has greater capacity is automatically regarded as better providing that it meets other engineering criteria such as reliability and cost. The versatility of the new information systems provides its own interest as a topic for study and development. Extensive effort is expended on ingenious data manipulation as an end in itself. This results in a radical change in the system development procedures. Instead of asking what the recipient needs the information technologist concentrates on what can be made available. The overall picture is now seen as a hierarchy of pictures at different levels with differing degrees and kinds of detail. The maintenance of all these pictures requires vast amounts of data. The system has to be fed continuously with new data mostly obtained by telecommunication.

The concept of the commander viewing the total situation is retained but, because of the quantity and variety of available data, he or someone else has a new set of tasks concerned with the selection of what might be relevant from vast data stores and available varieties of presentations.

The new systems are susceptible to the criticism that they have been over-engineered and under-designed. Good design implies fitness for purpose. The ultimate purpose of all non-trivial telecommunication is to convey information to a person who has the role of a decision maker. Oscar Wilde made the essential point a century ago in the earliest days of the telephone when he said that the value of the telephone is the value of the conversation between the people using it. In short, communication is a human problem for which the information technologists merely provide the vehicle. This is self-evident but it needs to be regularly reiterated because the sheer weight of the engineering effort and expertise tends to overwhelm the users and the people specialists such as ergonomists, organisation theorists and psychologists.

The human operator appears in command and control systems at the end, the beginning and sometimes at various intermediate stages. The commander at the end is central to the whole business because he is the primary reason for the existence of the system. He uses it to provide information which enables him to make decisions about dispositions and actions.

6.1.2 The commander

The level of complexity varies from a whole battlefield to one navigation officer trying to assess the position of his submarine in relation to other vessels. Whatever the level, a command and control system is essentially a picture builder. Data is put together from various sources and the output is an information presentation which is intended to aid decisions. The system may also disseminate commands for action and may be involved in the allocation of various resources for specific tasks.

Information collected and displayed for this purpose has a number of advantages:

1   It need not be confined to any particular geographical area. It is possible to change the scale and thus to present the detail of a small area or an overall view of a large one.
2   It need not be confined to immediate events; it is possible to present histories, trends and predictions.
3   It can be a physical or a functional picture or a mixture, that is it can vary in level of abstractness from topographical correspondence to the use of alpha-numerics and other specialised symbols.
4   It can be an amalgam of data from a variety of sources.

There are also a number of disadvantages:

1   The user is at the mercy of the designer who has set the limits of what is available.
2   The presentation is heavily biased toward the visual system; there will be auditory inputs via voice links, but these will not be directly related to the pictures. There are no correlated noises, smells and temperature changes.

3   Usually there is little indication of the relative validity of different aspects of the presented data; some of it may be exact, but other parts may be someone's guesses.
4   The instrumentation itself is not perfectly reliable; it may give some spurious indications or it may collapse temporarily.

Most of the disadvantages stem from the fact that there is no direct sensory contact with the real situation. This is irreplaceable in several respects:

(a) There is always slight doubt about a purely visual impression; this doubt is removed if it can be confirmed by auditory, tactile and olfactory impressions.
(b) There is always doubt about the instrumentation itself; it may be generating false data.
(c) Action in reality usually has immediate feedback; action through an interface may involve considerable lags. Furthermore, the feedback through an interface may indicate merely that the information system has accepted a command; it does not necessarily follow that action has occurred.
(d) Direct contact always provides subtle cues which are not available second-hand through reports and instrumentation, e.g. the way in which a subordinate responds to a question as opposed to the content of the response.
(e) The impact of direct observation has an emotional as well as a factual content which may be important, e.g. observing casualties is qualitatively different from receiving lists of losses.

For all these reasons the optimal system is one in which displayed information is used to complement directly sensed information. Current fashion is to underestimate the value of the latter. As displays get more elaborate and attempt to be comprehensive there is a danger that the user will regard the whole thing as a computer-based game. There is data on the screen which can be manipulated without regard to the fact that it represents happenings out there. Command and control displays are never more than windows on reality, and it is essential to look through them rather than at them.

Fortunately, the commander is never alone with his displays. There are other people he can converse with, some in the same room and some accessible by telephone and radio. The VDU-based displays in particular are only one part of his total information source, and should be designed as such. Screens are still much too small to form a model that can be shared by several people holding a discussion. For this purpose maps and other schematic displays are much more satisfactory.

6.1.3 The picture builders
Several kinds of professionals influence the pictures which finally appear before the commander. The world is full of potential information and there has to be selection and filtering.

Much of this filtering is done implicitly at the design stage. The engineer designs his sensing array and communication network according to what can be detected and transmitted by available mechanisms. This is unlikely to coincide with what is ideally needed.

The array of sensing devices can sometimes be changed in orientation, and more usually it can be sampled at the discretion of another kind of operator — the field observer or system manipulator. His role is crucial; essentially it is to convert impressions into facts by allowing entry into the system. Any mistakes he makes are very difficult to remove at any later stage of information processing. In principle his selection is governed by what the commander needs, but he may not have sufficient expertise or awareness of the overall situation to make an optimal series of choices. It is important to recognise that he also is a decision maker requiring his own information sources about what to do and why. If his instructions are to detect or report everything that is available then the task of selection and filtering is pushed further into the system; it must still be carried out by someone.

Thus there is invariably a need for some human intervention at the input end of command and control systems, for several reasons:

(a) Much of the data is probabilistic in that the information is an inference, perhaps from several different sources, which require collation and rejection of whatever seems spurious. For example, the position of an enemy unit may be a function of a trend over time and of current sightings by different devices and from different reference points.

(b) Some of the available data may be deliberately misleading in that the enemy is attempting to feed the system with false information. This requires an assessment in terms of validity and veracity which only a skilled human operator can conduct, e.g. intercepted messages within the enemy systems.

(c) Some data has to be entered into the system via keyboards. This task is less skilled than interpretation of ambiguous data but it is just as crucial for effective system performance (Alderman et al., 1980). Keyboards are also used for control of information presentation, and again effective performance is essential.

Whether or not further human intervention is required within the network is largely a function of the total size. For a very large system a hierarchy of pictures may be appropriate, with human assessments taking place at various levels. If the human action includes intervention as well as monitoring then of course the nature and timing of these interventions must be incorporated for consideration at higher levels. That is, the commander will want to know not only what situations face his subordinates but also what they have done or propose to do about them.

So far this paper has emphasised the general importance of consideration of the human operator in relation to information systems. Such activities can be considered systematically within the design process using two approaches — decision aiding and the contribution from psychology. These are complementary rather than competing methodologies; each can make its own contribution.

6.1.4 Decision aiding

6.1.4.1 Background: Over the past twenty years there has been extensive research and development work in the context of the question: 'We have the technology; what shall we do with it?' This has resulted in much new jargon: expert systems, knowledge engineering, artificial intelligence, information management, intelligent knowledge-based systems, intelligent support systems, decision analysis systems and so on. Because of the obvious need there has been a wide range of human engineering work, and there are several conferences each year under one or more of the above headings. The journals and books are correspondingly loaded with this new material. There is an appearance of rapid progress, but this is largely an illusion for several reasons:

(a) Understanding human behaviour in this situation has proved highly intractable in the sense that models of man as a decision maker are not easy to develop. For example, the classic view that a decision is a selection between alternative actions ignores the aspect that the skilled operator is often more distinguishable by his timing than by his choice (Singleton, 1981). That is, decisions are about precisely when to do it as well as what to do.

(b) Psychological concepts such as intelligence, knowledge, problem solving and thinking appear to conceal many interacting variables and processes. For example, computer support is too easily assumed to take the form of knowledge content in an information sense, but it has long been established that 'knowing that' is quite different from 'knowing how'. More recently Kolers and Palef (1976) have demonstrated that 'knowing not' also appears to be a positive human characteristic.

(c) Intelligence is itself an ambiguous term. It may refer to the quality of human intellect or to information which is of military value. Terminology such as 'intelligent systems' is usually intended to refer to the first meaning, but this is dubious if taken too literally. The human operator is intelligent; he needs complementary not competitive activity within his decision aids. The second meaning is more appropriate for information technology.

(d) The replacement of proven command and control systems must necessarily proceed cautiously and slowly. Massive injections of new technology may appear sound theoretically but may cause great difficulties when untidy, unforeseen, ambiguous situations have to be coped with (e.g. Fialka, 1981).

(e) System components such as microcomputers can change rapidly, but they meet a considerable inertia in having to await the development of exploiting skills in users coping with real problems.

(f) The implicit assumption that man is a rational diagnostician and strategist may be tenable in dealing with a technology-based system such as a chemical processing plant, but this assumption becomes much more dubious when dealing with commercial or military systems. The successful manager or general does not necessarily think like a scientist.

This is not to denigrate the research and development that has been and is proceeding. It is worth while to sample this activity, although the scale is such that some subdivisions are needed to make sense of it. There seem to be four main approaches:

270 The man–machine interface

Design approach The designer constructively manipulates the available technology and uses himself and his colleagues as the evaluators.

Behavioural approach Recent developments in psychological theory, particularly in decision theory, are considered in the context of their relationship to the new technology.

Computer-based support Some established aspect of human performance is selected and specifically aided or compensated.

Task analysis Observable man–computer interaction is used as the basis for understanding how the human operator functions.

These are not mutually exclusive categories but they serve as a structure to comprehend the new literature.

6.1.4.2 The design approach: Most development work comes under this category. Man–machine interfaces for use in specific control rooms are produced by close liaison between designers and software specialists with some advice from human factors engineers. Presented information is divided into two main categories — status information and alarm information — on the principle that the operator needs to know what is happening in general and also with accentuation when some parameter has departed from an expected range, e.g. Herbert (1984). A review of naval work in this area is provided by Glenn *et al.* (1982). They and other authors have concluded that principles of allocation of function in systems need to be better developed, not only decision aspects (Price, 1985) but also on action aspects (Parsons and Kearsley, 1982). An alternative procedure is to identify a formal series of design steps required, with emphasis on the importance of considering the total system as opposed to support for separate tasks (Hopson and Zachary, 1983). A less ambitious approach is to start from the proposition that two main modes of display are feasible: alphanumerics and pictures. The question is then how to design such presentations with mutual supplementation and a match to the needs of the user (Tainsh, 1982).

Systems theory would suggest that the human operator is a particular kind of subsystem with specialist properties which contribute towards the system objectives. This leads directly to the concept of allocation of function between man and machine on the grounds that all man-made systems are man–machine systems. Even in the early days of systems thinking (Wulfeck and Zeitlin, 1962) it was appreciated that men and machines are inherently incomparable, although it was considered at the time that descriptors of human performance might be developed so that a series of elaborate design trade-offs would be possible using operational effectiveness as the criterion. This has not proved to be the case. More recently there have been many attempts to predict the probability of failures of system performance by incorporating human error data into fault trees (Swain and Guttman, 1983). Again the results are of dubious value.

6.1.4.3 The behavioural approach: Sage (1981) has provided a useful review of theories related to problem solving and decision making with more than 400 references and an unusual list of 27 sources of cognitive bias from perceptual anchoring to wishful thinking. Wohl (1981) reviews air force tactical management decision requirements, and lists four types of error based on an assumed sequential decision-making process of stimulus—hypothesis—options—response. There is always an element of self-fulfilling prophecy in designing systems based on hypotheses about how the human operator will behave; being intelligently adaptable, he adjusts his processes to correspond to the available processed information.

There has been a very large research effort on the general principles of interface design starting from the needs and characteristics of the user. An extensive bibliography of studies mainly published in the period 1979—84 is given in Chevelaz *et al.* (1985); this report also contains a useful annotated bibliography of 40 'seminal works'. The authors comment that 'in general the literature appears to be fragmented, situation specific, and lacking in cohesive theoretical structure'. The difficulty seems to be that so much psychological and behavioural research is laboratory based (Wickens, 1984). There is still a widespread belief that only experiments can yield evidence that is scientifically acceptable. It is at least equally tenable that practical and theoretical progress should come from generalisations based on systematically studied real or simulated situations incorporating skilled operators.

6.1.4.4 Computer-based support: The most straightforward approach is to use the computer to make more automatic the visual aids that are already used. For example, Moses and Vande Hei (1978) and Rebane *et al.* (1980) have used a computer graphics system which does rather more efficiently and more flexibly what used to be done with maps, overlays and charts. This approach has the advantage of a natural compatibility with current operator skills, but the disadvantage that the new technology may be under-utilized in not providing sufficient opportunities for new and different skills to develop.

Concentrating on the goal-directed nature of decision making, Pearl *et al.* (1982) have developed a support system that assists the user in structuring and using his own knowledge rather than providing him with displayed knowledge. This ensures that the operator will be systematic (perhaps too systematic) and will consider many action alternatives. It does not in itself use the computer's facility as a data store. This kind of decision analysis aid can be supplemented by other aids. A data store can be added and so also can operational research techniques such as dynamic programming (Smith and Lane, 1984). Considerable effort is currently being devoted to the design of expert systems in which are given indications not only of the answer but of how it was arrived at (Swartout, 1983).

Currently it is accepted that the human operator functions by comparing the real problem situation with his internal model of it. It follows that another useful step forward might be to model the system in the computer and allow the operator to observe a comparison of the model's prediction of what should be happening with data as to what is actually happening (Tzelgov *et al.*, 1985). This also utilises another behavioural attribute: it is established that the human operator is much more efficient as a comparative device than as an assessor of an absolute position. Not only can the computer model predict the future; it might also generate the possible causes of discrepancies between actual and predicted developments (Herbert and Williams, 1985). Thus it compensates for a third human weakness: the limited ability to enumerate all the possible causes or remedies. Studies are proceeding on the possibility of developing an influence modelling and assessment system (IMAS) (Embrey and Humphreys, 1984) which explores causes and associated consequences.

6.1.4.5 Task analysis: Accepting that computer support is useful, the future systems will inevitably have tasks or part-tasks proceeding in parallel in the computer and in the operator with extensive dialogue via the interface. Such systems have proved to be a useful research tool in finding out how the operator proceeds, because what happens within a flexible computer system and what happens across the interface is readily observable and from this data what happens within the operator can be inferred, e.g. Tainsh (1985). In this way the computer functions as an excellent communication bridge between the engineer, the human factors specialist and the user. It is well established by experience that better communication aids are needed to further interdisciplinary research and development and to match more precisely the needs of the user (Licklider, 1982).

Task analysis in this context needs to be related to the information presented and more particularly to the man—computer interaction. The task description can take the form of a diagrammatic listing of computer tasks, human tasks and their interaction; this is called by Tainsh (1985) a job process chart. Other ways of presenting data include tables relating decision-aiding techniques to decision situations, and more tentative Venn-type mapping of functional situations (Hopson and Zachary, 1983). A more personnel-based skills analysis will incorporate other factors such as the available human resources — that is, persons who have been through particular selection and training procedures and can be expected to assume particular levels of responsibility.

Task analysis is related to the design of procedures which is also important, partly as a support for the training procedures and partly to ensure reasonable standardisation in system operation (Sidorsky *et al.*, 1984).

6.1.5 The contribution from psychology
There are natural tendencies to underestimate the complexity of human performance. Simplifying assumptions make for easier analysis techniques, but this can be counter-productive. To take the extreme case, if the operator is treated as a stimulus/response device and the interfaces and training are designed from this viewpoint, then the operator has little option but to behave in this way; the prophecy is self-fulfilling. There is also an implicit belief that operators are less skilled than designers — hence the approach of designing something which is 'foolproof'. This belief is much less tenable for advanced systems where the operator may have accumulated years of experience, giving him an insight into specific

system performance which is quite different from and in some respects superior to that of the designer.

In studying behaviour as it is rather than as it logically ought to be, one is always faced with two fundamental aspects of human performance. First, the operator learns continuously: the way he behaves on first acquaintance with the system will be different from his behaviour after some months of experience, and this will be different again from behaviour after some years of experience. To complicate the picture further, in conditions of high stress he sometimes regresses to earlier forms of behaviour. Secondly, he usually performs on at least two different levels. He responds and simultaneously monitors his own responses. There may even be higher levels where for example he makes allowances for his own biases. This self-monitoring enables him to be very effective in error correction (if the designer has provided such opportunity) but it complicates matters in that, for example, he may deliberately make an error just to check that the system will behave as he expects it to.

Together these factors make for very large differences within and between individuals which need to be taken account of in any task and skill descriptions. Finally, the variety of information he can use, the mental models he can build, and the range of strategies he can adopt also add to the complexity of operator performance.

Human tasks of receiving and transmitting information, decision making and risk taking have been extensively studied by psychologists. Knowledge is acquired and consolidated mainly by the use of laboratory-based experiments. It is rare that the results of these experiments are directly relevant to real problems; their contribution is to the development of principles or theories of human behaviour, and occasionally these theories have some meaning in the context of real tasks. The process from experiment through theory to practice (with feedback and iterations) takes decades. Thus, the useful psychology at present dates mostly from the 1950s and 1960s. Even after a long period of gestation care is needed because academic psychologists are notorious for not defining their terminology precisely. For example, the acceptance of information relates to the concept of perception, but even after a century of experiment and theory it is still not clear precisely where the borderlines are in the sequence sensation–perception–decision or indeed whether they exist. It appears that these processes overlap and overlap also with other processes such as thinking and problem solving.

However, perceptual theory as developed by Gibson (1950, 1966, 1979) provides considerable insight into the characteristics and limitations of the human operator seeking and accepting new information. This theory emphasises elaborate processes and structures which depend at the periphery on the performance of perceptual systems, that is sense organs performing in concert to provide mutually reinforcing data and internally on the modelling of the environment by continuously modifying existing impressions.

Concepts of thinking owe a great deal to Bruner (Bruner *et al.*, 1967; Anglin, 1974), Bartlett (1932, 1958) and Neisser (1967, 1976). Thinking is considered to be an active problem-solving process, characterised by continuous development, in touch with reality by the utilisation of feedback but going beyond the evidence by interpolation and extrapolation. Sensory data is not used directly but is fitted into an already existing internal model of the situation which is kept continuously up to date. Any response is a response to this model rather than directly to reality.

Learning theory has made only a minimal contribution to applied problems — even those of training. The conditioning approach (Hull, 1943) had a long and unproductive life, and it is only recently that some *rapprochement* between learning, training and cognition has been attempted (Lesgold et al., 1978; Thomas and Harri-Augstein, 1985).

In the post-war period there has been a parallel movement depending on concepts first developed for engineered mechanisms and applied to behaviour by analogy (Weiner, 1948; Shannon and Weaver, 1959; Sheridan and Ferrell, 1974). The three main theories in this category are information theory, servo theory and decision theory. Each has contributed some basic ideas to psychology, e.g. channel capacity from information theory, feedback from servo theory and utility from decision theory. Each has generated a vast literature which has added little to the basic ideas because the analogy with human behaviour is a limited one. For example, information theory assumes that the range of alternatives is known, but usually the most difficult issue faced by a human problem solver is the specification of the alternatives. Servo theory is a convenient descriptor of linear stationary systems, but the human operator is neither linear nor stationary; he can change his strategy abruptly and he learns continuously. Decision theory and its offshoots such as signal detection theory have been curiously unproductive in the sense that there is little that the human factors specialist can use. The principles of vigilance as enunciated by Mackworth (1950) provide a useful description of human behaviour in situations where 'men are required over long periods of time to watch or listen for hardly distinguishable signals occurring irregularly at unpredictable intervals', but extensive experimentation and the application of signal detection theory again seem to have generated little further knowledge of use in practical situations.

Thus, for the present, the human factors or ergonomics specialist relies on three attributes which enable him to make a unique contribution to system design:

(a) A general sensitivity to the importance of the human operator and his advantages and limitations as a system component
(b) Expertise in how to find out about some specific aspect of human performance
(c) A theory of human skill which incorporates ideas about learning and performance in controlling complex systems.

The first category includes the sometimes obvious and still vitally necessary reminders to others in the form of consistently asking 'what about the operator?', and pointing out simple errors which are still made such as ambient lighting that reflects from the surface of visual display units. This expertise is contained in textbooks and manuals about human performance, some general (e.g. McCormick and Sanders, 1982), some specific to particular kinds of performance (e.g. Easterby and Zwaga, 1984) and some specific to particular industries (e.g. EPRI, 1982).

The second category includes expertise on the design of experiments and also field studies such as task analysis and error analysis. There are many books about the design of experiments relevant to systems (e.g. Moraal and Kraiss, 1981), but few as yet about field study procedures. The one which most nearly approximates to a comprehensive survey is Rasmussen and Rouse (1981).

Theories of human skill are dealt with in Singleton (1978–83) and Rogers and Sloboda (1983). The latter extends to language skills, which are increasingly important in relation to computer-driven displays; these skills are described in more detail in Kolers et al. (1980). The controller is a person with a repertoire of skills at his disposal which enables him to function competently at the two extremes of following standard procedures and diagnosing unexpected situations. For the standard procedure he is familiar first with how to locate it quickly and unambiguously within the large library of operating instructions, and thereafter to follow it accurately but not mechanically. That is, for each action he will check that the system responds as expected, and he will have the confidence to cope with any discrepancies that may arise due to errors in the instructions or any failures in the system reaction.

His diagnostic role is crucial and depends on continuous learning before and during his period of responsibility for system performance. The assumption that he responds to his model of the situation rather than to the situation itself is a useful explanation of at least two phenomena. On the positive side he can escape from the tyranny of reacting immediately; his behaviour is not tied closely to specific inputs and his outputs are a function of the developing history of the event rather than its immediate manifestation. On the negative side he can be wrong in that his model does not sufficiently match reality, and this can lead to major errors. Minor errors occur all the time because he generates hypotheses (that is, extrapolations from the model of the current situation), checks these by tentative responses and modifies them when the system response is not in line with his prediction. Even so the model can depart from reality because it steers his selection of significant data and provides a filter through which he rejects data that is not, for the moment, regarded as relevant. In older psychological terminology his separation of figure from ground is governed by the model which sets the criteria of relevance.

Command and control systems have specific problems in that the system user is a particular kind of person in terms of the way he was selected and trained as a military officer. His cognitive style will differ depending on whether he is of junior, intermediate or senior rank (Page, 1981). The junior officer is an action man rather than a thinker; his success depends on intervening immediately and drastically with little time for formal analysis. In the intermediate ranks this dashing style gives way to a more cautious logical approach with considerations of past trends and future consequences. The senior officer has to take a wider view, considering, for example, political and mass media reaction to his activities; because of the multiplicity of relevant factors he tends to make decisions more intuitively again. Clearly the information needs for these contrasting but complementary styles of operation are very different. The design of decision aids should take account of these differences as well as the task requirements.

6.1.6 Conclusion

It seems that men and machines are not usefully regarded as equivalent system components in any performance sense. This is partly because their attributes are complementary rather than competitive and partly because the human components have a unique role in monitoring and guiding progress towards overall objectives.

It is questionable whether the user of the information system should be regarded as part of the system; rather he is the customer of the system output. The latter view emphasises the importance of designing the whole system including the output man—machine interface to suit his requirements.

The design of man—machine interfaces is always critical for the reason already mentioned, namely that the interface must facilitate communication between components with fundamentally different characteristics. The human operator side is highly flexible, adaptable by the use of training techniques and easily reprogrammable in terms of revised instructions, but nevertheless there are fundamental limitations in sensory and motor performance and in information handling.

Human operators within the information network have similar roles to those in other kinds of information networks such as those used in commerce. The central human factors problem is neither allocation of function nor interface design but an intermediate one in which consideration is given to the position and role of human intervention within the network. This is best considered as a problem of tasks, with allocation of tasks to operators dependent on factors such as load and required level of skill. The required procedures are those of task analysis and skills analysis.

Given these and other complexities it is becoming more and more necessary to regard the design of an information system not so much as a one-off which is completed and handed over for operational use, but rather as an evolutionary system with the potentiality for adaptive changes in the light of increasing experience.

6.1.7 References

ALDERMAN, I. N., EHRENREICH, S. L. and BINDEWALD, R. (1980) 'Recent ARI research on the data entry process in battlefield automated systems', US Army Research Institute for the Behavioural and Social Sciences, RR 1270
ANGLIN, J. M. (ed.) (1974) *J. S. Bruner: Beyond the Information Given*, George Allen & Unwin, London
BARTLETT, F. C. (1932) *Remembering*, Cambridge University Press, Cambridge
BARTLETT, F. C. (1958) *Thinking*, George Allen & Unwin, London
BRUNER, J. S., GOODNOW, J. J. and AUSTIN, G. A. (1967) *A Study of Thinking*, Wiley, New York
CHEVALAZ, G., TIJERINA, L., MODISETTE, L. and HERSCHLER, D. (1985) 'Final report on menu and display literature search to US Army Human Engineering Laboratory', Battelle, Columbus, Ohio
EASTERBY, R. and ZWAGA, H. (eds) (1984) *Information Design*, Wiley, New York

EMBREY, D. and HUMPHREYS, P. (1984) 'Support for decision making and problem solving in abnormal conditions in nuclear power plants', London School of Economics and Political Science, Decision Analysis Unit, research paper
EPRI (1982) 'Human engineering guide for enhancing nuclear control rooms', EPRI NP-2411, Palo Alto
FIALKA, J. J. (1981) 'The Pentagon's exercise "Proud Spirit"': little cause for pride', *Parameters* (Journal of the US Army College), **XI**(1), pp. 39–41
GIBSON, J. J. (1950) *The Perception of the Visual World*, Houghton-Mifflin, Boston
GIBSON, J. J. (1966) *The Senses Considered as Perceptual Systems*, Houghton-Mifflin, Boston
GIBSON, J. J. (1979) *The Ecological Approach to Visual Perception*, Houghton-Mifflin, Boston
GLENN, F. A., BENNETT, J. M. and ZACHARAY, W. W. (1982) 'Design of air strike planning aids', Office of Naval Research Technical report 1454-B, Arlington, PA
HERBERT, W. R. (1984) 'A review of on-line diagnostic aids for nuclear power plant operators', *Nuclear Energy*, **23**(4), pp. 259–64
HERBERT, W. R. and WILLIAMS, G. H. (1985) 'An examination of qualitative plant modelling as a basis for knowledge-based operator aids in nuclear power stations', CEGB, Leatherhead, research paper
HOPSON, J. and ZACHARY, W. (1983) 'Experiences in aiding airborne decision-making', Society of Automotive Engineers, 9104 4388–4401
HULL, C. L. (1943) *Principles of Behaviour*, Appleton Century, New York
KOLERS, P. A. and PALEF, S. R. (1976) 'Knowing not', *Memory & Cognition*, 4(5), pp. 553–8
KOLERS, P. A., WROLSTAD, M. E. and BOUMA, H. (eds) (1980) *Processing of Visible Language 2*, Plenum, New York
LICKLIDER, J. C. R. (1982) 'On communication between systems scientists and behavioural scientists', NATO, Brussels, annex 11 to AC/137-D/653
LESGOLD, A. M., PELLEGRINO, J. W., FOKKEMA, S. D. and GLASER, R. (eds) (1978) *Cognitive Psychology and Instruction*, Plenum, New York
MACKWORTH, N. H. (1950) *Researches on the Measurement of Human Performance*, HMSO, London
MCCORMICK, E. J. and SANDERS, M. S. (1982) *Human Factors in Engineering and Design*, McGraw-Hill, New York
MORAAL, J. and KRAISS, K.-F. (eds) (1981) *Manned Systems Design*, Plenum, New York
MOSES, F. L. and VANDE HEI, R. P. (1978) 'A computer graphic-based aid for analyzing tactical sightings of enemy forces', US Army Research Institute for the Behavioural and Social Sciences, TP 287
NEISSER, U. (1967) *Cognitive Psychology*, Appleton-Century-Crofts, New York
NEISSER, U. (1976) *Cognition and Reality*, Freeman, New York
PAGE, M. R. H. (1981) 'Management of military organisation', in Singleton, W. T. (ed.) *Management Skills*, MTP Press, Lancaster
PARSONS, H. M. and KEARSLEY, G. P. (1982) 'Robotics and human factors', *Human Factors*, **24**(5), pp. 535–52
PEARL, J., LEAL, A. and SALEH, J. (1982) 'Goddess: a goal-directed decision structuring system', *IEEE Trans.*, **PAMI-4**(3), pp. 250–62
PRICE, H. E. (1985) 'The allocation of functions in systems', *Human Factors*, **27**(1), pp. 33–45
RASMUSSEN, J. and ROUSE, W. B. (eds) (1981) *Human Detection and Diagnosis of System Failures*, Plenum, New York
REBANE, G. J., WALSH, D. H., MOSES, F. L., SCHECHTERMAN, M. D. and LEVI, L. R. (1980) 'Dynamic displays for technical planning', US Army Research Institute for the Behavioural and Social Sciences, RR 1247
ROGERS, D. R. and SLOBODA, J. A. (eds) (1983) *The Acquisition of Symbolic Skills*, Plenum, London

SAGE, A. P. (1981) 'Behavioural and organizational considerations in the design of information systems and processes for planning and decision support', *IEEE Trans.*, **SMC-11**(9), pp. 640–78

SHANNON, C. E. and WEAVER, W. (1959) *The Mathematical Theory of Communication*, Illinois Press, Urbana

SHERIDAN, T. B. and FERRELL, W. R. (1974) *Man–Machine Systems*, MIT Press, Cambridge, Mass.

SIDORSKY, R. C., PARRISH, R. N., GATES, J. L. and MUNGER, S. J. (1984) 'Design guidelines for user transactions with battlefield automated systems: prototype for a handbook', US Army Research Institute for the Behavioural and Social Sciences, RP 84-08

SINGLETON, W. T. (ed.) (1978–83) *The Study of Real Skills., Vol 1–4*, MTP Press, Lancaster

SINGLETON, W. T. (ed.) (1981) *The Study of Real Skills. Vol. 3: Management Skills*, MTP Press, Lancaster

SMITH, S. B. and LANE, R. S. (1984) 'An intelligent operator aid for dynamic route planning', *Expert Systems*, **1**(2), pp. 143–8

SWAIN, A. D. and GUTTMAN, H. E. (1983) *Handbook of Human Reliability Analysis with Emphasis on Nuclear Power Plant Applications*, NUREG/CR-1278

SWARTOUT, W. R. (1983) 'EXPLAIN: a system for creating and explaining expert consulting programmes', *Artificial Intelligence*, **21**, pp. 285–335

TAINSH, M. A. (1982) 'On man–computer dialogues with alpha-numeric status displays for naval command systems', *Ergonomics*, **25**(8), pp. 683–703

TAINSH, M. A. (1985) 'Job process charts and man–computer interaction within naval command systems', *Ergonomics*, **28**(3), pp. 555–6

THOMAS, L. F. and HARRI-AUGSTEIN, E. S. (1985) *Self-organised Learning*, Routledge & Kegan Paul, London

TZELGOV, J., TSACH, U. and SHERIDAN, T. B. (1985) 'Effects of indicating failure odds and smoother outputs on human failure detection in dynamic systems', *Ergonomics*, **28**(2), pp. 449–62

WEINER, N. (1948) *Cybernetics*, Plenum, New York

WICKENS, C. (1984) *Engineering Psychology and Human Performance*, Merrill, Columbus, Ohio

WOHL, J. G. (1981) 'Force management decision requirements for Air Force tactical command and control', *IEEE Trans.*, **SMC-11**(9), pp. 618–39

WULFECK, J. W. and ZEITLIN, L. R. (1962) 'Human capabilities and limitations', in Gagne, R. M. (ed.) *Psychological Principles in System Performance*, Holt, Rinehart and Winston, New York

Chapter 6.2
An engineering standard for a systematic approach to the design of user–computer interfaces

A. Gardner
(HUSAT Research Centre)

6.2.1 Introduction

A Naval Engineering Standard (NES) on human factors in the design of military computer-based systems is being prepared by HUSAT under contract to the UK Ministry of Defence. The draft will be field tested during 1986–7 and revised before publication. This paper is intended to give a foretaste of the possible contents of the standard.

The standard is being written to present an overall methodology of human factors in the sense of a coherent collection of methods and principles which ought to have a wide relevance to many kinds of systems design. However, its prime focus of attention is C^2 systems within the UK Royal Navy and the advice on design solutions suitable for these specific applications. For example, general advice on task allocation and task design is supplemented by specific advice on the selection of dialogue styles and interface devices. In these respects, the NES is intended to complement the many existing handbooks on human factors which focus on general, physical ergonomics (e.g. UK MoD (Navy) Defence Standard 00-25) and general user–computer interface design (e.g. Smith and Mosier, 1984).

6.2.1.1 User-centred design: The overall design methodology in the standard is called user-centred design. It is user centred in three ways:

(a) All computer-based systems exist for *human purposes*. The standard gives advice on how to identify the users' purposes and how to represent the information gathered so that it is actionable by designers.
(b) The criteria on which systems are to be judged are the requirements of those who must use it. The standard gives advice on how to establish *testable system goals from the point of view of the users*.
(c) The design team must have full *user involvement* in all aspects of systems

development. The standard gives advice on how to involve users in design decision making.

User-centred design always assumes that the users' views are important. It does not assume that the users are always right. There is a strong emphasis on objective quality control and empirical testing in all parts of the standard to help counteract personal biases on the part of individual members of the design team.

User-centred design is a methodology for creating user–computer systems which give *effective performance* and are *acceptable to users*. These two goals are interrelated since effective performance helps to make systems more acceptable to users (especially when their lives might depend on it) and acceptability to users helps to make systems more effective. The standard gives advice on systems design directed at both goals. The goals are not treated separately.

It is recognised that there are other important system goals (e.g. costs, quantities and delivery dates) but these are outside the scope of this standard. It is recognised, too, that the methodology is neither foolproof nor complete. The standard does not remove the need for the design team to show judgement and creativity in arriving at a satisfactory product.

6.2.1.2 Relationship to other methodologies: User-centred design is complementary to established methodologies for systems development (e.g. CORE, DIADEM, MASCOT). Other methodologies concentrate on computer factors, whereas the standard concentrates on human factors. Neither approach is complete without the other. Similarly, user-centred design will impact upon project management but it is not a methodology for project management.

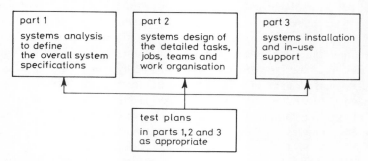

Fig. 6.2.1 *Organisation of the standard*

6.2.1.3 Organisation of the standard: The standard is organised into three parts (or volumes) covering three major stages in systems development (see Fig. 6.2.1). The human factors decisions which have to be made at each stage are explained and advice is given on how to make them. Advice on test plans and prototyping are included in each part as necessary.

This contribution deals specifically with part 2 — 'human factors in systems design'. This means that it deals not only with the design of user–computer inter-

Engineering standard to the design of user-computer interfaces 281

faces but also with the design of tasks, jobs, teams and work organisations; see Fig. 6.2.2. As shown in the figure, systems design is conceived of as a transform of inputs into outputs with associated tests. The main features of this process will be described here, with some features treated in greater depth than others as being more directly relevant to the aims of this contribution.

6.2.2 Inputs to systems design

A central concern of systems analysis is to determine the information shown in Fig. 6.2.3. Each action ('What is done?' or 'What is to be done?') is related to all other actions by the reciprocal questions 'When' and 'What next?' (giving the time line relationships) and 'Why?' and 'How?' (giving the purpose and method relationships).

If these relationships are pursued systematically the result is a database of

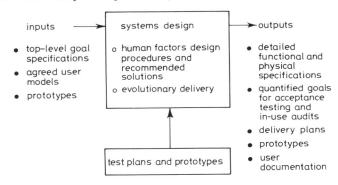

Fig. 6.2.2   Outline contents of part 2

Fig. 6.2.3   Basic questions of systems analysis

actions and relationships (sometimes called objects and attributes). This database can then be represented diagrammatically using formal conventions common in structured programming. For example, Fig. 6.2.4 shows certain planning functions in C^2 represented in the format of CORE action diagrams (Stevens and Whitehead, 1985). Briefly, the boxes are processes (the actions) linked by message flows (giving time relationships of 'When?' and 'What next?') and nested within other processes (giving the purpose and method relationships of 'Why?' and 'How?').

Note that the extract of Fig. 6.2.4 is used here merely as an illustration. The accuracy or adequacy of the model shown is not important for the purposes of this contribution. Also the use of CORE in the standard is not to be taken to mean that

282 Engineering standard to the design of user–computer interfaces

analysts are to be constrained to use CORE. Other notations could be used just as effectively.

Figure 6.2.4 is an example of what in the standard is called an *agreed user model*. It is a user model in that it specifies the functionality that users want from the system to be designed. It is agreed in the sense that it has to be authorised by the user authority as a correct description of the system to be designed.

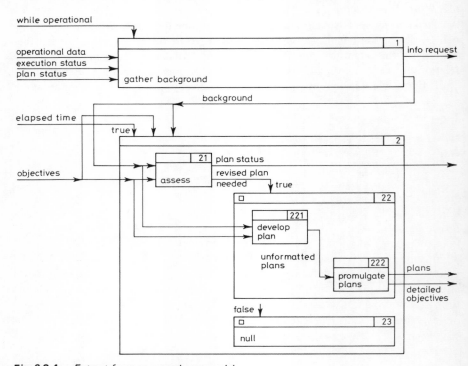

Fig. 6.2.4 *Extract from an agreed user model*

An agreed user model like that shown in Fig. 6.2.4 has to be supplemented by other information collected during systems analysis, since the action diagram notation cannot represent all the things that the design team needs to know. For example, designers also need to know the details shown in Fig. 6.2.5. These are displayed as annotations to the action diagrams or recorded in associated documents.

More completely, the agreed user models provide a description of the duties and activities to be allocated, along with information on how they are allocated in current systems, how well they are being performed and to what extent the current allocations are acceptable to users.

The agreed user models and the *top-level goal specifications* also include information on constraints affecting the new system. For example, they define: the target users in quantity and quality; the development and delivery deadlines; the costs

Engineering standard to the design of user–computer interfaces

and resources; the performance effectiveness to be attained; the policy with regard to the extent of automation; and the degree of technological risk acceptable. *Prototypes*, which for these purposes include operations research simulations, provide information on the dynamics of the system scenarios and, if sufficiently detailed and of the user-in-the-loop type, provide information on users' performance and preferences.

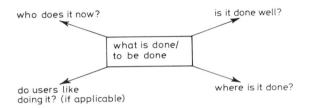

Fig. 6.2.5 *Some additional questions for systems analysis*

These inputs are unlikely to form a complete specification of the system to be designed. Many details will remain to be decided during task design, job design etc. In particular, design decisions taken in the latter stages may cause a major iteration, forcing one to change the earlier specification. For example, it may prove to be technically impossible to automate a set of critical functions or to have these done by users. This will force a major revision of the top-level goal specification.

In addition, it is unrealistic to believe that systems analysts can define the system's intended functionality without presuming something of the physical allocations and means of implementation (as implied, for instance, by the distinction between logical design or functional design during systems analysis and physical design during systems design). Systems analysts are encouraged to defer the physical design as long as possible, in order to encourage worthwhile innovation, but it cannot be ignored entirely.

Hence, it is recognised that the precise demarcation between systems analysis (part 1) and systems design (part 2) is not clear cut. Contractors are advised to use parts 1 and 2 in conjunction rather than in isolation.

6.2.3 Tasks

6.2.3.1 Concept of a task: The central human factors concept in systems design itself is the task. Tasks are defined in terms of:

Duties in the sense of operational goals, delegated responsibilities, resources etc.
Activities in the sense of processes and messages (data and control). Activities may be externally observable (e.g. actions) or internal and unobservable (e.g. thinking). Activities are the functional primitives of human factors design.

Both have to be specified for a full description of a task. Duties and activities are identified during systems analysis and are represented in the agreed user models. In the CORE notation (Fig. 6.2.4) activities are the primitive processes (e.g. 21) nested within duties (e.g. process 2). Activities describe 'How?' and duties describe 'Why?'. It is recognised that the systems analysts cannot define tasks in every detail during systems analysis. Some features will need to be revised or completed during systems design. The agreed user models are a starting point.

6.2.3.2 Task allocation: As shown in Fig. 6.2.6, task allocation is the design process whereby the human factors goals and recommended solutions are synthesised into detailed functional and physical specifications in the light of the characteristics of the target users.

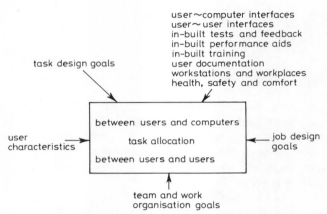

Fig. 6.2.6 *Outline of sections on human factors design procedures and recommended solutions*

Duties are always allocated to users since ultimate responsibility for the system lies with the users. Allocating duties between users and other users involves the designers in job, team, and work organisation design. Point defence against missiles is the classic example of a duty which relies heavily on automation to achieve the rapid response needed. Even so, point defence is the responsibility of a specific user who has responsibility for ensuring that the automated functions are available and, most importantly, has right of veto.

Allocating activities between users and computers involves the designers in task design. These are the primary decisions affecting the system.

User–computer interfaces, user–user interfaces etc. (as in Fig. 6.2.6) are added to the system to facilitate the performance of the users' duties and activities. This point is stressed because, in the past, it was common for human factors to be thought of as concerned solely with interfaces, workstations and workplaces, and health, safety and comfort. The standard recognises the importance of these topics and provides detailed advice on them. However, we take the position that these topics have been over-emphasized, in the past, and seek to redress the balance by placing greater emphasis on the more global issues of systems analysis and specifications, task and job design (etc.), and systems installation and in-use support.

Engineering standard to the design of user–computer interfaces 285

6.2.3.3 Methodology for task allocation: Task allocation can be approached as an iterative series of design steps as shown in Fig. 6.2.7. Experienced designers will be able to undertake task allocation concurrently rather than serially. For example, the initial provisional allocations will be done with an eye to the implications for the additional features (e.g. interfaces and performance feedback) and the designs will be checked as they are being devised rather than when complete. Designers who are less expereinced in human factors, however, are advised to tackle the problems serially as in Fig. 6.2.7 and accept that they will need to revise their designs many times.

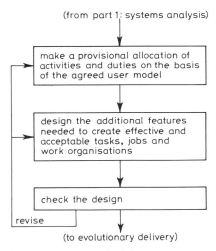

Fig. 6.2.7 *Outline of task allocation methodology*

6.2.3.4 Initial task allocations: If task allocation is treated as an iterative design process then it does not matter what rules are used for the *initial* allocations, since these allocations can be revised at a later date. It is recognised that designers tend to start from whatever features they are most familiar with and work outwards from there. Nevertheless, this standard makes a strong recommendation that *designers should consider using the procedures given below as an initial rule set*. These procedures are not mandatory but are based on sound practical experience.

Task allocation has to be done so as to create effective and acceptable tasks, jobs and work organisations. Design goals for each of these are provided in their respective sections in the standard. The following four rules are a subset and have been selected for use in initial allocation because they are actionable early in the system design stage:

1 *Let users do the things they are good at and like doing*   Make this allocation on the basis of user performance and opinions with current or comparable systems or with prototypes. If none of these is available, designers will have to make an initial allocation on the basis of user preference and expert judgment.

2 *Allocate new duties and activities to users in a way that is consistent with existing duties and activities*   Make sure that:

   (a) New duties and activities are similar in content to existing duties and activities
   (b) They are contained within the same or closely related duties
   (c) There is no negative transfer between the old and the new.

3 *Design for the lowest common denominator*   Make this allocation on the basis of the performance of the least capable one-third of target users on current or comparable systems or with prototypes. Make sure that:

   (a) Activities involve a low level of abstraction and are uncomplicated
   (b) No more than three activities are nested concurrently.

4 *Design for least effort* such that users are not straining (mentally or physically) under normal 'exercise' conditions. Make this allocation on the basis of user opinions and expert judgments in the light of performance with current or compable systems or with prototypes. Make sure that activities are not widely separated physically such that users would have to hurry between workplace and workplace.

These four rules will enable designers to resolve much of the initial uncertainty about the allocation of duties and activities. It is recognised that these initial allocations may need to be revised when consideration is given to other rules, as suggested in Fig. 6.2.7. Note that these rules are actionable only in so far as the agreed user models contain the necessary information. This means that certain prerequisite questions have to be asked during systems amalysis, as shown in Fig. 6.2.5.

The emerging design must be recorded by revising and annotating the agreed user models, as described in the task allocation charts described in the following section.

6.2.3.5 Task allocation charts: Figure 6.2.8 shows the duties and activities in Fig. 6.2.4 expressed as a task allocation chart. Processes 1 and 2 have been allocated to users 1 and 2 respectively. Process 222 has been allocated to a computer under the direct control of user 2.

The task allocation chart is simply a recasting of the agreed user models into a format which reveals more clearly the allocation of functions to users and computers. In these charts, functional chains of command are indicated by the ways in which lower-level tasks are nested within high-level tasks.

6.2.4 User–computer interfaces

User–computer interfaces are the logical functions and physical devices that enable users and computers to communicate. User–computer interfaces translate messages between the preferred task language of the users and the preferred operating language of the computers: controls translate from users to computers, and displays translate from computers to users.

Engineering standard to the design of user–computer interfaces 287

Figure 6.2.9 illustrates the additions of a user–computer interface (process 221/222) to facilitate communication between user 2 and the computer, and the addition of a user–user interface (process 1/2) to facilitate communication between user 1 and user 2.

Task allocation charts can become very complicated, especially if the tasks involve several mutually exclusive processes (e.g. process 22 and 23 in Fig. 6.2.9) and if several additional features are included. Charts are to be extended over several pages as necessary in the interests of clarity.

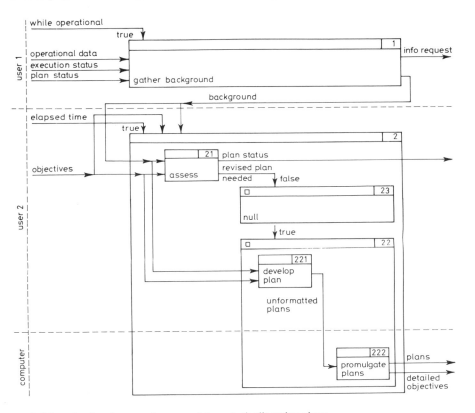

Fig. 6.2.8 *Casting the agreed user model as a task allocation chart*

The disadvantage of charts which spread on to several sheets of paper (or screens if computerised) is that it is hard to grasp the whole and to appreciate the detail. One way of possibly overcoming this is to describe the task activities as lines of psuedo-code in a structured programming format. To some extent the choice of format reflects personal preference. Non-programming users find the diagrams easier to follow than the psuedo-code, so it might be necessary for designers working with psuedo-code to convert to diagrams for presentations to users.

288 Engineering standard to the design of user–computer interfaces

6.2.4.1 Methodology for designing user–computer interfaces: User–computer interfaces are additional features added to the system to facilitate the performance of the users' duties and activities. The methodology for designing user–computer interfaces, therefore, starts with a consideration of the users' duties and activities

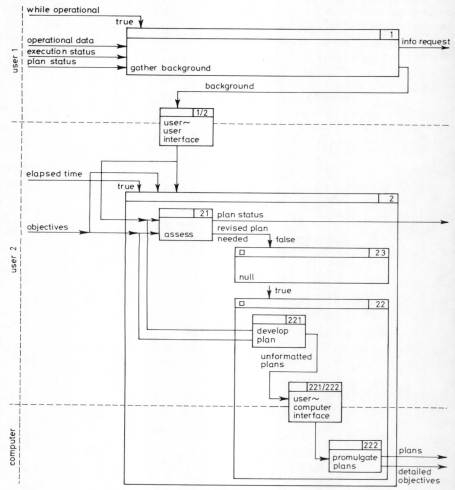

Fig. 6.2.9 *Example of additional features added to a task allocation chart*

and then determines how best to achieve these ends bearing in mind the users' characteristics. Figure 6.2.10 shows the methodology in outline.

The methodology is presented here as a serial sequence of design decisions. Experienced designers will tackle these problems concurrently along with other additional features.

6.2.4.2 User–computer interface protocol: Designers will be familiar already with the need to design protocols for communications such as NES 1026 and 1028 (Section 5.4) and the ISO/OSI seven-layer model (ISO, 1982) for computer–computer interfaces. This standard approaches the design of user–computer interfaces in the same way. The steps in Fig. 6.2.10 reflect the recommended protocol for user–computer interfaces outlined in Table 6.2.1.

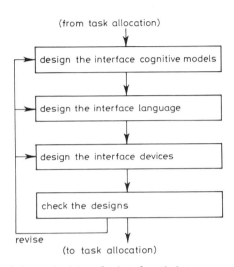

Fig. 6.2.10 *Outline of the methodology for interface design*

6.2.5 Interface cognitive models

Interfaces convey messages to and from users about their duties and activities. The user–computer interface protocol requires designers to define first the functionality of the interfaces and secondly the imagery to be used to convey the message.

The *functionality* of the interfaces is defined in terms of:

(a) Which duties and activities? (the subject matter)
(b) Who from? (the source) Who to? (the destination)
(c) Broadcast? (for all to receive) Point-to-point? (private communication)

Table 6.2.1 *Outline of user–computer interface protocol*

Name	Concerned with	See
Application tasks	System functionality and the imagery of the messages	Interface cognitive models
Task language	Message dialogues, grammar, and vocabulary	Interface languages
User devices	Physical aspects of controls and displays	Interface devices

(d) Sent with prior arrangement? (connected) Sent without prior arrangement? (no connection)
(e) Receipt acknowledged?

These details define the context of the message. Wherever possible they are to be included with the message without requiring the users to supply information already held by the computer.

The *imagery* to be used to convey the messages must be defined such that there is a good fit between the imagery of the messages and the imagery of the users' cognitive models. These terms are explained below. Good imagery and good cognitive models help to make the tasks more concrete and more resistant to stress and disaster. They also help to make the tasks easier to learn and easier to remember.

Many messages can be represented in a variety of images (e.g. set membership can be described verbally, algebraically and diagrammatically) and different users may prefer to think in different images. Designers are to consider the possibility of providing users with a choice of imagery — especially for critical tasks.

6.2.5.1 Transparency: Interfaces should be transparent in the sense that they enable the users to work through them to perform their necessary duties and activities without appearing in any way as a barrier. A transparent interface is like a well cleaned window: it is there but it can be looked through without hindrance. The user—computer interfaces differ from a well cleaned window, however, in that the users do not look through to the physical world directly; they work with a computer representation of the world rather than with the world itself. Hence, the main prerequisite for transparency is that designers can create an adequate representation or image of the worlds that users must work with. In practice, this means that the interfaces must help users to create cognitive models that are understandable to them and accurate enough for the system's purposes. For example, a three-dimensional (3-D) task ought to be tackled with a 3-D cognitive model and, hence users ought to be provided with 3-D controls and displays or linked 2-D devices which the users can easily integrate into a 3-D mental picture (as in approach radars which give a pair of displays showing plan and elevation).

6.2.5.1.1 Transparency to the users' external world
Some interfaces exist to enable users to interact with the world external to the system via the sensors (e.g. radars and sonars) and the effectors (e.g. missiles and external communications). The following provides an example of the interface design problems and their solution. It is taken from the domain of navigation using passive sensors which provide details of a contact's bearing (direction from own ship) but not its range, speed or heading. To obtain such information, the operator has to employ complicated mathematics which results in numerical estimates of the most likely range etc. and the margins of error associated with these estimates. To a mathematician, the uncertainties involved are expressed most

Engineering standard to the design of user–computer interfaces 291

succinctly in terms of means and standard deviations. The following shows an extract from a display in an old system. The solutions are presented to the commanding officer in these mathematical terms, e.g. 'range' is the estimated mean range and 'sigrge' is the standard deviation (sigma) of the estimated mean range:

tr no	014		
brg	A	sigbrg	a
range	B	sigrge	b
course	C	sigco	c
speed	D	sigspd	d

The design problem here is that although the display is sufficiently accurate for navigation purposes, it is not intelligible to the least common denominator of user. Statistical concepts are notoriously difficult for most users to understand. And in this case the commanding officers were being required to combine mentally four separate distributions of uncertainty to determine where the contact might be.

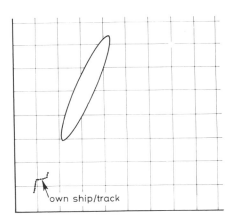

Fig. 6.2.11 *Improved display*

The design solution was to re-examine the display requirements by asking what was the real duty that the system was designed to undertake. In this example, the operator was being tasked by the commanding officer with the order 'Show me where you think the contact might be, based on the best data currently available'. The design solution, therefore, was to answer this question directly using imagery which the ultimate recipient (the commanding officer) might best understand. Since the commanding officer was already using map-like displays for planning, the navigation output was converted to this format. Figure 6.2.11 shows the new version. The track of own ship is shown in the bottom corner of a labelled plan display (LPD). The navigation uncertainties have been combined and presented as an area of the LPD (in this case an ellipse) within which the contact is likely to be located with some ascribed probability (80%, 90% or whatever accuracy is required). This new display is more transparent to the external world in the sense that it transforms the sensor data into an image that better fits the user's cognitive model and the system's needs (the numerical version is available as an option).

6.2.5.1.2 Transparency to other users' subsystems

Some interfaces exist to enable users to interact with other users' subsystems (e.g. to send and receive orders and data).

Some of these may be implemented as user—user interfaces rather than as user—computer interfaces, but in terms of designing their functional transparency the two can be treated in the same way. The decision to implement the interface directly or via a computer is a later stage in the design process. The first step is to decide upon its functionality.

The world that has to be represented, in this case, consists of processes and messages making the system. Typically, this can be shown by some form of flow diagram and status display at a level of detail appropriate to the particular user's needs. Examples are mimic diagrams in machinery control and functional block diagrams in electronics maintenance. Where there is a need to represent the system as a set of duties and activities, designers might consider the possibility of using the task allocation charts themselves in the form of an interactive map of system functions.

6.2.5.1.3 Transparency to the internal world of the users' computers

Some interfaces exist to enable users to look inside their own computers (e.g. check the contents of a file; debug a faulty program; use the word processing software). This type of transparency is sometimes referred to as allowing users to open up the computer process. It is the sense in which transparency is used, commonly, by designers of intelligent knowledge-based systems (**IKBS**) when they refer, for example, to the computer being able to explain its reasoning having made a decision. The design requirement is similar to that for transparency to other users' subsystems: the typical representation will be as a flow diagram status display of processes and messages (e.g. task allocation charts; lines of code in a structured program; sequence of rules in a production stack).

6.2.5.2 Virtual machines

: A complementary way of thinking about the interface functions is to think of the interface as presenting a virtual machine to the user (in the sense that the external world, the other users' subsystems and the users' own computers, are known to the users only in terms of models and the detail provided by the interfaces). This will be determined, usually, during systems analysis when decisions are made about what is to be the activity level of the agreed user models. During interface design, however, this is to be re-examined. It will be found, often, that the interface implications of a particular choice of activity level (or a particular initial task allocation) lead to an unworkable virtual machine being presented to the users. Often, this can only be discovered by testing with prototypes.

6.2.5.3 Visual metaphors: It has been stated that a good interface will have a good fit with the users' cognitive model of the tasks, world etc. Equally obviously, not every facet or detail can be represented at the interface, or certainly not all at the same time (i.e. the interface presents a virtual picture to the users). It follows that designers have to make decisions as to what they include or leave out.

For example, it has become common for office automation computers to represent activities as objects in the form of visual metaphors (e.g. symbolic telephones, documents, wastepaper baskets etc. appearing on an office desk). The objects are made explicit but their functions are left implicit. This form of interface is successful only to the extent that users *already possess* a good cognitive model of an office desk and the objects/activities that it presumes. Such models are highly task specific. There are no agreed visual metaphors or equivalent cognitive models that can be applied to all naval C^2 tasks but designers are encouraged to look for visual metaphors wherever possible (e.g. Fig. 6.2.11).

6.2.5.4 Multiple levels of interfaces: The preceding discussion has emphasised the need to match the interfaces to the duties and activities of the task allocation charts. Designers must bear in mind the possibility that these duties and activities may have to be reallocated while in use and that each new allocation may necessitate a change in the interfaces.

In practice, a reallocation will mean either that the breadth of a user's duties has to be changed or that the level of detail has to be changed. Every single duty could be reallocated to become an activity with its own user–computer interface if need be. This might prove to be impossibly costly, so designers should consult users to determine the most likely reallocations (especially of critical tasks) and design multiple interfaces accordingly. These are to be added to the task allocation charts as variations.

6.2.6 Interface languages

The user–computer interface protocol requires designers to specify the interface languages in terms of:

Vocabulary (characters, words etc.)
Grammar (sentences and paragraphs)
Dialogue (control, style and sequencing).

These are explained in the subsections below.

6.2.6.1 Vocabulary: The vocabulary of the interface is the whole range of words, abbreviations, punctuations, numbers, mathematical symbols, map symbols, lines, points, colours etc. that are the primitives of the messages and of the cognitive models. By definition, these primitives are used in combination to create a message. For convenience, all vocabulary primitives are referred to as *characters* or *words*.

There is a strong operational imperative to standardise on the vocabulary of military systems in the interests of consistency. Designers must show that the proposed vocabulary is adequate to express all the messages required by the system. Synonyms may be needed to accommodate different cognitive models.

Each user subsystem requires a different subset of the total vocabulary because of different task demands. Hence each interface may be fitted with only a limited subset of the whole. But, wherever possible, all interfaces are to be fitted for the total vocabulary so as to allow for flexibility. For example, a user's displays and controls may be configured to deal with the words, characters, symbols etc. of a particular task but are to be capable of being reconfigured to deal with other vocabularies for other tasks.

6.2.6.1.1 Abbreviations
Abbreviations are a special form of words. The standard gives detailed advice on how and when to construct abbreviations. Briefly:

(a) Do not abbreviate unless the abbreviation is significantly shorter or more meaningful than the full term. Use the full term or the accepted standard abbreviation wherever possible.
(b) Let the length of the abbreviation depend upon the length of the term being abbreviated. Do not restrict all abbreviations to the same length.
(c) Do not try to form all abbreviations by using only a single method of formulation.
(d) Check the proposed abbreviations with users for acceptability and discriminability.
(e) Avoid the use of non-standard abbreviations in outputs to other subsystems. Limitations in the communication links may force the message to be transmitted in abbreviated form. This is to be expanded on receipt into terms that the recipient can readily understand.

6.2.6.2 Grammar: The grammar of the interface is the arrangement of the vocabulary primitives into meaningful message units. Examples of message units are: a sentence in written English; a diagram in graphics; an equation in mathematics. For convenience, all such functional equivalents are referred to as *sentences* in this standard. Combinations of sentences are referred to as *paragraphs*.

The use of the terms 'sentence' and 'paragraph' does not mean, of course, that written English is the only language to be used in interfaces. The languages implemented are to be appropriate to the users' tasks in every instance.

The minimum sentence is to consist of:

(a) A *prompt*, which has the function of signalling that the user or computer is ready to send or receive a message. For example, the appearance on the screen of a menu signals that the computer is ready to action the user's choice.
(b) The text of the message, made up of combinations of *actions*, e.g. verbs and arithmetic operators; *objects*, e.g. noun phrases, numbers and symbols; *conjunctions*, e.g. 'and', 'with' and brackets (); and *punctuations*, e.g. comma and stop.

Engineering standard to the design of user–computer interfaces 295

(c) A *terminator*, which has the function of signalling the end of the message, e.g. pressing an 'enter' key; pressing the = key on a calculator.

Interfaces are to be designed so that users can send and receive messages in whole sentences or paragraphs. This is to emphasise the functional completeness of the messages with respect to the interface cognitive models; to give guidance to users as to the legal grammar for the benefit of the computers; and to counteract the limitations of short- and long-term memory.

One-word sentences may be needed (e.g. 'abort') which combine the functions of prompt, text and terminator. Such sentences should be kept to a minimum.

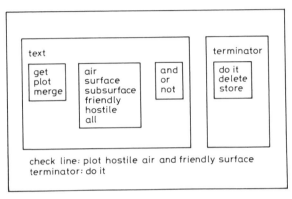

Fig. 6.2.12 *Example of a whole sentence interface*

Figure 6.2.12 gives an example of the use of screen windows and menus to display the grammar of a single sentence. The window contents change depending upon the legal options. The check line builds up as the user constructs the message by picking words from the windows. Finally, the user sends the message by choosing an appropriate terminator.

The standard will give detailed advice on the design of menus, windows etc.

Figure 6.2.13 gives an example of an interface dealing with a paragraph of sentences. The screen is used to set up a training simulator for aircraft control. Each of the parameters shown (bearing, range etc.) is quantified by the user in form filling mode. Each parameter represents a shortened sentence which can be specified in any order and changed as often as required. When the user is satisfied with the details, the command 'proceed' is keyed. This is a message terminator which means 'proceed (to set up the starting characteristics of the fighter aircraft as specified here)'. The command 'proceed' terminates a paragraph of functionally related single sentences.

This guideline does not mean that interfaces are to be capable of the full range of sentences implied by 'natural language'. Restricted grammars are more practical and, usually, will be sufficient. For example, the grammar might allow shortened sentences in which certain words are implied rather than stated explicitly (as in Fig. 6.2.13).

Designers are to devise a simple and consistent grammar for use throughout the system. This is to simplify learning and facilitate flexibility of allocation.

6.2.6.3 Dialogue: The dialogue of the interface is the set of rules governing the dynamics of the messages exchanged between users and computers. The standard requires designers to specify dialogues in terms of:

(a) The rules for initiating a message (see Section 6.2.6.3.1)
(b) The rules for guiding a user in the construction of the messages (see Section 6.2.6.3.2)
(c) The rules for guiding a user in the sequencing of the messages (see Section 6.2.6.3.3).

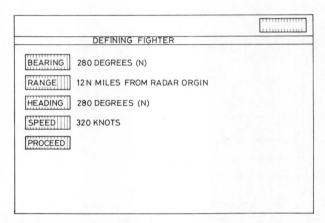

Fig. 6.2.13 *Example of a paragraph of sentences*

6.2.6.3 Mixed initiative dialogues
In many past systems, dialogues were initiated either always by the user (user driven) or always by the computer (computer driven). In future, all systems are to be 'mixed initiative' in that they are to be a mix of user driven and computer driven *as appropriate* to the tasks. Designers are to examine each message in the task allocation charts and determine who (user and/or computer) is to be in control. The task allocation charts are to be annotated accordingly.

6.2.6.3.2 Mixed styles of dialogue
Designers are to determine also the extent to which users need to be guided in how to construct their messages. There are three possibilities:

Freestyle dialogues typified by 'command language' interfaces. In this case, the user is given a screen prompt (e.g. a special symbol meaning that the computer is ready to receive a message); a screen area or field within which to construct the message; and a terminator (e.g. an 'enter' key). The on-screen guidance is minimal.

Engineering standard to the design of user—computer interfaces 297

Structured dialogues typified by 'menu-based' interfaces. In this case, the user is given a screen prompt (e.g. the appearance on the screen of the menus); an indication of the legal vocabulary and the sentence and paragraph grammars (e.g. Figs. 6.2.12 and 6.2.13); and a terminator (e.g. labelled keys that differentiate between different terminator functions). The on-screen guidance is considerable. Menus are not limited to alphanumeric messages (e.g. direct manipulation of symbols is a graphical form of menu) and are not limited to screen displays (e.g. some menu items may be provided by special keys placed alongside the main screens or on the keyboard; global commands like 'abort' and 'enter' are commonly handled in this way).

Hybrid dialogues typified by 'question and answer' and 'form filling' interfaces. These combine features of both freestyle and structured dialogues.

All three styles of dialogue will be needed in naval systems. Table 6.2.2 shows how they are to be used. Structured and hybrid dialogues are uniquely useful for communicating at the sentence and paragraph levels of detail. Freestyle dialogues are uniquely useful for creating new 'macros' — messages that have not been anticipated by designers — and 'command stacks' providing short cuts through menus.

All user—computer interfaces are to be provided with facilities to enable them to be operated in both freestyle and structured/hybrid styles of dialogue as far as is practicable. A designer will be required to obtain the agreement of the customer in advance to any instance where it is considered impracticable to implement both styles of dialogue.

The reasons for insisting on both styles of dialogue are to provide users with a

Table 6.2.2 *Recommended dialogue styles and their uses*

Dialogue style of the interface	Language level of the interface	
	Character and word	Sentence and paragraph
Freestyle	Command language interfaces for new messages and short cuts through menus	Do not use
Hybrid	Do not use	Question and answer and/or form filling interfaces for messages (e.g. parameters and data) which it is not practicable to provide via menus
Structured	Do not use	Menu interfaces are to be the normal form of interaction. All naval systems are to use menus if physically possible

298 Engineering standard to the design of user–computer interfaces

choice when operating the interface (e.g. experienced and inexperienced users may choose different styles) and to avoid premature decisions by designers as to which form of interface is best when, in practice, both are useful (structured and hybrid dialogues provide guidance; freestyle dialogues provide flexibility).

It is essential, however, that all the dialogues are totally compatible (again, as far as is practicable). This can be achieved by ensuring that the vocabulary and grammar are identical (e.g. the same words are used and have the same meaning in all styles of dialogue). This will simplify learning of the system and will reduce user errors.

6.2.6.3.3 Dialogue flow

The interfaces are to be designed to make the dialogue flow clear to the users. Dialogue flow is used in two senses:

(a) Flow *within* tasks (i.e. around the activities in a duty)
(b) Flow *between* tasks (i.e. around the duties in a job).

The recommended procedure for indicating the flow in structured dialogues is:

(a) Each *duty* is to be designed as a logical screen of displays and controls (by analogy a 'page' worth). The aim is to enable users to have a distinctive screen for each and every duty so that they (and the computer) will always know which duty they are working on and everything they need will be found within the screen. This is not always physically possible: a single logical screen may have to be implemented as several physical screens (or pages).

(b) Each *activity* is to be designed as a logical window within the logical screen. Each activity is to include details of all the computer inputs and outputs relevant to that activity. These may be implemented as fixed windows which are always 'open' or as multiple windows which can be 'closed' and/or overlapped. Inputs and outputs which have to be viewed in conjunction will be implemented as fixed windows: those which can be viewed sequentially or in various combinations will be implemented as multiple windows.

The result will be a screen with reserved window areas as shown in Fig. 6.2.14. The reserved screen areas are to be used as follows:

(a) This duty *name*. The top left corner of the screen is to be reserved for the full name of the duty to which the screen refers. If the duty extends to more than one physical screen then the subparts are to be named also (e.g. planning page 2). The name is to appear as a fully open window.

(b) The top middle of the screen is to be reserved for *external control inputs* (e.g. change in background circumstances). Control messages are to be written to this area using the conventions for alarms and warnings.

(c) The top right of the screen is to be reserved for *close duty* command. This is to appear as a half open window which goes fully open (thick border) before the screen changes to park the duty (see (e) below).

(d) The biggest area of the screen in the middle left is to be reserved for *open activity windows* (i.e. activities that are being worked on).

Engineering standard to the design of user–computer interfaces 299

(e) The middle right of the screen is to be reserved for windows for *other duties* (i.e. windows leading to other duties allocated to the user). These windows are to appear as 'closed' windows. These windows enable users to navigate between tasks.

(f) The bottom left of the screen is to be reserved for *closed activity windows*. These are windows relating to the activities within the same duty which are closed and parked. These windows allow users to navigate within tasks.

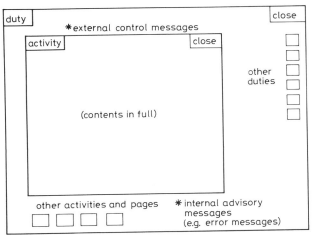

Fig. 6.2.14 *Example of screen layout showing windows for duties and activities*

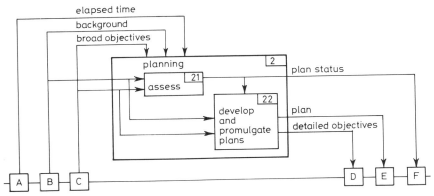

Fig. 6.2.15 *Extract from a task allocation chart*

(g) The bottom right of the screen is to be reserved for *internal advisory messages* (e.g. error messages referring to current activities and advice on processing delays). Messages are to be written to this area using the conventions for alarms and warnings.

Figures 6.2.15 and 6.2.16 show worked examples. The simplified duty 'planning' in

300 Engineering standard to the design of user–computer interfaces

Figure 6.2.15 is shown in Fig. 6.2.16 as a screen containing activity windows, control messages etc. as described in Fig. 6.2.14.

The fact that the task allocation charts and the logical screens and windows bear a fairly direct relationship one to the other raises the possibility of computer-aided interface design (at least partially).

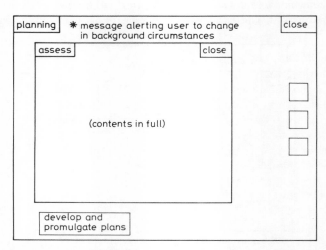

Fig. 6.2.16 *Example of planning screen*

6.2.6.3.4 Help facilities

Computer facilities that are provided to guide users in the construction and sequencing of messages are commonly known as help facilities. Many existing systems have facilities limited to special pages that help users to understand the vocabulary. Designers must provide much more help in future systems (e.g. for sentence construction and dialogue sequencing indicated by the menu layout; inbuilt tests and feedback; performance aids and training aids). These sorts of help are more extensive and pervasive than the existing sorts of help and have to be fully integrated into the primary tasks. The idea of *separate* help facilities and pages is no longer useful. Help commands may be part of the system, but the meaning of such commands must be made clear to the users. Global help commands are not to be used without qualifying commands (e.g. help with vocabulary; help with sequencing; help with learning etc.) unless the context provides all the guidance necessary.

6.2.7 Interface devices

Interface devices are the physical ways of implementing the necessary interface cognitive models and interface languages. The design of the cognitive models and languages ought to be prior to the design of the devices (although revisions may be needed due to device limitations).

The standard makes recommendations about devices only in terms of their

Engineering standard to the design of user–computer interfaces 301

functions and placement relative to the users. The internal characteristics of the devices (e.g. the electronics and the associated programming) are not covered. Designers are not constrained by the standard other than with regard to the human factors implications of the devices.

6.2.7.1 Principles of interface device design: Briefly, interface devices are to be chosen and placed in the workstation according to the following principles:

(a) Controls are to be operated by hands and fingers. Voice recognition (by computer) is not yet at a stage where it can be recommended other than for special applications.
(b) Displays are to be visual or auditory (i.e. aimed at the eyes or ears).
(c) Visual displays and associated controls are to be placed within the user's main cone of vision and within easy reach. Frequently used and critical visual displays and associated controls are to be especially easy to see and reach.
(d) If several displays and/or controls have to be operated in conjunction they are to be placed so that this can be done easily.
(e) Controls are to be spaced or otherwise protected to prevent accidental operation (e.g. of adjacent keys).
(f) Do not require the user to operate more than two controls at the same time.

6.2.7.2 Interface device recommendations: The overall principles (given above) and the detailed facts about particular devices (given in the standard) lead to the recommendations shown in Table 6.2.3. This shows that the various recommended

Table 6.2.3 *Recommended interface devices and their uses*

	Dialogue style and language level of the interface		
	Freestyle at character and word level	Hybrid at sentence and paragraph level	Structured at sentence and paragraph level
Displays		Screens* Indicator lamps ⟵——— Headphones ———⟶ Loudspeakers Telephones	
Controls	QWERTY keys* ———⟶ Number pad*	⟵———	Touch screens* Cursor keys* Tracker ball*
	⟵———————————	Fixed function keys (abort*, enter*) Variable function keys ———⟶ Graphics pad Digitiser	

* Essential. Others are dependent upon the tasks

dialogue styles (Table 6.2.2) are to be implemented using a variety of controls and displays. Some are regarded as essential for all workstations and tasks: others are optional depending upon the tasks.

These recommendations may be implemented by workstations which are fitted with the essential devices and fitted for the others.

The standard gives detailed advice on the physical characteristics of controls and displays and advice on their layout in the workstations.

6.2.7.3 Checking the interface designs: Advice on checking the interface designs is given in the standard but will not be covered in this contribution. Designers will find impact estimation especially useful as a preliminary to prototyping.

6.2.8 Outputs from systems design
The main outputs from systems design are the *detailed functional and physical specifications* passed to the manufacturer. The standard does not lay down a standard format for these. The specifications, however, are likely to include written descriptions of the system such as task allocation charts, user–computer interface device recommendations and dialogue flows. In addition, the specifications may include dynamic demonstrations using prototypes.

Almost as important are the *quantified goals* which describe what is to be attained by the system and how it is to be measured. These will be used during acceptance testing after installation and for the in-use audits.

Delivery plans have implications for users. Evolutionary delivery includes the concept of short cycles of install–test–revise in which users have a special role. In essence, evolutionary delivery means identifying subsets of the total system that are useful in their own right and are safe to deliver early (relative to high-risk subsets which take longer to develop). It is a form of phased release of the product.

The *prototypes* which are output from systems design are those developed primarily as design aids for use within the systems design stage. These may have an extended life as training simulators in routine use and/or as prototypes for system modifications. They may have a use, too, in the systems analysis stage of the next generation of product.

It is unlikely that all the *user documentation* (e.g. operator and maintainer manuals) will be produced before the manufacturing stage. The output from systems design will usually be specifications for user documentation and some more or less complete manuals. The increasing use of computer aids to design (e.g. so-called programming support environments) ought to make it easier to produce better documentation earlier in the development cycle than in the past.

6.2.9 Final comments
This contribution has tried to give a brief idea of the type of advice which might appear in an engineering standard for a systematic approach to the design of human–computer interfaces. The paper has concentrated on interface issues; this is a pity in some respects, since it runs the risk of making the design of interfaces

Engineering standard to the design of user–computer interfaces 303

seem like the primary concern of the human factors designer. To repeat a point made earlier: interfaces are obviously important, but they are not more important than other aspects of the system. The highest priority is to spend more time on a thorough systems analysis leading to a better agreed user model and a better initial task allocation.

6.2.10 References

ISO (1982) 'Open systems interconnection (OSI) ISO/DIS 7498', ISO/TC97
MoD (1983) 'Human factors for designers of equipment', Defence Standard 00-25
NES (1984) 'Standard for inter-system communications protocols', NES 1028
PARRISH, R. N. *et al.* (1983) 'Development of design guidelines and criteria for user/operator transactions with battlefield automated systems', Phase III final report, Volume II. WF-82-AD-00. Prepared by the Synectics Corporation for the US Army Research Institute for the Behavioural and Social Sciences
SMITH, S. L. and MOSIER, J. N. (1984) 'Design guidelines for user–system interface software', The Mitre Corporation, Bedford, MA, report ESD-TR-84-190
STEVENS, M. and WHITEHEAD, K. (1985) 'The analyst – a workstation for analysis and design', in conference proceedings of MILCOMP 85, published by Microwave Exhibitions and Publishers Ltd, Tunbridge Wells

Copyright © Controller HMSO, London 1987. This work has been carried out with the support of the Procurement Executive, Ministry of Defence.

7: Advanced processing

Chapter 7.1

Expert systems in C^2 systems

C.J. Harris
(Southampton University)*

7.1.1 Introduction

In 1979 the US Department of Defense sponsored a series of command and control (C^2) colloquia which concluded that there is no adequate foundation for a theory of C^2 and hence no principles for overall system design and evaluation (see also Chapter 1). Despite the massive investment by NATO in C^2 systems research development and implementation in the intervening period, a fundamental methodology for the design of C^2 has not emerged. This is not surprising, since high-level C^2 systems are essentially an extension of basic human decision processes by means of procedures, organisations, equipments, situation assessment (threat assessment) and resource allocation. C^2 systems are among the largest and most complex real-time resource management systems known to man; their effectiveness is severely limited by lack of speed, data saturation and the cognitive limits of the human decision maker.

Control theory, large-scale systems theory, state variable ideas and fuzzy set theories have all been applied to human decision making in command systems (Morgan,, 1985; Levis and Boettcher, 1983; Leedom, 1979). All offer only a partial model of decision making, and all fail to recognise that in C^2 systems this process is not simply the selection of the perceived best option but the creation, evaluation and refinement of what the situation is and what (if anything) can be done about it. (See also Singleton's contribution on the man-machine aspects of C^2 in Section 6.1.)

The majority of existing C^2 systems are action information systems: that is, they concentrate on the acquisition of information, on data processing and communications and on physical MMI aspects rather than on underlying analysis of the tactical decision processes themselves. The actual process of command

---

* Formerly at the Royal Military College of Science, Shrivenham.

and control continues to remain with the decision maker or operator, in spite of the tendency of modern battlefield command and control systems to become increasingly sophisticated and complex though:

1. The increasing speed, effectiveness and mobility of modern weapon systems (calls for dispersion and mobility of C^2 centres).
2. The increasing number, diversity and capability of sensors which, together with intelligence, is leading to an information explosion problem. This data must be interpreted (filtered, correlated) and fused in real time to be effective (see also the contributions on data fusion by Wilson in Section 7.2 and Lakin and Miles in Section 7.3.)
3. The increasing number of battlefield sensors, which require complex adaptive asset management.

It is clear tha the rate, complexity, dimensionality and uncertainty of events and the information about them in a crisis is rapidly increasing. This, coupled with the need for real-time decision making, will result in high returns for decision-aided technology for C^2 systems.

7.1.2 C^2 complexity and artificial intelligence

The majority of C^2 systems are complex, ill defined, generally stochastic and unbounded, and are characterised by non-unique solutions (if they exist at all), frequently with no stopping rule or solution in finite time (Partridge, 1981). It is worth restating (see Chapter 1) that a real-time decision support/planning problems in a dynamically stochastic environment is a *hypothesis*. At the highest C^2 level, man's role is as a planner and decision maker. However, he is under stress and is prone to several perceptual decision/planning idiosyncrasies such as conservatism, data saturation, self-fulfilling predictions, habit, overconfidence, lack of confidence, gambler's fallacy and panic (see Chapter 1 and Sage, 1981 for full discussion of these attributes). To alleviate this situation, artificial intelligence (AI) or intelligent knowledge-based systems (IKBS) can be used as decision/ planning aiding techniques (Nilsson, 1980; Masai et al., 1983) that allow the user to concentrate on the options offered and to determine the most appropriate response. AI techniques and the emerging technology of concurrent processors now offer the possibility of real-time intelligent decision aids for C^2 applications. AI provides the tools to utilise an expert's (commander or user) knowledge for effective use in a real-time computer-based control structure. This offers the potential of an efficient use of a knowledge base (KB) of production rules for managing system processing resources, the establishment and evaluation of a current situation, and dynamic problem-solving strategies. A novel and significant feature of such an approach is that it enables the user to ask for justification of the decisions offered (Friedland, 1981).

7.1.3 Knowledge-based systems in C^2 systems

In a typical military command and control application, the knowledge-based system (KBS) would act as a special purpose intelligent controller in support of the commander (not to supplant him). The KBS will accept data from a variety of sensors, databases and intelligence sources, will correlate, combine or fuse them, will liaise with the knowledge base (KB) via optimal search or inference rules, and will then perform plausible or rational decision tasks from which modification of the existing KB would occur consistent with the new data.

An appropriate model of human decision making is that which has been used by Wohl (1981) in tactical air defence systems — the SHORE (Stimulus; Hypothesis; Options; Response) paradigm, which in the C^2 context this could be interpreted as shown in Fig. 7.1.1 (compare also with Fig. 7.3.1 of Middleton's contribution on Tactical Decision Making). Each of these functions involve complex inferential operations on the current sensor data and past data using a knowledge base of military equipments (weapons, stores, sensors, etc.), resources (men, ...) and operational tactics.

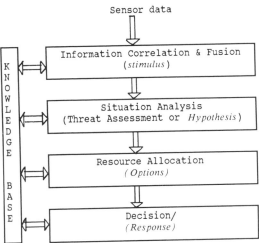

Fig. 7.1.1   SHORE C^2 paradigm

It follows that an intelligent C^2 tactical decision/planning system needs to be able to perform a series or levels of tasks:

(i) Data acquisition from sensors or intelligence.
(ii) Sensor data interpretation and data fusion (see Wilson and Lakin and Miles, Sections 7.2 and 7.3).
(iii) Situations or threat assessment.
(iv) Decision plans or goal generation for desired outcomes — resource allocations.
(v) Implementation of plans and monitoring of response/actions to ascertain success through sensors or intelligence [process returns to (i)].

At the highest level the decision maker/commander has a set of *a priori* goals or plans and the consequent actions, but C^2 expert systems must respond to an uncertain continuously changing dynamic environment. New data/intelligence may affect the decision process by changing the goals to be achieved and the manner in which they are to be achieved. In high information acquisition a balance has to be sought between full data assimilation (and consequent data saturation) with no planning solution, and an incorrect solution based upon an incomplete database. C^2 systems demand real-time solutions which, by their complexity, limit the time available for decision making. Time criticality directs the C^2 system to concentrate upon those actions which are currently most important; however, it must not ignore the low-priority or unpredictable solutions that have been so successful in past military campaigns. Essentially the C^2 system must restrict the solution space if real-time solutions are to be feasible. The risk of failing to generate any real-time solution suggests that the control system methodology accepts only partial goal satisfaction. It therefore follows that, with data saturation, real-time decision making or planning should be based upon an incomplete database. There are therefore two supplementary problems (a) recognising that there is an incomplete database (b) overcoming incompleteness. Doyle (1983) and Todd (1984) have suggested that a solution to the time-critical incomplete information problem can be made through non-monotonic reasoning with plausible or default assumptions. In this mechanism a default value is preferred to cost evaluation through lack of data or time. However, a correction mechanism is necessary to incorporate new data or the results of inconsistencies revealed through inferences. McAllester's (1980) work on reason maintenance systems has considered the update of databases subject to inconsistencies of information by considering the rationale of decisions in the context of other decisions. The application of this methodology and the theory of reasoned assumptions (Doyle, 1983) within the context of artificial intelligence applied to C^2 systems has yet to be assessed.

The majority of artificial intelligence systems have tended to be static inference systems, such as the adviser type methods based upon EMYCIN (Van Melle, 1979) and SAGE (1983) kernels, or systems which have dynamic networks to support their reasoning. The dynamic AI networks tend to utilise semantic nets, frames and blackboard systems. These systems support architectures that are capable of supporting continuous data inputs and planning for an arbitrary range of problems. Semantic nets and frames are particularly appropriate for detailed knowledge representations but, unlike blackboard systems, are less able to deal with globally accessible areas for a solution space. The main activities of the higher levels of a C^2 system are situation assessment and resource allocation/planning, for which the blackboard expert systems architecture has been advocated by several authors (Bell, 1984; Lesser and Erman, 1977; see also Sections 7.3 and 7.4).

A blackboard architecture is based upon a group of independent experts or knowledge sources, who are ignorant of each other's expertise. They surround a 'blackboard' or global data area. As new data is written on to the blackboard, each expert examines the blackboard to see if his expertise/knowledge has any contribution to make through a new hypothesis. This in turn may be used by the other knowledge sources to jointly contribute to the problem solution. The blackboard

Expert systems in C^2 systems 311

is both a representation of the current state of knowledge and a means for communications among the experts.

The speech recognition system Hearsay-II is the archetypal blackboard expert system; however, it only solves a single problem by concentrating upon segments of continuous signals. A blackboard model of co-operating knowledge sources in air defense is the ADX (Bell, 1984; see also Section 7.4). Like Hearsay, ADX attempts to keep the knowledge sources separate so as to allow easy systems modification, but, unlike Hearsay, ADX deals with multiple discrete problems over a long observation period. Therefore ADX has to adopt a simple static priority scheme, but yet is sufficiently general to allow Hearsay-II algorithms to be implemented on top of it if required. There are implementation problems associated with ADX such as continuous reasoning, time and consistency of data etc.

In Section 7.4, Middleton describes investigations which have been carried out into applying IKBS techniques to air defense C^2 tasks — in particular, the high-level tasks of real-time threat assessment. A threat assessment system must deal with data in real-time and in the order in which it arrives; therefore a special expert system framework is introduced. In the initial study Middleton has utilised a blackboard system based upon an ADX framework built in the POPLOG environment. To evaluate this against a variety of scenarios, a simulation facility based upon a simple object-oriented programming language has been used (Middleton, 1984). However, problems associated with the blackboard architecture have resulted in the development of an ACTOR-type object-oriented language BLOBS (Dickinson, 1985) being used to support continuous reasoning expert systems as well as simulation. From this, Middleton reconfigures the threat assessment expert system into a network of objects or blobs!

At a lower level of the C^2 systems architecture, the multiple expert architecture (MXA) of Rice (1984) provides a rule-based system framework which allows the knowledge engineer to develop an efficient and integrated real-time blackboard system, and is claimed to be the most highly developed operational blackboard system in Euorpe. The MXA language owes much to Pascal; it has a highly legible source code which makes it amenable to serious software engineering, and the added attribute of friendly rule writing. MXA was initially developed as a tool in the creation prototype systems for naval radar signal processing and tactical picture compilation. It consists of a compiler for translating a rule-based knowledge representation language, and an executive to which the twice-compiled knowledge is linked. The executive provides the blackboard environment for rule invocation and assists the development and debugging of rules.The MXA provides a problem-solving framework which can accommodate combinations of heuristic rules, statistical principles or mathematical formulae. It is these attributes that make MXA so attractive to Lakin and Miles (Section 7.3) for an IKBS solution to intelligent data fusion in the naval environment, which must cope with a large and diverse quantity of real-time input sensor data and non-real-time encyclopaedic information, achieve fast reaction times and relieve the operator of exhausting routine low-level monitoring tasks. The specific C^2 objective of Lakin and

Miles is to produce, as a result of intelligent data fusion and consequent parameter estimation, a set of conclusions for each perceived vehicle from which a consistent view of the tactical position of the resources deployed may be ascertained.

A C^2 expert system which uses blackboarding and back chaining paradigms is the project on multi-sensor integration data processing and target recognition in the underwater sonar environment (ASDIC: Hopford and Lang, 1984). ASDIC takes multisonar inputs, intelligence reports and libraries of acoustic signatures and compiles a tactical command picture. Similar work in the USA on a fixed base heuristic adaptive surveillance project (HSAP: Nii *et al.*, 1982) for an all-source classifier has been implemented and extended to the development of a classification and information management system (CLAIMS) which operates as a multilevel dynamic real-time sonar situation analyser.

The blackboarding expert systems approach is clearly appropriate to C^2 subsystems that involve a strong planning element. Therefore systems such as aircraft and communications, which require a high degree of real-time self-testing and diagnostic facilities that utilise multisensor inputs, suggest the use of blackboarding expert systems; examples include prototype intelligent helicopter fault management systems (Reynolds and Craighill, 1984).

Backchaining Bayesian inference expert systems have attracted considerable interest in C^2 systems, such as an IKBS to provide advice on electronic warfare plan evaluation (Roberts, 1984). SAGE (1983) was selected as the expert system framework as it provides a control structure and high-level language similar to that used by the PROSPECTOR system (Reboth, 1981) — including the knowledge acquisition system for developing the inference net. To compensate for the inadequate MMI to this system, the relational database management system RAPPORT (1982) of stored plan details of equipments and their selected frequencies, and algorithmic and interactive graphic interfaces to support the operator in plan creation, have been constructed for the adviser. The interaction between the operator and the knowledge base adviser is reduced to a small number of questions requiring simple answers, thus improving consistency of input evidence as well as reducing long and potentially erroneous MMI interactions.

7.1.4 A prototype structure for an overall knowledge-based C^2 system

A C^2 KBS is generally composed of three major components: a knowledge base (KB), a situational database (SDB) and a control system (CS).

Fig. 7.1.2 illustrates the general KBS configuration for a C^2 system.

7.1.4.1 The knowledge base: The KB is that part of the system which contains a map of the C^2 expertise which operates on known, conjectured or hypothesised cause and effect relationships, implications, heuristics, inferences and operational plants. Typically a knowledge-based system (KBS) contains rules and procedures. In C^2 applications the conventional KBS based upon the construction rules of if/then clauses is inadequate, since situation evaluation frequently involves uncovering situations that satisfy complex evolving sets of constraints rather than simply evaluating a set of fixed truth conditions.

The rule format proposed for situation analysis is based upon predicted calculus which enables us to represent knowledge in a descriptive manner. By using this general form we achieve both rule standardisation for C^2 systems involving thousands of rules, and effective representation. This approach places all quantified variables in front of all predicates; this enables maximal processing efficiency since the problem is then in the well known generate and test format. The situation

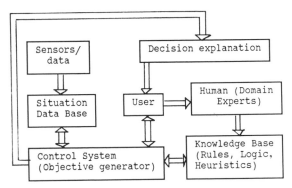

Fig. 7.1.2   Generalised C^2 KBS

analysis rule is therefore in the shape of a number of (situation) conditions; if these are satisfied, an indicated action takes place. Such a rule-based system requires interpretation of *confidence limits* used in the conditions, such as 'near', 'range', 'numbers'; this enables greater flexibility in system assessment. For example:

if	enemy heliborne gun ships are sensed 'near' some location L_1
and	enemy MBT squadron ('numbers') is at L_2
and	the 'range' between L_1 and L_2 is less than L_3
and	there is a defensible position at L_4
unless	the range between L_4 and L_1 (or L_2) is greater than L_5
then	the MBT squadron is going to attack

7.1.4.2 The situational database: The situational database is the basic set of information required by the system to define entities, relationships between objects, operational relationships, constraints etc.. It contains either situation related information — that is, the systems understanding of what constitutes the situation to be analysed, such as the location of resources (enemy's and own), their role in the situation and predictions of possible consequent actions — or background information. As the current situation develops, inferences drawn from the situation (through a current best estimator) are updated, confirmed or countered. A blackboarding expert system as discussed above could well operate effectively as a situational database. A belief maintenance mechanism is utilised throughout to ensure that all logic paths and the resulting conclusions are reassessed against new sensor information. At any time during the reassessment, users can request information concerning any entity in the situation database. Whereas the knowledge base stores encyclopaedic data on the user's resources (weapon systems, sensors, aircraft, ships, men, etc.), a relational database defines the functional relationship (or cause and effect) between these data elements and the initial plans (objectives) that follow inferentially from the functional relationship between data. And as such the relational database may be extended by the user through embedded rules that allow recursive interactive dialogue to generate hierarchical relationships between primitives such as physical objects, people, units etc.

7.1.4.3 Knowledge-based control system design: The control system actions (see Fig. 7.1.3) are to process input information from sensors, and to interact with the knowledge base rules to produce the situation analysis and to provide

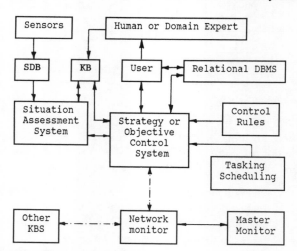

Fig. 7.1.3 *Distributed C^2 KBS with master monitor*

user input/output information (including explanations), all of which must be processed in real-time. One such implementation of this could be the data fusion IKBS system of Lakin and Miles (Section 7.3) applied to a wider class of problems than the naval environment. To achieve real-time operations using conventional von Neumann computer architectures, task scheduling based upon priorities is necessary. Satisfaction of all goals or plans in real-time is generally not feasible; then the choice of what tasks have greatest urgency and priority is significant. Partial goal fulfilment and the associated representation and reasoning about the effects of goal failure has attracted little attention. As C^2 systems require a large knowledge base, there are two primary constraints — information storage limitations and execution time. Distributed knowledge-based systems as well as parallel processors (Broomhead, 1984) offer the option of several independent systems operating on various aspects of the overall problem communicating via a network. For example, one knowledge-based system could be devoted to sensor gathering, data fusion and the situational data system; another could be devoted to high-priority tasks such as crisis threat assessment; and another could be devoted to low-priority (but equally essential) tasks such as logistic analysis and allocation. Within each knowledge-based system the control systems must optimise the rule base, that is, it must be able to reduce the number of rules utilised to the minimum required to analyse the current situation. With a multiple tasking knowledge-based system and adaptive control system functionality, an overall control system structure or master monitor is necessary to ensure that the tasks, priorities and execution order are optimised.

7.1.4.4 C^2 control system structure: Fig. 7.1.3 illustrates a generalised C^2 knowledge-based system control system structure. The centralised control system or master monitor provides concurrent overall subsystem knowledge-based system management through a communications network. It accesses other knowledge-based systems, situational database systems, users and strategy monitors; It also can provide dynamic allocation to each subsystem of new tasks — reallocation being based upon the current situation and predicted state space through current best estimators.

The strategy or objective control system provides, in order of priority, the tasks or rules determined by the current situation that require processing in the situation assessment system. The solution procedures are based upon human expert guidelines and general analytical or logical processes (for example, fuzzy logic); these procedures provide links between the tasks and suggest what should be done next. Solution time is dictated by complexity rather than the size or dimension of the rule base.

The control rules contained in the strategy control system form another knowledge base and provide prime procedures for what should be carried out next, and as such can be used to specify analysis strategies and management of general processor resources. An adaptive priority/urgency scheme should be adopted that updates *a priori* goals/plans on the basis of the current situational database. An important architectural attribute would be an interrupt system that enabled the system to adopt new goals without misinterpretation of decisions determined. Todd (1984) has suggested the adoption of graceful degradation for efficient resource management via fall-back positions in order of increasing utility, in which the set of achieved objectives and resulting plans grows incrementally until either the full solution is produced or the system is interrupted. Each intermediate solution is sufficiently complete to allow implementation.

Situation assessment or rule processing is essentially stochastic and can be executed through a probabilistic data structure priority queue. The rules are probabilistically selected from the queue, based on the relative priority of the subject contained within the rule. These priorities act as dynamically varying operating constraints upon the control system, since they can be altered by the control system itself via the control rules in the strategy control system. This situation analysis methodology enables the high-speed (independent of number of rules) synthesis of multiple processing goals and the adaptive ability to cope with misleading sensor data.

7.1.4.5 C^2 control system operation: New sensor or intelligence data is filtered and added to the database. The significance of the data is determined by data fusion and correlation with other data, and then by application of interpretation rules in the knowledge base. The practical implementation of data fusion could well be through the blackboarding methodology adopted by Lakin and Miles (Section 7.3). The significance of the new data is achieved through a progressive forward and backward chaining process to identify the most significant or the highest priority situation data. Essentially this is a higher-level situation assessment procedure such as that considered by Middleton (Section 7.4). Following data acquisition and situation interpretation, suggested actions, strategies or conclusions are given to the user via an MMI, together with explanations. The decision justification is as a consequence of complex logical inference chains in the knowledge base under the control of the strategy control system. The atribute of explaining the reasoning behind advice/status/decision recommendations given to the user by the knowledge-based system is essentially to the human decision maker, since he can directly assess the justification for recommendations before implementing the actions. The centralised master monitor system allows the individual knowledge-based system to communicate with other dedicated knowledge-based systems and their data sources via a communications and sensor network. The overall methodology proposed in this section is to provide a broad framework of command and control knowledge-based system engines, each with an ability to self-tune to specific situation assessments and objectives.

To implement such an overall C^2 expert systems strategy is currently not feasible. However, the knowledge-based subsystems for data fusion and situation assessment given in Sections 7.3 and 7.4 demonstrate that considerable progress has been made towards complete C^2 computer-based decision aiding.

7.1.5 Alternative approaches to C^2 architectures
We have already noted that C^2 systems are chracterised by a dynamic environment with ill defined representations and with uncertainty of current and future states. The domain of interest at the level of the planner/commander does not provide an adequate basis for generating the plan. It is usually too large, complex and stochastic for explicit predicate calculus representation. The methodology adopted by expert system approaches to C^2 assessment and planning is to emulate the expert's reasoning and restrict the solution domain to that perceived by the expert rather than to perform exhaustive or global searches of the solution space. Practical experience in the development of C^2 expert systems (Hopford and Lang, 1984) illustrates the difficulty of eliciting knowledge from experts who frequently

disagree among themselves! An alternative methodology (Markosian and Rockmore, 1984) is to allow for the existence of non-expert solutions by narrowing and partitioning the solution space into a set of skeletal plans based upon *a priori* scripts (Schank and Abelson, 1977) that capture the heuristics used by expert planners. Although these skeletal plans do not produce optimal solutions, they produce at least one skeletal plan for every given problem. A skeletal plan is initially matched to the domain state, and its current domain state effectiveness is measured; further refinements of skeletal plans then take place. At each iteration, evaluation for effectiveness and satisfaction of constraints is carried out. Current work in this direction includes extension of the basic framework to include more flexible controls, which enable opportunistic planning and stochastic scripts within skeletal plans to generate more exact plan evaluation.

Artificial intelligence is not the only approach to solving real-time dynamic C^2 system problems. Indeed Sutherland (1986), in a strongly critical essay on the contribution of artificial intelligence to decision aiding, argues that AI has not yet performed the inference operations that stochastic decisions demand, or transcended the more traditional decision disciplines. Wilson (Section 7.2) considers the use of Bayesian/Shafer-Dempster probability/likelihood methods in the solution of the data fusion problem addressed by Lakin and Miles (Section 7.3) via expert systems. However, Wilson concludes that IKBS methods are very effective in utilising a wide variety of information (including prior knowledge) to generate optimal evaluations of complex command and control situations.

7.1.6 References

BELL, M. (1984) 'The ADX – an expert system framework for air defence', Proceedings IKBS defence seminar, RSRE, Malvern, UK

BROOMHEAD, D. S. (1984) 'The application of the INMOS transputer and OCCAM to IKBS', Proceedings IKBS defence seminar, RSRE, Malvern, UK

DICKINSON, I. J. (1985) *BLOBS User Manual,* Cambridge Consultants Ltd.

DOYLE, J. (1983) 'Some theories of reasoned assumptions – an assay in rational psychology', Carnegie-Mellon, University, report CMU-CS-83-125

FRIEDLAND, P. E. (1981) 'Acquisition procedural knowledge from domain experts', Proceedings 7th IJCAI, pp. 856–92

HOPFORD, R. P. and LANG, J. (1984) 'The ASDIC project: lessons learnt while setting up an IKBS research programm', Proceedings ZKBS defence seminar, RSRE, Malvern, UK

LEEDOM, D. R. (1979) 'Representing human thought and response in military conflict simulation models', Symposium on modelling simulation of avionic systems and C^3, AGARD NATO conference proceedings 268

LESSER, V. R. and ERMAN, L. D. (1977) 'A retrospective view of the Hearsay-II architecture', Proceedings 5th IJCAI, pp. 790–800

LEVIS, A. H. and BOETTCHER, K. L. (1983) 'Decision making organisation with acyclical information structure', *IEEE Trans.,* **SMC-13**(3), pp. 384–91

MARKOSIAN, L. Z. and ROCKMORE, A. J. (1984) 'A paradigm for dynamic plan generation and execution monitoring', Proceedings IKBS defence seminar, RSRE, Malvern, UK

MASAI, S., McDERMOTH, J. and SOBEL, S. (1983) 'Decision making in time critical situations', Proceedings, 8th IJCAI, Karlsruhe, pp. 233–5

McALLESTER, D. A. (1980) 'An outlook on truth maintenance', MIT AI Laboratory memo 551

MIDDLETON, S. (1984) 'Actor based simulation of air defence scenarios in POPLOG', Proceedings IKBS defence seminar, RSRE, Malvern, UK
MORGAN, P. D. (1985) 'Modelling the decision maker within a command system', IEE international conference on advances in C^3, IEE conference publication 247, pp. 65–9
NII, H. P., FEIGENBAUM, E. A., ANTON, J. J. and ROCKMORE, A. J. (1982) 'Signal to symbol transformation: HASP/SIAP case study', *AI Magazine*, **3**(2), pp. 23–35
NILSSON, N. J. (1980) *Principles of Artificial Intelligence,* Tioga, Palo Alto, Calif
PARTRIDGE, D. C. (1981) 'Computational theorising as a tool for solving wicked problems', *IEEE Trans.*, **SMC-11**(4), pp. 318–21
RAPPORT-3 Fortran User Manual (1982) Logica Ltd, New Street, London
REBOTH, R. (1981) 'Knowledge engineering techniques and tools for the PROSPECTOR environment', technical note 243, AI Centre, SRI International
RICE, J. P. (1984) 'MXA – a framework for the development of blackboard systems', Proceedings IKBS defence seminar, RSRE, Malvern, UK
ROBERTS, A. (1984) 'A knowledge based system for EW tasking plan evaluation', Proceedings IKBS defence seminar, RSRE, Malvern, UK
SAGE, A. P. (1981) 'Behavioural and organisational considerations in the design of information systems and processing for planning and decision support', *IEEE Trans.*, **SMC-11**(9)
SAGE expert system (1983) Language specification (SAG 02), SPL Int., Abingdon, UK
REYNOLDS, and CRAIGHILL, (1984)
SCHANK, R. and ABELSON, P., (1977) *Scripts, Plans, Goals and Understanding,* Wiley
SUTHERLAND, J. W. (1986) 'Assessing the artificial intelligence contribution to decision technology', *IEEE Trans*, **SMC-16**(1), pp. 3–21
TODD, S. J. (1984) 'Research issues in KB for C^2', Proceedings IKBS defence seminar, RSRE, Malvern, UK
VAN MELLE, M. (1979) 'A domain independent production rule system for consultation programmes', Proceedings 6th IJCAI, pp. 923–5
WOHL, J. G. (1981) 'Force management eefision requirements for air force tactical command and control', *IEEE Trans.*, **SMC-11**(9)

7.1.7 Bibliography

AIELLO, N., NII, H. P. and WHITE, W. C. (1981) 'Joy of AGE-ing: heuristic programming project', Stanford University
ARDEN, B. W. (1980) *What can be Automated? The Computer Science and Engineering Research Study,* MIT Press
BARSTOW, D. R. (1979) *Knowledge Based Program Construction,* Elsevier, North-Holland, NY
BELZER, R., ERMAN, L. D., LONDON, P. and WILLIAMS, R. (1980) 'Hearsay-III: a domain independent framework for expert systems', *PROC. AAAI*
CHARRIER, G. and GIRARD, E. (1981) 'Contribution of AI to tactical command systems', NATO AFCEA symposium on NATO C^3 – the challenge of the 80s, Brussels
CLOCKSIN, W. and MELLISH, C. (1981) *Programming in Prolog,* Springer Verlag, Berlin
DAVIS, M. H. (1985) 'Artificial intelligence in on line control applications – an air traffic control example', IEE international conference on advances in C^3 systems, IEE conference publication 247, pp. 224–33
DAVIS, R. (1978) 'Knowledge acquisition in rule-based systems', in *Pattern Directed Inference Systems* (eds. D. A. Waterman and F. Hayes-Roth), Academic Press, NY
DAVIS, R. and LENAT, D. (1982) *Knowledge Based Systems in Artificial Intelligence,* McGraw-Hill, NY
DOYLE, J. (1979) 'A truth maintenance system', MIT AI Laboratory memo 521
DUDA, R. O. (1978) 'Development of the PROSPECTOR consultation system for mineral exploration', AI Centre, SRI International
ENGLEMAN, C. (1979) 'KNOBS: an experimental knowledge based tactical air mission plann-

ing system and a rule based aircraft identification simulation facility', Proceedings 5th IJCAI
ERMAN, L. D. (1980) 'The Hearsay-LL speech understanding system: integrating knowledge to resolve uncertainty', *Computing Surveys*, 12(2), pp. 213–53
ERMAN, L. D., HAYES-ROTH, F., LESSER, V. R. and REDDY, D. R. (1980) 'The Hearsay-II speech understanding system: integrating knowledge to resolve uncertainty', *Computing Surveys*, 12(2)
ERMAN, L. D., LONDON, P. E. and FICKAS, S. F. (1981) 'The design and example use of Hearsay-III', Proceedings 7th IJCAI, pp. 409–15
FEIGENBAUM, E. A. (1977) 'The art of AI. I: themes and case studies of knowledge engineering', Proceedings 5th IJCAI, pp. 1014–29
FENNELL, R. D. and LESSER, V. R. (1977) 'Parallelism in artificial intelligence problem solving: a case study of Hearsay-II', *IEEE Trans.* C-26(2), pp. 98–111
FORGY, C. L. and McDERMOTT, J. (1977) 'OPS: a domain independent production system language', Proceedings 5th IJCAI, pp. 3–9
FREIHERR, G. (1980) 'The seeds of artificial intelligence', Division of Research Resources NIH, Bethesda, MD, report 80-2071
GADSDEN, J. A., LAKIN, W. L. and ROBERTS, A. (1984) 'A knowledge based system for EW tasking plan evaluation', ASWE memo XCC 84003
GEVARTER, W. B. (1983) 'Expert systems: limited but powerful', *IEEE Spectrum*, August
HAAS, N. and HENDRIX, C. G. (1980) 'An approach to acquiring and applying knowledge', Stanford Research Institute, technical note 227
HENDRIX, G. G., SACERDOTI, E. D., SAGALOWICZ, D. and SLOCUM, J. (1978) 'Developing a natural language interface to complex data', *ACM Trans. Database Systems*, 3(2), pp. 105–47
HEWITT, C. (1976) 'Viewing control structures as patterns of passing messages', MIT AI Laboratory, memo 410
KEIRSEY, D. M. (1980) 'Natural language processing applied to Navy tactical messages', Naval Ocean Systems Centre, San Diego, Tactical C^2 Division, Technical report 234, NOSC code 824
KLAHR, P. (1982) 'SWIRL: simulating warfare in the Ross language', Rand Corporation note N-1885-AF
McALLESTER, D. A. (1982) 'Reasoning utility package user's manual – version one' MIT AI Laboratory memo 667
McARTHUR, D. and KLAHR, P. (1982) 'The ROSS language manual', Rand Corporation note N-1854
MICHALSKI, R. S. (1983) 'Integrating multiple knowledge representations and learning capabilities in an expert system: the ADVISE system', Proceedings 8th IJCAI, pp. 256–8
MINSKY, M. (1975) 'A framework for representing knowledge', in *The Psychology of Computer Vision* (ed. Winston), McGraw-Hill, NY
NAU, D. S. (1983) 'Expert computer systems', *IEEE Trans.*, C-16(2), pp. 63–84
NEWELL, A. and SIMON, H. A. (1972) *Human Problem Solving*, Prentice-Hall, Englewood Cliffs, NJ
NII, H. P. (1980) 'An introduction to knowledge engineering blackboard model and AGE; heuristic programming project', Stanford University, report HPP-80-29
NILSSON, N. J. (1971) *Problem Solving Methods in Artificial Intelligence*, McGraw-Hill, NY
POPLOG: a multi-purpose multi-language program development environment, System Designers, March 1984
ROBERTS, A. (1984) 'The use of a relational database in support of knowledge based systems', ASWE memo XCC 84004
SACERDOTI, E. (1977) *A Structure for Plans and Behaviour*, Elsevier, North Holland, NY
SHAPIRO, E. Y. (1981) 'An algorithm that infers theories from facts', Proceedings 7th IJCAI, p. 446

SIKLOSSY, L. and ROACH, J. (1974) 'Collaborative problem-solving between optimistic and pessimistic problem solvers', Proceedings IFIP Congress, North Holland, pp. 814–17

SMITH, D. E. (1980) 'A frame based production system architecture', *Proc. AAAI*

STEFIK, M. (1979) 'An examination of a frame structured representation system', Proceedings 6th IJCAI, pp. 845–852

STEFIK, M. (1981) 'Planning with constraints', *Artificial Intelligence,* **16**

STEFIK, M. (1983) 'Knowledge programming in loops', *AI Magazine,* **3**(13)

TODD, S. J. and FATMI, H. A. (1983) 'A cybernetic approach to intelligence based systems', IEEE International conference on systems, man and cybernetics, Bombay, India

VERE, S. (1982) 'Planning in time: windows and durations for activities and goals', NASA Jet Propulsions Laboratory report JPLD-527

WATERMAN, D. A. and HAYES-ROTH, F. (1982) 'An investigation of tools for building expert systems', Rand Corporation report R-2818-NSF

WEISS, S. J., KULIKOWSKI, A. and SAFIR, A. (1978) 'A model-based method for computer aided decision making', *Artificial Intelligence* **11**, pp. 145–72

WEIZENBAUM, J. (1976) *Computer Power and Human Reason,* Freeman

WILKINS, D. E. (1983) 'Representation in a domain independent planner', Proceedings 8th IJCAI, vol. 2, pp. 733–40

Chapter 7.2

Some aspects of data fusion

G.B. Wilson
(Ferranti Computer Systems)

7.2.1 Introduction
Commanders and their staffs, who are both users and active elements of command and control systems, need to have access to a wide range of information to carry out their duties. This information relates to their own forces and the enemy's as well as to the operating environment, and one of their primary requirements is that they must be able to obtain knowledge of the current activity in their areas of interest. As the range and effectiveness of modern weapons and the pace of warfare have increased, so the areas of interest of C^2 systems have grown, and ever shorter delays in transmitting information to the user have become tolerable. It is to meet these requirements that the last few years have seen a considerable increase in both the number and variety of sensor systems and the communications needed to carry the information from remote sensors to the command and control centres. The increasing complexity and capabilities of C^2 systems is probably more apparent in these improvements to the reporting systems than in any other aspect.

However, it is not sufficient to just provide more and more sensors with greater individual capabilities because the increasing volume of information brings its own problems — the so-called information explosion — and it is these that have led to growing interest in techniques for data fusion.

It is an obvious fact, but one which frequently seems to be overlooked in discussions, that the actual activity in an area does not depend on either the number or the quality of the sensors. For example, it is obvious that the number of aircraft operating in the area of interest of an air defense C^2 system is not increased by doubling the number of sensors. If there are 100 aircraft in the area concerned then the system users should be provided with 100 comprehensive and accurate reports — one about each of the aircraft. Of course if there are 20 sensors there could conceivably be 20 reports on each aircraft to start with, and the problem to be faced

322 *Some aspects of data fusion*

is how to reduce 2000 possible reports to just the 100 with which the system should finish up.

It is to this process that the blanket description of *data fusion* has been applied, and the use of this single description seems to suggest that there is a single universally applicable technique suitable for every possible kind of information. It also seems to imply that 'fusion' — whatever it means — is the only way of reducing the volume of information to the correct number of comprehensive reports that we want.

The purpose of the following discussion is to consider briefly the various types of sensor information with which a C^2 system is likely to be concerned and to discuss some of the techniques needed to process that information.

7.2.2 Positional information

7.2.2.1 The positional frame of reference: In the introduction it was stated that the users of the C^2 system must be able to obtain knowledge of the current activity in their areas of interest without attempting to define what is meant by activity. What we are concerned with here is where objects are, what they are and what they are doing — in other words, position, identity and behaviour. It is worth considering each of these in turn to see if they are really different and so needing different treatments.

The first thing which is clear is that the C^2 system has to operate in the four-dimensional world of space and time, so one of our primary concerns must be with information about the position of objects in our area of interest. In fact, position is of overriding importance because identity and behaviour mean little unless they can be associated with position. If we report that we have identified an enemy aircraft we are asked, 'Where is it?' If we report that a ship seems about to carry about an attack we are asked 'What ship? Where?' It would appear then that the first task with which we are faced is how to deal with positional information, in which we should sensibly include speed and direction as well as position itself. And of course since we are almost certainly concerned with a dynamic environment we need to take time into account as well.

7.2.2.2 The positional problems: If all the sensors used to obtain positional information about targets were absolutely accurate and if all of them used a common co-ordinate system for reporting, there would be no problem in combining information from any number of them. All the sensors detecting a target would report an object at exactly the same position and with exactly the same velocity so that the situation pictures from the individual sensors would register exactly and we could easily construct a single, comprehensive, accurate picture. In real life, inaccuracies in sensor position and alignment (particularly with mobile sensors), measurement inaccuracies and variations in the propagation of the atmosphere all result in variations between the sensor reports. Then there is the additional problem that most sensors measure target position in terms of range and bearing from their own position — that is to say in polar co-ordinates. A bistatic ESM

Some aspects of data fusion 323

system, consisting of two sensor stations working together to fix target position by triangulation, provides information in the form of angles from the baseline joining the two sensors. In practice C^2 systems all use some form of geographical grid which is essentially Cartesian, and sensor reports have to be converted into this common system form. Co-ordinate conversion is not always such a trivial process as it might appear to be. The need to represent the curved surface of the earth on flat maps can itself lead to serious errors in co-ordinate conversion. Furthermore, in a dynamic environment successive reports from a single sensor on a single target are likely to refer to different positions and every report has to undergo co-ordinate conversion leading overall to a sizeable computational load.

This brief discussion has identified the first problem to be tackled: deciding whether two or more reports actually refer to the same target. An operator using a PPI display does this by making a visual comparsion of the information provided by two or more sensors. He looks at the plots and tracks on the screen and by comparing position and velocity, perhaps over several minutes, makes a judgment based on his experience and, (one suspects) a measure of intuition. If we want to do this job automatically we need to formalise and quantify the factors involved.

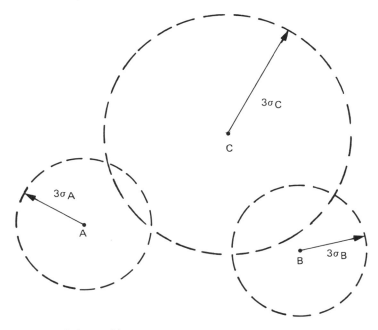

Fig. 7.2.1   *A correlation problem*

7.2.2.3 Correlation: The total performance of a sensor including all the inaccuracies mentioned above can, at least theoretically, be expressed in dimensional terms such as variances in range and bearing measurements. This means it should be possible to use accepted mathematical correlation techniques to establish

324   Some aspects of data fusion

whether two or more reports match each other. This, like co-ordinate conversion, in theory seems easy but in practice proves to be otherwise.

Consider the simple situation illustrated in Fig. 7.2.1. Suppose that a radar is reporting two targets at positions A and B and that there is a circular, normal distribution of errors with the 3σ circles round the positions A and B being as shown. Suppose that another sensor of a different type, say an ESM system, with a different performance is reporting a target at position C, with its 3σ circle also as shown. Clearly there is some correlation between the reported positions at A and C as indicated by the overlap in the respective 3σ circles. Similarly there is a measure of correlation between B and C. This means there are three possibilities: either the target reported at C is the same as that reported at A, or it is the same as that reported at B or it is a third completely separate target. The area of the intersection between two circles can be taken as representing the degree of correlation, which means that the correlation between C and either A or B is small. The most probable situation is therefore that C is a third completely separate target.

In this simple example two-dimensional position only has been used, and in a real-life static environment with a sparse target density this could be sufficient. However, in practice and especially in a densely populated dynamic environment it is better to test velocity as well as position. In a system using a Cartesian grid co-ordinate system the correlation test could cover six variables x, y, z, $\dot{x}$, $\dot{y}$, $\dot{z}$, although it is frequently unnecessary even in air defense systems to use z and $\dot{z}$,

It is worth noting in passing that sensors rarely have circular error distributions such as are used for illustration in Fig. 7.2.1. Radars are usually more accuate in range than in bearing, whereas with a bistatic passive sensor system the error distribution pattern depends on a target position relative to the system baseline. The actual error distribution for our simple example is therefore more likely to resemble that shown in Fig. 7.2.2. This presents an even more complicated correlation problem since the variances in x, y, $\dot{x}$, and $\dot{y}$ all depend on the bearing of the target relative to the sensor.

This discussion on correlation has been included not only because it is such an important process in picture compilation but also because it is not a trivial exercise. Nevertheless it has to be appreciated that establishing a correlation between two reports does not combine the information; it merely determines whether the two reports refer to the same target. If we decided that the target reported at C and that reported at B in Fig. 7.2.1 are the same, perhaps by correlation of velocity, we still have two reported positions for the target. Correlation has not helped to determine the true target position, or even to make a better estimate of it.

7.2.2.4 Fusing positional data: The next logical step would seem to be to combine or fuse these separate reports to give us a best estimate, and this does not appear to be too difficult. To continue with the earlier example; a decision that B and C represent the same target means that the true target position actually lies within the area of intersection of the respective 3σ circles. This in fact represents graphi-

Some aspects of data fusion 325

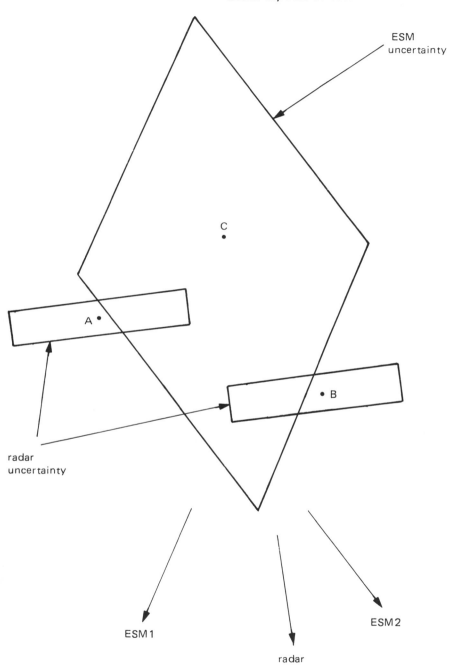

Fig. 7.2.2 *Practical sensor uncertainties*

cally the process of taking a weighted average of the two reports taking account of their relative accuracies. It has also the advantage that even the weighted average still retains a measure of uncertainty. In fact it is the exception rather than the rule for this type of fusion to be carried out, particularly in highly dynamic scenarios. In most current systems one report from a pair or group is taken as the representative or 'system' report so that other reports, at least as far as positional information is concerned, are in practice filtered out. One of the obvious reasons for taking this approach is that it serves the immediate purpose, and if the accuracy of a single sensor is sufficient to satisfy the system requirements then it is sensible to use it. It also needs to be borne in mind that sensors will typically update positional information several times a minute. If we decide that reports from two sensors correlate we do not need to retest for correlation at every update; indeed if we are absolutely confident in the correlation decision we do not need to make any retest of its validity. However, if we decide to combine the information we must repeat this process at every update of sensor input if the best estimate of position is also to be kept updated. The size of the processing load which this represents needs to be taken into account, particularly if the benefits are minimal, as they frequently are.

Although there are systems where it may be beneficial or even necessary to fuse separate sensor reports to achieve the desired accuracy, it is generally true that the processing of positional information at this stage is limited to correlation. This accomplishes one of the prime requirements, which is that of ensuring that there is a single system report on the position of each target.

Another important reason for correlation is that it is a prerequisite for the fusion of identity and behavioural information. It was noted earlier that these types of information are of little practical use unless they can be associated with position. Clearly the individual sensors providing such information, say identity, need to make such an association and only when this has been done is it possible to combine the various items of information. The positional correlation process enables us to determine that non-positional information from different sensors relates to the same target and hence, as already stated, it is a prerequisite for fusion.

7.2.3 Identity information

7.2.3.1 The nature of identity information
The correlation of positional information as described above is a mathematical process that deals with dimensional data. If we say that the measured range of an object is x miles and the measurement standard deviation is y miles it has a precise meaning. However, identity is a non-dimensional attribute and cannot be treated in the same way. A statement that the identity of an object is Q with a standard deviation of r has no meaning because there are no dimensions. But if a sensor merely makes a statement that an object's identity is Q we have no means of knowing how accurate the assessment is. The situation is even worse if two sensors make

different statements — for instance, one that the object's identity is Q and the other that it is S. There is no way of deciding, from the information given here, which is correct. Note the implicit assumption here that only one report is correct, whereas when we were discussing positional information it was accepted that two or more sensors could report a single object at different positions but both reports could be 'correct' within the measurement accuracies of the sensors. We also need to note that whereas the object of correlating positional information was not necessarily to obtain the best possible estimate of position, the object of processing identity information is almost certainly to obtain the best possible estimate of identity.

Clearly if the sensors can only provide the bald conflicting statements given above — 'the identity is Q' and 'the identity is S' — we can do very little except admit we do not know which is correct! However, if we had some idea of the accuracy or quality of the individual assessments we could be better placed to make a decision.

7.2.3.2 Identity probabilities: One method of indicating the accuracy of the assessment is to add to it a statement of the probability that it is correct. Suppose in the example already used that sensor 1 declares that there is a 60% probability that the object is Q (denoted for the present by $P_{Q1} = 0.6$) and sensor 2 declares a 50% probability that it is S, i.e. ($P_{S2} = 0.5$). The weight of evidence seems to be slightly in favour of the true identity being Q, but it is clear that the information is still not complete. If we are certain that an object is present and there is a probability of 0.6 that it is Q then there is also a probability of 0.4 that it is something else — but what? We would like to know if sensor 1 attaches any probability to the object being S and if sensor 2 gives any probability that it is Q. We require the sensors to give us a complete set of probabilities, and it is easy to see that to satisfy this condition the probabilities must sum to unity, i.e.

$$P_{Q1} + P_{S1} + \ldots = \Sigma P_1 = 1$$

Similarly for other sensors, so that

$$\Sigma P_1 = \Sigma P_2 = \Sigma P_n = 1$$

Suppose now that the two sensors give the complete sets of data as follows:

$P_{Q1} = 0.6$ $P_{S1} = 0.3$ $P_{T1} = 0.1$

$P_{Q2} = 0.1$ $P_{S2} = 0.5$ $P_{T2} = 0.2$ $P_{V2} = 0.2$

Note that since $\Sigma P_1 = \Sigma P_2 = 1$ all possibilities have been included and hence by implication $P_{V1} = 0$.

There now seems to be more weight of opinion in favour of the true identity being S, since this has a fair measure of support from both sensors. Furthermore, athough sensor 1 gives a probability of 0.6 that the identity is Q, sensor 2 gives a probability of 0.9 that it is *not* Q. Similarly, although sensor 2 gives a probability of 0.2 that the identity is V, sensor 1 states that there is zero probability that it is V ($P_{V1} = 0$).

328 Some aspects of data fusion

There are two questions to be asked: How do the sensors derive the probabilities? How do we combine the information from two or more sensors to provide a single assessment?

7.2.3.3 The probability matrix: One possible way of deriving the probabilities is to use the concept of *forward conditional probability,* which can be defined, in this instance, as the probability that the sensor will make a certain declaration, say D_i, given that a specific object, say O_N, is present. Since this is a statistical quantity it should be possible to obtain it provided we can allow the sensor to observe the object and make its declaration often enough to give a representative sample. This probability is denoted as $P(D_i|O_N)$ and clearly there is a value of $P(D_i|O_N)$ for every possible declaration that the sensor can make. If there are n possible declarations then we can see that

$$\sum_{i=1}^{i=n} P(D_i|O_N) = 1$$

For every other object that the sensor can separately distinguish there will be a similar set of probabilities, e.g. $P(D_i|O_M)$, where again

$$\sum_{i=1}^{i=n} P(D_i|O_M) = 1$$

These probabilities can be assembled in the form of a probability matrix as shown in Table 7.2.1.

It is important to note that the matrix is an assembly of columns each of which lists the probabilities of specific declarations resulting from the presence of the object in question. Clearly the number of elements in the column must equal the number of possible declarations since the elements must sum to unity. Zero elements arise in the matrix when the sensor cannot make a particular declaration as a result of a particular object being present. At this stage it can be assumed that the number of columns is determined solely by the number of objects for which observations are required.

Table 7.2.1 *Prior probability matrix format*

Declaration	Object					
	O_1	O_2	...	O_N	...	O_X
D_1	$P(D_1|O_1)$	$P(D_1|O_2)$	...	$P(D_1|O_N)$	...	$P(D_1|O_X)$
D_2	$P(D_2|O_1)$	$P(D_2|O_2)$	...	$P(D_2|O_N)$	...	$P(D_2|O_X)$
D_3	$P(D_3|O_1)$	$P(D_3|O_2)$	...	$P(D_3|O_N)$	...	$P(D_3|O_X)$
$\vdots$	$\vdots$	$\vdots$		$\vdots$		$\vdots$
D_i	$P(D_i|O_1)$	$P(D_i|O_2)$	...	$P(D_i|O_N)$	...	$P(D_i|O_X)$
$\vdots$	$\vdots$	$\vdots$		$\vdots$		$\vdots$
D_n	$P(D_n|O_1)$	$P(D_n|O_2)$	...	$P(D_n|O_N)$	...	$P(D_n|O_X)$

Table 7.2.2 *Prior probability matrix*

Declaration	Object			
	O_1	O_2	O_3	O_4
D_1	0.5	0.1	0.3	0.2
D_2	0.3	0.8	0.4	0.1
D_3	0.2	0.1	0.3	0.7

If there is a second sensor, which makes declarations $d_1, d_2, \ldots, d_n$ given the presence of the same objects $O_1, O_2, \ldots, O_N$, then it is possible to set up a probability matrix in a similar way to that described earlier where the elements consist of the probabilities $P(d_j|O_N)$.

If we now take the case where the object O_N is present then the probability that sensor 1 will make the declaration D_i and that in the same situation sensor 2 will make the declaration d_j is $P(D_i|O_N)P(d_j|O_N)$. It has to be noted that the declarations need not be the same but they must be made in relation to the same object. Also if the number of possible declarations by sensor 1 is n and by sensor 2 is m then the number of combined probability elements will be nm.

It will be recalled that the question was raised earlier as to the method of combining sensor declarations. This shows that it can be carried out by the statistical process that is applied to any other pair of probabilities. However, the use of the simple multiplication rule can only be justified if the declarations are statistically independent, i.e. the probability of sensor 2 making the declaration d_j is not affected by the fact that sensor 1 has made the declaration D_i.

7.2.3.4 Posterior probabilities: The probability matrix discussed above is concerned with, and constructed from, forward conditional or prior probabilities, i.e. $P(D_i|O_N)$. However, in trying to identify an unknown target we are given the sensor declaration and have to derive the probability that the target is present, i.e. $P(O_N|D_i)$. The problem now to be faced is how to convert $P(D_i|O_N)$ to $P(O_N|D_i)$.

The process can be illustrated by a simple example. Table 7.2.2 is a sensor probability matrix of prior probabilities. If we have a scenario in which there are 100 of each object then we can say that statistically the declaration D_2 will be made 30 times as a result of the O_1 objects present, 80 times as a result of the O_2 objects present and so on. So for this scenario the declaration D_2 will be made 160 times and, given a particular instance of this declaration being made, the probability that it results from O_1 being present is $30/160 = 0.19$. For the given matrix and scenario this represents the desired posterior probability $P(O_1|D_2)$. If we carry out this process for the complete matrix the result is as shown in Table 7.2.3, where the probabilities are given to the nearest 1%.

It can be seen that the elements in this matrix could be obtained simply by normalising the elements in each row of the prior probability matrix. However, this can only be done directly if the numbers of each object present are equal, as in the sample scenario. For a different scenario, say, where the numbers of ob-

330 Some aspects of data fusion

Table 7.2.3 *Posterior probability matrix 1*

Object	Quantity	Declaration		
		D_1	D_2	D_3
O_1	100	0.45	0.19	0.15
O_2	100	0.09	0.50	0.08
O_3	100	0.27	0.25	0.23
O_4	100	0.18	0.06	0.54

jects present are $O_1 - 100$, $O_2 - 200$, $O_3 - 50$ and $O_4 - 100$, the posterior probabilities can be calculated in the same way. Using the same prior probability matrix as before (Table 7.2.2) it can be seen that the declaration D_2 will now be made 220 times, although of these the number resulting from the presence of the O_1 objects is still 30 times. So the probability that D_2 is made as a result of O_1 being present is now $30/220 = 0.14$. The complete posterior probability matrix for this object mix is shown in Table 7.2.4.

The way in which the matrices in Tables 7.2.3 and 7.2.4 were derived raises a further problem which can be illustrated by expanding on the example. It will be recalled that the sensor probability matrix was made up of columns each of which referred to an object which we wish to be able to identify, in the example $O_1 - O_4$. However, it is possible that other objects could result in the declarations $D_1 - D_3$ being made, although we may not wish to identify these objects. Suppose in the sample already used that there is another object O_5 whose presence always results in the declaration D_2, i.e. $P(D_2|O_5) = 1$ (it follows that $P(D_1|O_5) = P(D_3|O_5) = 0$). If in the object mix used to derive Table 7.2.3 there are also 100 of O_5, then the total number of declarations D_2 made is now 260 and the probabilities in the corresponding row of the matrix become

$$P(O_1|D_2) = 0.12 \quad P(O_2|D_2) = 0.31 \quad P(O_3|D_2) = 0.15$$

$$P(O_4|D_2) = 0.04 \quad P(O_5|D_2) = 0.38$$

This shows that the posterior probability matrix must include all objects which can result in any of the declarations being made by the sensor. This is true whether or not we wish to separately identify those objects, because it is necessary to be able to ascertain that it is *not* one of those we are interested in — in the example, not O_1, O_2, O_3 or O_4. In this case O_5 could cover all others. However, it follows

Table 7.2.4 *Posterior probability matrix 2*

Object	Quantity	Declaration		
		D_1	D_2	D_3
O_1	100	0.48	0.14	0.16
O_2	200	0.19	0.73	0.16
O_3	50	0.14	0.09	0.12
O_4	100	0.19	0.05	0.56

Some aspects of data fusion 331

that if the posterior probability matrix has to include all possible objects then so must the sensor prior probability matrix. It may be difficult to derive or obtain the sensor probabilities for such an omnibus 'object'.

7.2.3.5 Bayes' transform: The method used to derive the posterior probability matrices in Tables 7.2.3 and 7.2.4 is, in practice, an application of Bayes' theorem, which is stated as follows:

$$P(O_N|D_i) = P(D_i|O_N)P(O_N) \bigg/ \sum_{K=1}^{K=x} P(D_i|O_K)P(O_K)$$

This can be regarded either as using the prior probabilities to modify sensor declarations or conversely as using sensor declarations to modify the prior probabilities of objects being present. However, in either case it is a necessary condition that the object class is complete and that the objects are mutually exclusive.

The disadvantage of Bayesian transformation is that definite values must be allotted to each element in the sensor probability matrix and to the prior probabilities of each object being present. In practical terms this means that there is no way of allowing for uncertainty either in the object mix ratio or in sensor performance.

7.2.3.6 Uncertainty: In Bayesian-based fusion the sensor declaration, as described above, consists of one row of a probability matrix. Because such a declaration is made up of a set of probabilities or likelihoods it is commonly known as a *likelihood vector*.

It was seen that one of the difficulties with the Bayesian approach to fusion is that definite probabilities have to be assigned to each element of the matrix and its constituent likelihood vectors. In practical terms the likelihood vector produced by a sensor can give no indication of the quality of the input information. Probabilities of 0.5 assigned to each of two possibilities could result either from consideration of very high quality input data or by the assumption of equal random probabilities in the absence of any input data whatever. Similarly in the application of Bayes' formula to convert the prior probabilities to posterior probabilities the mix ratio of target objects has to be stated in definite terms.

In both these cases there is clearly a need to be able to express uncertainty, and a generalisation of the Bayesian process proposed by Dempster (1968) and Shafer (1976) allows this to be done. The method involves the use of belief functions in which belief in a proposition, e.g. the presence of an object, is expressed in terms of support for it, its plausibility (not necessarily supported) and disbelief in it. Plausibility plus disbelief together add up to unity, and the difference between plausibility and support represents the degree of uncertainty. A fundamental difference between the Shafer-Dempster approach and the Bayesian probabilistic technique described above is that Shafer-Dempster does not in principle rely on the forward conditional probability concept; nor does it require the prior probability ratio, the object mix used in the Bayesian transform.

However, it is difficult to see how a sensor can assign a measure of support to a belief without relying on previous knowledge of performance. It is a fundamental fact of life that the process of identifying an object must always depend on prior knowledge. At a personal level we identify other individuals by recognising specific attributes — aspects of their appearance, the sound of their voices, or even handwriting. Similarly identification of objects in a C^2 system's environment relies on prior knowledge of specific attributes of those objects. Identity information can only be derived on the basis of prior knowledge, whether the combining or fusing process is to be Bayesian or Shafer-Dempster.

7.2.3.7 A comparison of Bayes and Shafer-Dempster: In a C^2 system the most basic assessment of identity is that of allegiance. It could be argued that the categories of allegiance in the military environment can be restricted to friend or enemy, but practical considerations require the inclusion of neutral among the possibilities. The inclusion of neutrality as an allegiance is also significant in the Bayesian process since it ensures that the object class, i.e. the possible allegiance categories, is complete. In other words in a military sense all objects of interest are either friendly, enemy or neutral and there are no other possibilities.

A sensor designed to provide allegiance to a Bayesian system would be required to divide a unit of probability between the three possible allegiances. A sensor working into a Shafer-Dempster system could assign degrees of belief to each of the three possibilities but an element of uncertainty could also be included to make a complete 'unit of belief'.

IFF is a sensor specifically intended to provide allegiance information, so it is worth considering its application as an input source for either Bayesian or Shafer-Dempster processing.

7.2.3.8 Processing of IFF information: The procedure for using IFF to assess allegiance can be divided into a number of distinct stages, the first of which is the interrogation of the target and reception of its response, if any. The interrogation and response are each coded and either or both codes are normally changed at regular intervals, typically 30 minutes. A response may be correctly coded or incorrectly coded, and a third possibility is that no response is received. The second stage of processing is to decide which of these three categories of reply has actually been received by the interrogator. The third and more difficult stage is to determine what inference can be drawn from the received response.

The method of drawing this inference depends on whether the assessment of allegiance is required in a Bayesian or Shafer-Dempster format. In this simple concept of IFF operation we have defined three possible allegiances and three possible responses. It is therefore possible to define the forward conditional probability matrix format shown in Table 7.2.5.

Consider first of all the elements in the first column. The probability that a

Some aspects of data fusion   333

Table 7.2.5   IFF probability matrix

Response	Allegiance Friend (F)	Enemy (E)	Neutral (N)
Correct (C)	$P(C|F)$	$P(C|E)$	$P(C|N)$
Incorrect (I)	$P(I|F)$	$P(I|E)$	$P(I|N)$
No reply (0)	$P(0|F)$	$P(0|E)$	$P(0|N)$

friendly platform will give a correct response to an interrogation, i.e. $P(C|F)$, depends on a number of factors; it must be fitted with a transponder which is serviceable, switched on and set to the correct codes. The interrogator and transponder aerials must not be screened from each other and the communication conditions must be such that neither the interrogation nor the response codes can be misread. The probability that an incorrect response will be received, i.e. $P(I|F)$, depends on similar factors except that either an incorrect code is set or there is sufficient distortion to result in the misreading of a code. If these probabilities can be determined then the probability of no reply being received, i.e. $P(0|F)$, can easily be derived since the elements of the column must sum to unity. It seems feasible that these probabilities could be derived from statistical records, but it is nevertheless clear that this would be a complex and lengthy task.

The elements of the third column are easier to derive since it can reasonably be assumed that neutral platforms will not be attempting to respond to IFF interrogations. This means that the probabilities in this column are as follows: $P(C|N) = P(I|N) = 0; P(0|N) = 1$.

The second column is potentially the most difficult since the probabilities depend on whether the enemy is 'spoofing', i.e. attempting deception response. If it is assumed that the enemy is not spoofing then the probabilities are the same as those for the neutral allegiance column, i.e. $P(C|E) = P(I|E) = 0$. $P(0|E) = 1$. However, if it is assumed that the enemy will be attempting to spoof then the factors which would enable him to do it successfully have to be considered. Fit and serviceability are relevant, as with friendly platforms, but his ability to determine the correct code settings must also be taken into account. Furthermore, it is unlikely that the probabilities for the 'enemy' column of Table 7.2.5 can be obtained from statistical records since a potential enemy is unlikely, to say the least, to divulge his spoofing tactics before conflict.

This discussion has in effect restated, in the context of a probability matrix, recognised deficiencies in the current IFF system. The next generation of IFF will be sufficiently secure to ensure that enemy spoofing is practically impossible and automatic code changing will eliminate the possibility of incorrect code settings. However, the possibility of no response being received from a friendly platform because of transponder failure, mutual screening or poor propagation will remain. This means that whereas a correct response will positively identify a friend a 'no reply' will still represent a considerable problem. This is especially so when

334 Some aspects of data fusion

account is taken of the fact that the forward conditional probabilities discussed have to be modified by the prior probabilities of friends, enemies and neutrals being present in order to obtain the required posterior probabilities, i.e. $P(F|0)$, $P(E|0)$ and $P(N|0)$.

A Shafer-Dempster approach simplifies some aspects of this problem but is unable to resolve others. The aim with this technique is to divide, as far as possible, the unit of belief between three propositions: that the platform being interrogated is friendly, enemy or neutral. As far as IFF is concerned support for each of these propositions can only be assigned on the basis of the received reply. This means that the 'no reply', which is most difficult to deal with by Bayesian techniques, is easiest to resolve by Shafer-Dempster. Because there is no reply then no support can be assigned to any of the propositions and the whole of the unit of belief is allocated to uncertainty.

However, in the case of current IFF, which is susceptible to spoofing, assigning support to one of the propositions on the basis of, say, a correct reply remains a problem. It depends in turn on the support assigned to a separate proposition that the enemy is spoofing. The assignment of numerical values in this case is no easier in Shafer-Dempster than it is in the Bayesian approach. The latter is firmly based on the use of prior probabilities, although these may, in many circumstances, be difficult to obtain. In the Shafer-Dempster approach use may be made of prior probabilities, but the additional problem is introduced of how much weight is to be attached to them.

When the next generation IFF, already mentioned, is introduced then Shafer-Dempster will offer an elegant method of assessment. When a correct response is received the complete unit of belief will be assigned in support of the proposition that the platform being interrogated is a friend, i.e. there is 100% belief that it is a friend. On the other hand when no response is received the entire unit of belief must be assigned to uncertainty, i.e. there is no direct evidence of its allegiance.

7.2.3.9 Prior probabilities in Shafer-Dempster: It was noted earlier that the Shafer-Dempster technique does not rely on prior probabilities in the way that the Bayesian approach does. However, this does not mean that use cannot be made of such probabilities. When trying to assess a platform's allegiance, the degree of support, based on prior knowledge, for the presence or absence of platforms of a particular allegiance represents a valid input to the Shafer-Dempster process. An advantage over the Bayesian approach is that there is likely to be much less uncertainty in prior estimates of friendly presence than that for neutral or, particularly, enemy. These varying degrees of uncertainty can be reflected in the amount of support assigned to the three allegiances. The principal problem, as with all applications of Shafer-Dempster to identity processing, is how to assign levels of support in a consistent, repeatable manner.

7.2.3.10 The fundamental problem of identity procesing: Two approaches to the problem of combining or fusing identity information have been discussed

Some aspects of data fusion 335

here in sufficient detail only to demonstrate that both techniques have practical application difficulties. The fundamental reason for this was covered briefly at the start of the discussion of identity processing. It is that identity is a non-dimensional or qualitative attribute whereas the fusion processes described are dimensionally or quantitatively based. Development of consistent and repeatable methods of deriving the necessary numerical inputs are prerequisites for successful application of both processes. The difficulty in using sensor outputs to provide these inputs lies in deciding not what the sensor output is but what it means in terms of identifying the object which is being reported on. In the example of IFF quoted earlier it is comparatively easy to decide that a response has been received to an interrogation. However, if it is suspected that enemy platforms are spoofing then it is difficult to determine what the response tells us about the platforms' identity.

The fundamental problem in processing identity information is not a lack of techniques to fuse data from a number of sources. It lies in the difficulty of expressing identity information in quantitative terms and in assigning accurately representative numerical values to sensor outputs.

7.2.4 Behaviour

7.2.4.1 Types of behaviour: The third type of information mentioned in the introduction was behaviour. When an object has been detected, its position has been determined and it has been identified, the user of the C^2 system wants to know what the object is doing — its behaviour. The method of describing position in dimensional terms is well understood and the problems of assessing identity in terms of allegiance or platform type have been discussed above. Behaviour is an even more abstract concept than identity and there is no single dimensional frame in which it can be described. In the C^2 system behaviour is important because it helps to determine both the current threat posed by a platform and its potential threat. Behaviour related to threat assessment can include elements of positional information: direction, speed, manoeuvre. It can also include operation of equipment: jamming, using radar or laser systems, opening weapon bay doors, releasing weapons. Not all of these aspects of behaviour are likely to be found at the same time and the same combination of behaviour elements may have different threat connotations depending on circumstances. For example an aircraft flying low and fast in a north-easterly direction with a target acquisition radar operating and its weapon bay doors open would be assessed as representing an immediate threat to a ship ten miles to its north-east. It is unlikely to represent the same threat to a ship ten miles to its south, although its behaviour is the same. This also re-emphasises the earlier statement that the primary importance of behaviour is directly related to its use in threat assessment.

It is the disparate nature of the elements of behaviour which make it impossible to define a system of dimensions for behavioural information. High-speed, low-level flight and operation of a radar are such different forms of behaviour that there is little to be gained in trying to define a common dimensional measure to describe them. What type of processing, if any, is applicable then to behavioural information?

7.2.4.2 Processes for behavioural information: There are three stages to the process. The first is to collate the information that is likely to be relevant to the user's task. This information may be unprocessed sensor information or it may be processed information such as identity of either platform or emitter. However, this is essentially a straightforward data formatting task and to that extent is a trivial function; the effectiveness of this stage lies in correctly defining the types of information that are relevant to the user's needs.

The second stage is to determine the implications of these collated items of information. In the example already quoted the implication of the aircraft having its weapon bay doors open and a target acquisition radar operating is that it is about to launch a weapon. In this case a process of deductive reasoning is involved in reaching such a conclusion. Because the relevance of behaviour is likely to be in its relation to position and identity the three must be considered together, and this is commonly and logically described as situation assessment.

The third stage of the process is making the threat assessment, and again this involves deductive reasoning. As already discussed, the same behaviour will represent different threats to different possible targets.

In current systems the analysis of behaviour and threat assessment are usually manual processes, although some short-range weapon systems have automatic launch initiation against targets exhibiting specified behaviour patterns. For example there are several ship's self-protection systems which automatically launch weapons against any platform directly approaching the ship within a certain range, below a certain height and above a certain speed. Such simplistic rules are not likely to be acceptable in many other circumstances such as in the confused environment of a land/air battle where friendly and enemy forces are very likely to be closely intermingled.

7.2.4.3 Automatic processing of behavioural information: Replacement, or augmentation, of such manual reasoning processes by automatic processes suggests a possible area for the application of IKBS, and the literature on different suggested approaches particularly to threat assessment is already extensive. Threat assessment is of course primarily concerned with enemy activity but this presupposes that the enemy forces can be identified. If they cannot, any platforms or forces which are as yet unidentified (unknown) must still be assessed as a potential threat.

7.2.4.4 Behaviour as a source of identity information: Although the primary purpose of behaviour analysis is to assist in threat analysis there is another significant use, and that is as another source of identity information. In the example quoted above of the automatic initiation of close-in weapon systems it can be seen that it is the behaviour of the low, fast approaching aircraft that leads it to be assessed as an immediate threat, with an inherently implied decision that it is an enemy.

Conversion of behavioural information ratio into a probabilistic form suitable for the kind of identity processing discussed earlier is generally not easy. One case in which it can be used effectively is in establishing a probabilistic relationship between an aircraft and a flight plan. There are statistics available on the pattern of deviation of aircraft from their filed flight plans and these can be used to assess the probability that an aircraft is in fact one for which a flight plan has been filed. In a similar way a relationship can be postulated between an aircraft track and a safe transit route. An aircraft that follows such a safe lane precisely is probably a friend — but how probably? It is the conversion of a qualitative evaluation of probability into a quantitative level which presents the difficulty. This is, of course, another example of the fundamental problem of identity processing which was discussed earlier.

7.2.4.5 Prior knowledge in behaviour analysis: The importance of prior knowledge in assessing identity was discussed earlier. Such knowledge also plays a vital part in behavioural analysis leading to situation assessment and threat evaluation. It is the range of prior knowledge related to the immense variety of position, identity and behaviour patterns that single out situation and threat assessment as being particularly suited to the application of IKBS. Another important factor, particularly in threat assessment, is that account has to be taken of the intentions of the forces concerned, and this implies an appreciation of tactics. These may range from the set textbook tactics of the staff college to the sometimes highly idiosyncratic tactics of individual commanders.

The extent to which IKB processes used in the way described here can be described as fusion processes may be debated. However, it cannot be doubted that they represent very effective methods of combining information from a variety of sources, including prior knowledge, in a synergistic way to enable the user to make the best possible assessment of the current situation in his area of interest. Sections 7.3 and 7.4 cover the use of IKBS in data fusion and threat assessment in some depth.

7.2.5 Conclusion

At the beginning of this contribution it was suggested that 'data fusion' was a term which could be misleading when used in relation to C^2 systems. A number of different processes have been discussed which are necessary in such systems to enable the user to determine the position, identity and behaviour of objects in this area of interest. In short, they are necessary for him to work out what is happening when his information comes from a variety of sources, some of which may be imprecise, intermittent or even contradictory.

The differences between such processes as correlation which establishes that two sets of data refer to the same object, and fusion which combines data which have already been correlated, are important because of the different logical and computational approaches necessary. However, the interrelationship of these processes in the overall operation of a C^2 system is also important in the progression from data collection by sensors through picture compilation and situation assessment to threat evaluation.

A study of current systems and those projected for the near future suggests that whereas a large degree of automation has been applied to the correlation process, although not always with complete success, the same cannot be said of the other processes. Identification of non-cooperating platforms remains a key problem area in picture compilation, and the difficult problems of automatic identity processing have been discussed above. Behavioural analysis, situation assessment and threat evaluation are still almost entirely manual processes although the pace of modern warfare places increasing loads on the operator and, as mentioned above, in some cases it has been necessary to automate, albeit using very simplistic criteria.

The operator load in these vital processes makes it essential to provide automated assistance that is accurate, consistent and reliable. Development of such assistance requires a clear understanding of the precise nature of the various relevant processes and the differences between them.

7.2.6 References

DEMPSTER, A.P. (1968) 'A generalisation of Bayesian inference', *Journal of Royal Statistical Society,* series B, **30**(2), pp. 205–47

SHAFER, G. (1976) *A Mathematical Theory of Evidence,* Princeton University Press

Chapter 7.3

An AI approach to data fusion and situation assessment

W.L. Lakin and J.A.H. Miles
(Admiralty Research Establishment)

7.3.1 Introduction

This contribution describes our research programme at ARE to introduce knowledge-based systems to problems in command and control. Although we are most interested in applications for the Royal Navy, the fundamental requirements are common to most command and control systems and we hope the generality of the concepts will be apparent. The purpose of command and control is to achieve some desired objectives using available resources to best effect. Because these resources usually involve many individuals, both men and machines, a considerable amount of communication and information handling, including intelligence, must take place — hence C^3I. The reason for the complexity of C^3I systems is that they have to tackle very large problems in a complex unpredictable environment. It is these characteristics which generate the need for so much human skill, judgment and intuition and which have caused computers to make such little inroad into the more intellectual processes in command and control. With the arrival of expert systems, however, there is renewed enthusiasm to encapsulate human expertise in computer software, and C^3I is an obvious field for attention.

To set the context for our research and the rest of this contribution, we show in Fig. 7.3.1 a simple layered view of C^3I and, alongside it, the related technology. We considered, early on, possible applications of expert systems in the higher levels of C^3I, i.e. decision making, but opted for a more bottom-up approach by trying to extend from algorithmic data processing into the first level of manual activities within current systems. We believe this approach is essential because it attacks one of the most critical areas of C^3I — that of bringing together the information and forming a coherent picture of what is happening. This must be a prerequisite for all higher-level activities and, from our experience, we know that it is where the first major shortcomings appear in current systems.

In the naval environment, which this contribution primarily addresses, this first step of forming a tactical picture involves detecting, locating, tracking and, if

340 An AI approach to data fusion and situation assessment

possible, classifying all objects which might conceivably contribute to the tactical situation. This implies virtually every object within sensor range or within the volume of interest to a single warship or to a group of co-operating maritime units, which may be dispersed over a wide area of ocean. The shear volume of data involved generates severe problems for the level of human and conventional computing resources which can sensibly be deployed. It is necessary to include, not only all the real-time sensor data, but also what might be termed secondary or non-real-time data so as to provide further evidence for classifying the objects, predicting their intentions and gaining a general appreciation of the tactical situation.

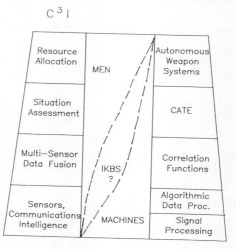

Fig. 7.3.1 C^3I *with related technology*

Here we have termed the bringing together of information into a coherent picture *multisensor data fusion*, though we are in fact including all sources of information, not just that from radio, acoustic or optical devices. There are for example human observers providing intelligence information and a background of encyclopaedic information and operational plans, all of which set the context for the more dynamic real-time sensor information. The task of combining such disparate data types has proved well beyond the capabilities of conventional computing methods, and has remained the province of the already overloaded human operator. To add to the problems, the overall data processing burden has to be undertaken in time scales which allow effective response to be taken against today's high-speed missile threat.

It is important to project the magnitude of the problems posed by these aspects of the naval command and control problem into the future. More powerful and more sophisticated sensors will generate an even greater volume of data to be processed, and new communications and intelligence systems will add further to the disparity of data types, particularly in the non-real-time categories. On the other hand, the need for increased control over emissions and the prospect of a severe ECM environment conspire to make what has hitherto constituted a disproportionate reliance on active sensors, particularly radar, no longer viable. This implies not only the more complex processing required by passive sensors, but also the need to glean as much information as possible from these data sources. Data fusion can to some extent compensate for the lack of positional information associated with such sources. Finally, it is projected that the number of operators assigned to these increasingly complex information processing activities will not become larger, but will in fact become significantly smaller.

Thus the goals of an intelligent data fusion system in a naval environment must be:

(a) To cope effectively not merely with the large quantity of input data which is available to a warship now, but that which will be available in increasing amounts in the future.
(b) To include and make full use of non-real-time and encyclopaedic information as well as real-time sensor data.
(c) To achieve the reaction times necessary to respond to rapidly changing and potentially hazardous situations, operating autonomously when the time penalty associated with manual intervention is unacceptable.
(d) To relieve the human operator of routine low-level tasks, thereby providing a rational basis for the reduced manning levels currently projected, and also allowing the man the time to exercise his higher-level decision-making capabilities.

The main purpose of this contribution is to report on a programme of research aimed at realising these objectives. Section 7.3.2 sets out the main characteristics of the naval data fusion problem and describes in more detail some of the information sources. Section 7.3.3 outlines the expert system approach that we have adopted; Section 7.3.4 explains the problem-solving strategy behind our 'stage 1' demonstrator, and Section 7.3.5 presents some of the results achieved so far. 'Stage 1' addresses primarily the problem of correlation, i.e. which items of evidence are associated with what real-world objects.

Section 7.3.6 extends data fusion into the domain of multiple receiving platforms, Section 7.3.7 examines the problem of combining the evidence for an object once the correct correlations have been made. Section 7.3.8 looks beyond data fusion to the higher-level problem of situation assessment, which forms the main thrust of our present research activity. Its purpose is to provide an interpretation of what the detailed 'world picture' means in tactical terms. Based on the experience accumulated over the course of the work, Section 7.3.9 seeks to identify the type of AI machine most suited to the data fusion problem. Finally, Section 7.3.10 provides a summary of observations and conclusions thus far and of our plans for the future.

7.3.2 Multisensor data fusion

The information on which a warship bases its tactical picture of the world is derived from a wide range of disparate sources such as own-ship's sensors, radio data links, signals, intelligence and tactical plans. The data is generally incomplete, inaccurate, ambiguous, conflicting and subject to deliberate interference and deception by the enemy. The techniques employed in the analysis, evaluation and correlation of this information include both algorithmic processing, operating largely on lower-level aspects of the problem such as vehicle detection and tracking, and heuristic reasoning for higher-level aspects such as the determination of identity and intent, based on behaviour patterns, intelligence and tactical appreciation and experience. It is the objective of this part of our research programme first to solve the problem of how to fuse all of this data so as to provide the best possible tactical picture, and secondly to encapsulate as much of this solution as possible in a computer.

The main characteristics of the problem are as follows:

(a) For a particular object in the tactical picture, each sensor (or more correctly each information source, since these include tactical plans, intelligence observations etc.) provides information on a subset of that object's properties. In other words, there is no single sensor that gives all the details to the accuracy and timeliness required. Therefore we need to combine information from a number of different sensors.

(b) If it is certain that the information from different sensors does refer to the same object, then it is not unduly difficult to combine this information. The difficulty arises because that certainty does not exist; hence the need arises to correlate information from different sensors.

(c) As well as the need to correlate information from different sensors, there is a need to correlate information received from the same sensor at different times. Correlating measurements or observations on the same object at different times is often achieved using simple association algorithms, provided the time intervals are short. Extended periods without contact on fast moving objects cannot be handled by such algorithms.

(d) One of the more important properties of an object is its position in space. Some sensors may provide poor positional information but good identity information, thus giving rise to considerable ambiguity. ESM, for example, gives good clues to identity, but only a measurement of bearing as regards position.

(e) Although many ambiguities can be resolved eventually, operational requirements generally demand that decisions are made immediately.

Current systems rely largely on manual correlation to form the tactical picture, but, even with moderately complex scenarios, the workload can easily overwhelm the limited number of operators available on a warship. This can lead to important data being overlooked. A computer program that could reliably carry out a large fraction of this task would greatly assist the operation and performance of the command and control function.

The 'world of interest' in the naval application consists of a volume of space

An AI approach to data fusion and situation assessment 343

which could be upwards of 100 kilometres in diameter and includes the airspace used by aircraft and the ocean used by submarines. The objects of interest could include all ships, aircraft and submarines. These objects are generally called platforms. The main platform parameters required in the model are position, velocity, identity, capability and mission.

Sources of evidence include:

Sensors
    Radar (short and long range, navigational, surveillance and tracking)
    IFF (an identification system)
    ECM (electronic counter-measures) detectors
    ESM (electronic support measures)
    Active sonar
    Passive sonar
Data links
    Ship to ship
    Aircraft to ship
    Shore to ship
Intelligence
    Electronic
    Communications
    Human

In general there will be many sensors of each type and many sensing platforms. As well as these sources there are several other categories of useful information:

Plan and command information
    Screen plans of ship and aircraft dispositions
    Flight plans for aircraft
    Aircraft control orders
    Weapon release orders
Environmental data
    Weather
    Oceanographic
    Geographic
    Political
    Shipping and air routes
Equipment database
    Objects (ships, submarines, aircraft, missiles, satellites etc.)
    Weapon systems
    Shore bases

A brief outline of each source or type of data follows.

7.3.2.1 Radar: Shipborne radars measure range, bearing and in some cases elevation and range-rate of targets. The measurements or plots as they are usually called, are processed by computer to obtain track data — that is, a series of estimated positions and velocities over the duration of each track. Tracks are allocated unique numbers so that new track data can be related to the set of tracks held. The tracking computer can also estimate the accuracy of the data it produces.

Radar tracks have the following characteristics:

(a) They consist of a series of estimates of position and velocity.
(b) They may also include estimates of accuracy.
(c) Most often they result from single objects, e.g. a ship or an aircraft.
(d) Sometimes they may result from several closely spaced objects.
(e) Some will be false alarms.
(f) It is possible for a track to change from one object to another.
(g) Tracks may result from any object that forms a radar target, e.g. lighthouse, oilrig, buoy, prominent coastline etc.

7.3.2.2 Identification friend or foe (IFF): The objective of IFF is to identify friendly platforms, typically aircraft. It is essentially a form of secondary radar which works as follows. A signal is transmitted by a ship, say, to interrogate any aircraft in the vicinity. On receiving the signal, any aircraft with an IFF transponder will transmit a coded reply. The ship examines any replies and those with the correct code can be assumed friendly. Civil aircraft use the same system for air traffic control purposes and include position information in the codes. For military use, a special code is used with values agreed between the parties beforehand.

IFF characteristics are:

(a) Only works with co-operating platforms.
(b) Provides position.
(c) Identifies friendly platforms and maybe neutrals.

7.3.2.3 Electronic counter-measures: Receivers built into radars can detect the presence of interfering signals (ECM) and provide direction (bearing) and signal strength data. This data is useful because it indicates where there is loss of radar cover and also indicates the presence of a hostile platform.

7.3.2.4 Electronic support measures (ESM): ESM sensors use wideband radio receivers to detect transmissions from equipments in the vicinity, particularly radars. Being passive, the ESM system can only measure the direction (bearing) of the transmission but, by analysing the received signal, may be able to identify the type of equipment. Parameters such as frequency, pulse length, pulse repetition frequency and scan rate are extracted from the received signal. Using a library of information about equipment characteristics, these parameters are then matched against all the known types in an attempt to establish the particular type. Knowing the type of equipment often enables the platform type to be inferred, for example a radar of type X may be known to be carried by an aircraft of type Y.

Detection of a particular emitter may last for some time so that many measurements may be taken. Although only bearing can be measured in the spatial

An AI approach to data fusion and situation assessment 345

dimensions it is at least possible to calculate a rate of change of bearing (bearing rate) and to form a track in one dimension.

ESM tracks have the following characteristics:

(a) They usually result from a single emitter (a platform with two emitters will probably result in two tracks).
(b) A complex emitter may result in more than one track.
(c) Bearing and bearing rate estimates are included.
(d) Emitter type identity may be produced.

7.3.2.5 Active and passive sonar: The two types of sonar are very roughly analogous to radar and ESM respectively, but they operate with sound below the sea surface and are primarily for the detection, localisation and classification of submarines. Sound waves in the sea are greatly affected by boundary layers, including the sea surface, seabed and water temperature layers in between. These layers make it difficult or impossible to detect submarines in some regions and, even if a detection is possible, the various paths that the sound could have travelled make it difficult to establish the submarine's exact position.

Active sonar relies on transmitting a sound signal and then listening for echoes. Using a directional array and measuring elapsed time between the transmission and echo, it can estimate both range and direction. It can detect submarines even when they are completely silent.

Passive sonar relies on the noise from the submarine to provide a signal for measurement. Only bearing can be measured but by analysing the sound, particularly its constituent frequencies, it is sometimes possible to recognise the characteristics of machinery on board the contact and hence to determine what type it is. Passive sonar is potentially of longer range than active sonar because of the one-way signal transmission, but of course the submarine must be noisy.

As well as detecting submarines, sonars will detect anything in the water which generates or reflects sound, such as surface vessels, whales, shoals of fish, wrecks and prominent seabed features. Unwanted detections must be eliminated using knowledge from other sources if detecting submarines is the primary task.

7.3.2.6 Data links: There are several types of data link in existence whose purpose is to transfer platform data in digital form between computers in ships, aircraft and shore bases. The message formats are standardised to maintain compatibility across different types of equipment. Data links can use a variety of radio channels for carrying the signals and the bandwidth of these channels will determine the amount of data received.

7.3.2.7 Intelligence: Intelligence usually refers to remotely sensed or observed events that could be of use to the naval task group. It could be electronically gathered data from hostile transmissions including communications which, for security reasons, are dealt with by the intelligence services, or it could be observations by human observers. The reports could contain information regarding platform movements, numbers of platforms, weapons carried etc. Because of the processing and communications involved, intelligence data may arrive some time after the original report was generated.

7.3.2.8 Plan and command information: A good deal of planning goes into any military operation and the plans that are drawn up can be of use when trying to interpret the real-time data produced by sensors. The identity, purpose and approximate position at any time of each own-force platform will be contained in some sort of plan. Screen plans give ship and aircraft dispositions for a naval task group and flight plans may be available for shore-based aircraft.

There is also a category of locally known information which stems from direct orders for objects under the control of a ship. Aircraft control orders and weapon release orders are examples. Again such data can be used to interpret the real-time sensor data.

7.3.2.9 Environmental data: Weather can affect the way equipments are used and their performance. Oceanographic data is used for sonar prediction and will affect the way submarines operate. Basic geography, coastlines, seabed contours and height contours provide context for inferring many properties of platforms, and the political map could be used to judge the general direction and objectives of forces. Shipping and air routes need to be included to explain high-density everyday traffic.

7.3.2.10 Equipment database: Information on the characteristics of all the objects involved in the world of interest can be used to assist in identifying platforms and to determine what equipment might be carried and hence their purpose, or what other platforms are likely to be associated with them. Knowledge of the enemy doctrine, procedures and tactics can be used to relate platform identity, behaviour and intent. Knowledge of the bases from which particular units operate can also help.

7.3.3 Expert system approach

Our initial approach to the multisensor data fusion problem was to search for suitable techniques from the world of artificial intelligence which (a) were well defined and (b) had been demonstrated on a similar type of problem. Perhaps the most well known and well defined part of the AI scene is expert systems.

The philosophy of an expert system is to produce a computer solution to a problem by capturing the expertise of a human expert. Generally the expertise is in the form of rules, and these rules form the knowledge base of the system. A controlling framework is used to allow the user to access the knowledge base in the manner of a consultation whereby the user may volunteer information or the machine may question the user until sufficient evidence is gathered to produce useful conclusions. The user may also ask the system to explain its reasoning so that he may understand how the conclusions were reached.

An AI approach to data fusion and situation assessment

We adopted this method of problem solving as it seemed to fit the multisensor data fusion problem, assuming that the human expertise exists. However, most of the well publicised expert systems are in quite different problem domains to multisensor data fusion — examples being medical diagnosis, fault diagnosis and the well known mineral prospecting expert system called Prospector (Duda, 1980). In this type of problem it can generally be assumed that all the symptoms belong to the same 'patient', whereas in multisensor data fusion there is the problem of finding out which evidence belongs to which patients and indeed how many patients there are present. Also the multisensor data fusion problem is a continuous real-time problem rather than a single-shot diagnosis. It was therefore necessary to find a different kind of framework to support the multisensor data fusion rules from the type commonly used and commercially available for diagnosis problems.

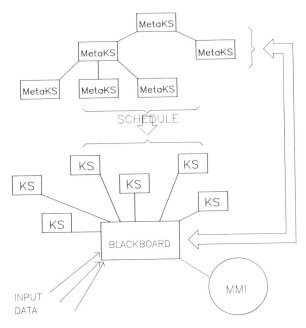

Fig. 7.3.2 *Blackboard system*

We finally settled on a scheme used by Stanford University and SCT to build a signal understanding system for submarine detection called HASP/SIAP (Nii *et al.*, 1982). The framework they used is commonly referred to as a blackboard system because the concept is analogous to a number of human experts who observe and modify the contents of a blackboard according to their individual specialisations. Their overall objective is to build up levels of knowledge on the blackboard, starting with the input data, until a solution to the problem emerges. Figure 7.3.2 illustrates the computer implementation of this scheme: the blackboard is a network of hypotheses and the experts are sets of rules called knowledge sources. There is also a control strategy for scheduling the knowledge sources in a timely manner according to the priorities of the knowledge required; this scheduling may also require expertise and be programmed by rules.

Software to support the blackboard architecture was not available at the start of our project; so we sponsored SPL International to design and build a general purpose blackboard framework which is now called MXA (multiple expert architecture) (Rice, 1984). The aim is to provide a problem-solving framework which can accommodate whatever mix of heuristic rules, statistical principles or mathematical formulae is appropriate.

7.3.3.1 MXA design concepts: MXA has a number of features in common with a Prospector-type expert system framework in that it supports:

(a) A language for expressing rules
(b) A hypothesis structure
(c) An inference engine, i.e. the knowledge source control program
(d) Explanation generation capabilities.

There are, however, significant aspects of the problem that are beyond the capabilities of a Prospector-type expert system. For example:

(a) The run-time hypothesis structure is dynamic.
(b) Hypotheses are needed to represent correlations of evidence as well as inference about meaning.
(c) The number of goals to be assessed is unbounded.
(d) The assessment is continuous.
(e) The system must run in real time with a variable and unpredictable workload.
(f) The data upon which the inferences are being drawn is varying (sometimes rapidly) with time.

One of the design objectives of MXA was to provide these facilities in a manner which is domain independent and which offers the system builder maximum flexibility. MXA is based on a blackboard architecture as described above. It supports a global data area upon which a dynamic structure of hypotheses is accessed and manipulated by knowledge sources (KSs). Scheduling is the process that determines which knowledge source is to be invoked next given the current state of the blackboard and any available input data. In MXA, the scheduling of knowledge sources is also provided by a rule-based approach. The rules for scheduling are held in meta-KSs which, like normal KSs, have access to the blackboard from which they can derive information pertinent to their scheduling decisions (Fig. 7.3.2). Meta-KSs may be invoked by other meta-KSs; thus an arbitrarily complex hierarchy of meta-KSs may be created to implement any desired scheduling strategy. This is obviously a very general approach, but is necessarily so in these early days of blackboard system technology.

7.3.3.2 MXA language: To enable the system builder to construct this domain-specific knowledge base, MXA provides, as one of its principal features, a language for expressing rules (Stammers, 1983). The MXA language is best understood as

An AI approach to data fusion and situation assessment   349

Pascal, upon which it is based, with extensions. The rules, expressed in this language, are converted by the MXA compiler into standard Pascal. A run-time environment is provided by the MXA executive (Fig. 7.3.3). Hypotheses are described by declarations, such that there is a class declaration for each type of hypothesis on the blackboard.

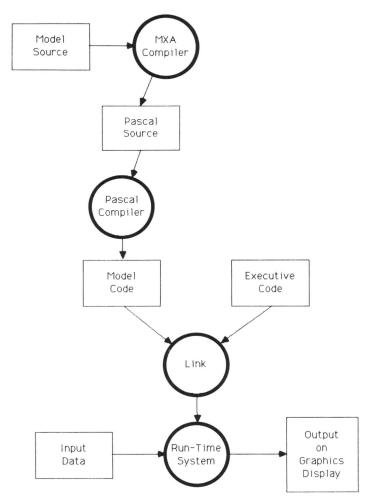

Fig. 7.3.3 *Multiple expert architecture (MXA)*

MXA was designed on the premise that rules would frequently be required to perform the same operation on a set of objects on the blackboard, or on a set of tuples of objects of arbitrary order. An example of a set of tuples of order 2 is:

'all radar-detection/sonar-detection pairs which are thought to correlate'

350 An AI approach to data fusion and situation assessment

For this reason, set selection and set manipulation were included as specific capabilities in the MXA language. Thus an MXA rule consists of three (main) parts:

1 Set description part (to identify the tuples upon which the rule is to operate)
2 Firing condition part
3 Action part

Functions are built into the MXA language to allow creation, deletion and manipulation of hypotheses and support links, and to facilitate navigation and interrogation of the blackboard.

7.3.4 Multisensor data fusion experiment: stage 1 demonstrator

A strategy to solve the multisensor data fusion problem has been evolved over the last two years and implemented using the facilities of MXA. In order to explain the reasoning behind this strategy, it is necessary to describe the assumptions made about the environment in which the system is expected to work, and particularly the characteristics of the input data.

The particular sources of information we are considering are shown in Fig. 7.3.4.

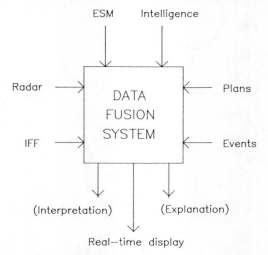

Fig. 7.3.4 *Multisensor data fusion experimental system*

These form a fairly general subset of the many specific sensor, communications and paperwork sources which would be found on a warship. Input data from the sources shown are assumed to have the following characteristics:

Individual vehicles It is assumed that the data will refer to individual vehicles, e.g. a ship, aircraft or helicopter, rather than to groups, e.g. the task force, a wave of aircraft. Groups will be dealt with in our next phase of implementation.

Track data The data will have the characteristics of what are often called 'tracks', i.e. a set of contacts over a period which refer to the same vehicle. This implies in

An AI approach to data fusion and situation assessment   351

some cases (e.g. radar) a considerable amount of preprocessing to associate individual measurements of a target, to reject spurious data, and to calculate course and speed. This is a reasonable assumption for modern sensors; observations from say an intelligence source can be similarly regarded, even if there is only one observation.

No duplication of sensors This means that if a ship is fitted with, for example, more than one radar, their outputs will be combined before entry into the data fusion system. Again this is a reasonable assumption as it is possible to use simple association algorithms to do this reliably without relating to any wider knowledge. It could be done in the data fusion system but would be an additional processing burden.

Timeliness of data For onboard sensors, we assume minimal delay between measurement and corresponding input to the data fusion system. For other data such as intelligence, delays between observations and data arriving are allowed for in the rules.

None of these assumptions is in fact critical to the strategy adopted, but they help to simplify the implementation and to clarify the following explanations.

The output of the current system is a real-time display of vehicle information in plan format similar to current command and control consoles. The other two outputs, interpretation and explanation, are shown bracketed in Fig. 7.3.4 as the knowledge sources to generate these are not yet written. They are the subject of our next phase of work.

The experimental set-up is shown in Fig. 7.3.5. It consists of:

(a) The MXA real-time inferencing and support structure.
(b) A knowledge base incorporating algorithms and expert rules for data fusion and referred to as the data fusion model.
(c) A colour graphics man—machine interface to display the compiled picture.
(d) A sensor data generation system (SDGS) which takes a simulated scenario description and generates from it synthetic sensor and other input data to test and exercise the data fusion system.
(e) The situation assessment knowledge base (under development), which will assist the operator in generating an informed assessment of what the compiled picture means in tactical terms.
(f) A facility for taking real data recorded at sea and presenting this to the experimental system. This is important not only to introduce an element of realism, but also to clarify the details of the problems encountered and to provide a mechanism for comparative assessment between the performance of the experimental system and that of the real operations. Our experimental radar automatic track extraction system (RATES; Shepherd *et al.*, 1982), used as a preprocessor for live or recorded radar video, forms a part of this facility.

Figure 7.3.6 illustrates some of the input data as a plan view of the world:

(a) Radar tracks are shown as lines from first detection to present position.

352 An AI approach to data fusion and situation assessment

(b) ESM contacts are shown as bearing lines along which electronic emissions have been detected, giving approximate direction but no idea of range.
(c) The sectors represent areas in which our own units have been ordered to operate or approximate positions of units reported as intelligence.

Figure 7.3.6 illustrates another important feature: radar tracks exhibit breaks when

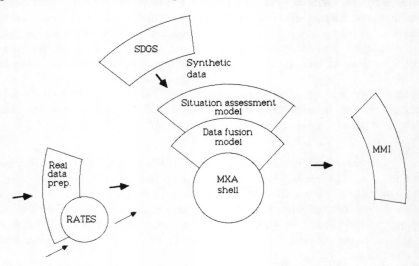

Fig. 7.3.5 *Experimental system*

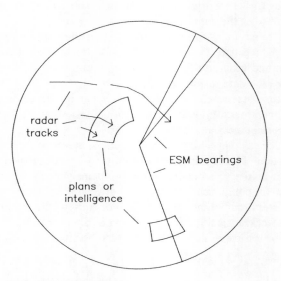

Fig. 7.3.6 *Example source data: plan view*

An AI approach to data fusion and situation assessment

the target goes out of range or into clutter, or the set is jammed. These breaks need to be repaired, which requires correlation in time of data from the same sensor as well as correlation between sensors.

This scenario is, of course, much simplified. In reality there could be many hundreds of track segments to deal with over say a one-hour period, and a very confusing picture can result. Hence it is important to tie together all data belonging to each real vehicle so that the picture resembles the real world as closely as possible.

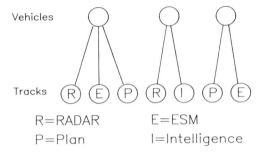

Fig. 7.3.7 *Ideal representation of data fusion. A track is set of data over time from any sensor about a single vehicle*

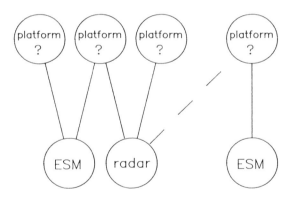

Fig. 7.3.8 *Correlation ambiguity*

Figure 7.3.7 represents what we are ideally seeking to achieve: the aggregation of information from tracks on the same vehicle from different sensors (using 'track' to mean a set of data over time from any sensor about a single vehicle). However, the problem of correlation ambiguities is illustrated in Fig. 7.3.8. Even if the radar and ESM contacts on the left of the figure are deemed by the correlation rules to be capable of correlation, there are still two possibilities:

(a) Either they are the same object – denoted by the middle platform hypothesis
(b) Or they are two different objects – denoted by the other two platform hypotheses.

354   An AI approach to data fusion and situation assessment

A single object detected by n sensors will generate $2^n - 1$ hypotheses. As each sensor detects not one but many objects, there is also further ambiguity as to which contacts go together, as illustrated by the ESM contact on the right of Fig. 7.3.8. The inherent inaccuracy of most of the sensor data implies loose correlation rules leading to large numbers of such ambiguities. The need to process several new contacts per second leads to unmanageable combinatorial difficulties with such an approach.

But there is a further problem, illustrated in Fig. 7.3.9. Having generated all possible hypotheses to explain the data, it is necessary to decide which ones to output as being the most likely. We attempted to do this by generating every possible consistent output set and then scoring these in some way. For example, taking two radar tracks either one of which could correlate with an ESM track, there are three possible self-consistent output sets or 'views' of the situation, these are:

Left-hand correlation is valid.
Right-hand correlation is valid.
Neither correlation is valid.

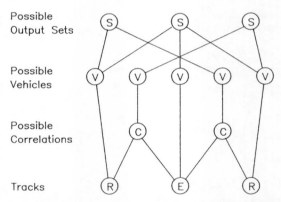

Fig. 7.3.9   *Correlation ambiguity example. Several sensors producing many tracks over a period of time ⇒ combinational explosion of correlations and output sets*

This approach is attractive since all consistent sets are represented and, given a satisfactory scoring system, the current best view is simply the view with the highest score. Unfortunately the enormous number of views resulting from any realistic scenario renders the approach impractical. For example, even for a single object detected by n sensors, the number of logically consistent views of the situation is given by the so-called Bell number $B(n)$, where:

$$B(n+1) = \sum_{i=0}^{n} \binom{n}{i} B(n-i) \qquad B(0) = 1$$

When $n = 10$, for instance, $B(10) = 116\,000$.

An AI approach to data fusion and situation assessment 355

We therefore adopted a less rigorous but more practical approach (see Fig. 7.3.10). This considers only pairwise correlations and involves three distinct rule-driven steps:

1. Assume all new contacts are separate and therefore each new contact implies a new vehicle.
2. Apply rules which create the possible pairwise correlations between each new track and existing tracks in the system. These correlations must be periodically checked to make sure they are still valid; those that fail the check are deleted.
3. Apply rules to confirm strong correlations and to deny others. Where alternatives are of similar strengths, wait for further evidence.

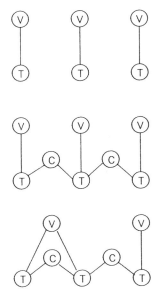

Fig. 7.3.10 *Current solution to data fusion problem: blackboard representation. See text for steps 1–3*

An example of this third type of rule is as follows. To confirm a correlation:

(a) Its likelihood must:

Exceed some absolute threshold; and
Significantly exceed other possibilities, i.e. be relatively unambiguous.

(b) The correlation must be part of an allowed set (see Fig. 7.3.11). For example, if A tentatively correlates with B, and B with C, these can only be confirmed if C also correlates with A. In other words, all tracks supporting the same vehicle must mutually correlate on a pairwise basis.

In addition to the rules for data fusion, the demonstrator incorporates graphics

356 An AI approach to data fusion and situation assessment

software to drive real-time pictorial and text displays. This graphics software allows limited interaction with the model at run-time including:

1. Selection of plan-display range scale
2. Selection of any combination of radar, ESM and plan input data or the correlated output to be displayed
3. Selection of radar track history to be displayed
4. Selection of text data to be displayed on any one track or vehicle
5. An expanded portion of the plan display in a separate window.

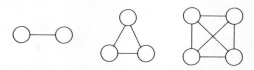

Fig. 7.3.11 *Allowed sets in correlation confirmation: see text for rules*

7.3.5 Stage 1 demonstrator: results

Figures 7.3.12 to 7.3.16 show the output from the stage 1 demonstrator at various points during a run with a test scenario. Figures 7.3.12 to 7.3.14 show examples of the input data for radar, ESM and plans respectively. It can be seen that the radar tracks have been allocated tentative vehicle types with appropriate symbols; this is based simply on rules that consider the maximum and minimum observed track velocity. The ESM bearing-only tracks show vehicle type and hostility (the original uses colour); it is assumed that these parameters have been derived from the emitter information. The plan data is a rather simple form of screen plan for a small group of ships. It is unrealistic in having two sectors overlapping.

Figure 7.3.15 shows the correlated picture. It can be seen that many of the ESM bearings have disappeared because they have been correlated with other data. In the case of a radar—ESM correlation, the ESM bearing is shown as a dashed line drawn radially from the track. For ESM—plan correlations the ESM bearing is shown as a dashed line across the plan (also shown dashed) to indicate that the vehicle is most likely to be somewhere along the line within the plan.

The two ships west-north-west of own ship are held on radar and there are also two friendly ship ESM bearings nearby but they have not correlated. This is a case where the correlations are ambiguous and the model cannot decide which of the ESM bearings goes with which radar track given the accuracy of the data. It is a design principle of the model that situations that are ambiguous are left uncorrelated until the ambiguity is adequately resolved, for example by the ships separating sufficiently in bearing. It may be noted that a similar situation exists over the two plans which also indicate the presence of two ships.

A number of steps could be taken to handle ambiguous situations like this, for example:

An AI approach to data fusion and situation assessment   357

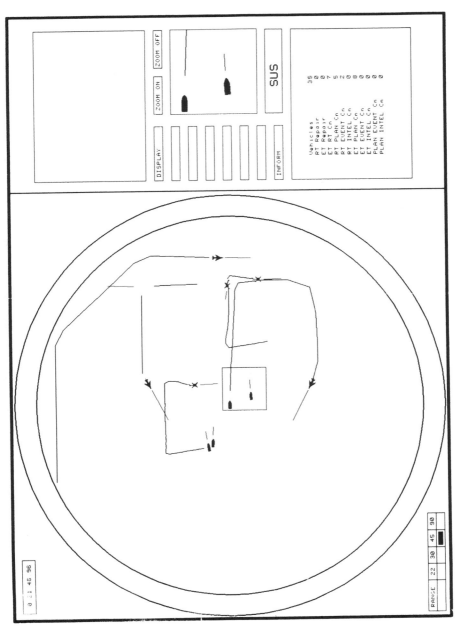

Fig. 7.3.12   Radar input data

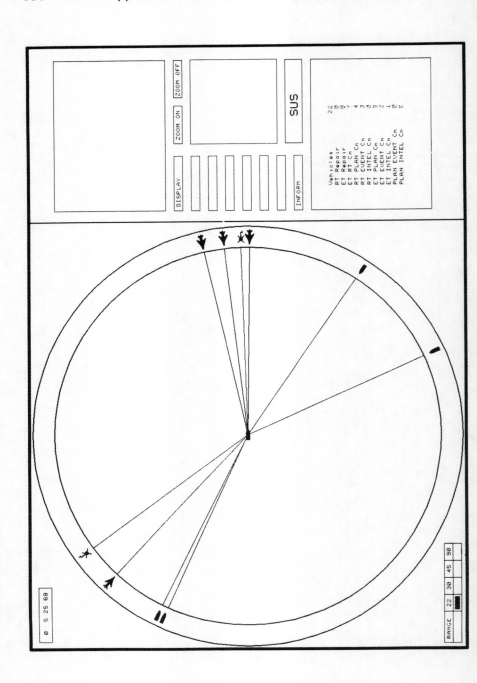

An AI approach to data fusion and situation assessment

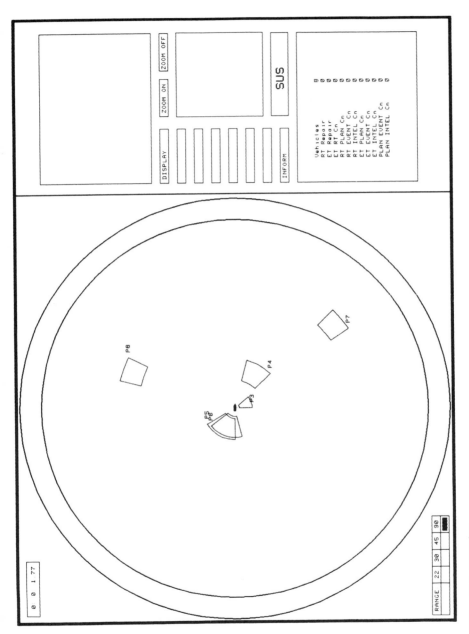

Fig. 7.3.14 Plain input data

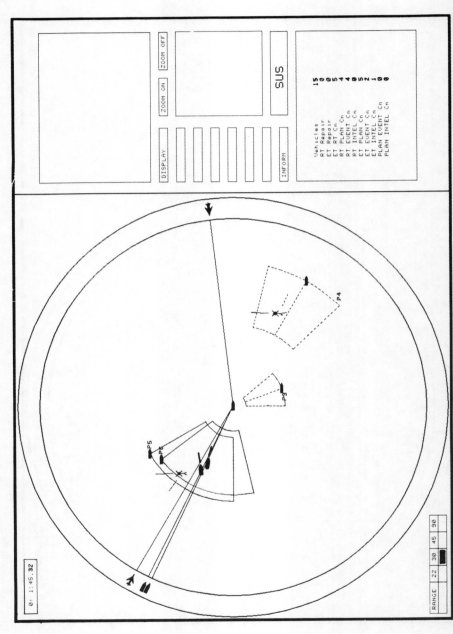

Fig. 7.3.15 Correlated picture

An AI approach to data fusion and situation assessment

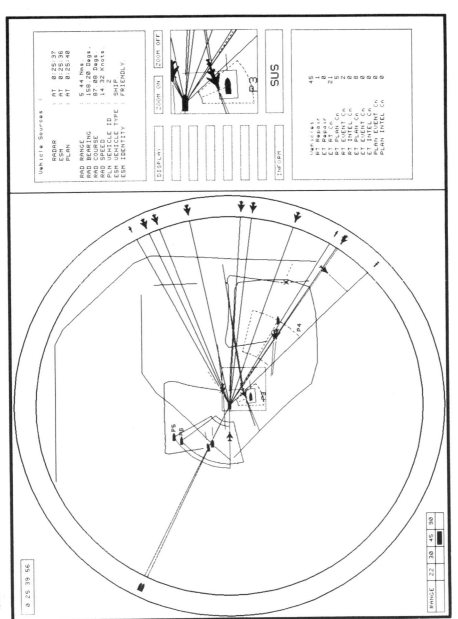

Fig. 7.3.16 Correlated picture – later in the scenario

1 The operator could be informed and asked if he had any further information.
2 The rules could be examined and refined to lessen the chance of ambiguities, perhaps using other information.
3 A strategy could be adopted whereby the situation was left to resolve itself when further information is received or when the situation changes (e.g. two close contacts may later separate). If it were of immediate importance to make a judgement, the best available choice would be made.
4 Instead of attempting to correlate individual tracks it might be better to correlate the data after grouping the tracks together.

In fact, most of these steps could be adopted to improve the performance of the data fusion system.

Only one hostile aircraft detected on ESM can be seen at this stage, but later in the scenario (depicted in Fig. 7.3.16) an attack formation arrives and fires its missiles. This picture gives an idea of the complexity of even a simple scenario when viewed in individual vehicle terms.

Figure 7.3.16 also shows the text display in operation. It shows the sources of data for the vehicle indicated by the small rectangle, in this case a ship. Radar, ESM and a plan sector have been correlated to produce this vehicle report. By a set of selection rules the model has derived the best set of parameters for the vehicle, i.e. position, course and speed from radar, vehicle type and hostility from ESM, and the name of the vehicle from the plan, which in this case was a screen plan of the task group.

In addition to the functionality issues discussed so far, the question of run-time performance is of paramount importance. The run-time performance of the experimental system is adequate only for simplistic scenarios such as the one described. The question of whether a general purpose framework can be optimised sufficiently to give real-time performance or whether an operational system would have to be hand crafted for efficiency or use special hardware is a subject of further research. Section 7.3.9 discusses possible hardware architectures which might provide both the representational power and the throughput needed.

7.3.6 Multiplatform data fusion

In a typical ASW scenario, there will be a number of surface ships with their air assets and perhaps submarines operating as a group. Assuming that data link and voice communications are available, it will be possible to collect sensor data from other platforms to form a tactical picture over a much wider area than that perceived directly by own ship. There is also the possibility of combining passive forms of data such as ESM or passive sonar to produce much more accurate track data, for example, by triangulation of bearing lines.

Stage 2 of this project will include enhancement of the data fusion system to demonstrate a multiplatform capability. What follows here is a discussion of how the implementation might be organised and the factors to be considered.

In addition to the sensor characteristics, multiplatform data fusion will also depend on the navigation accuracy of the sensor platforms and the characteristics of the data links or voice channels.

An AI approach to data fusion and situation assessment   363

7.3.6.1 Navigation accuracy: Accurate sensor platform position is essential for matching the spatial measurements made by sensors on different platforms. Either accurate navigation can be assumed, e.g. using Navstar GPS, or some form of accurate relative position estimation will have to be done for each platform prior to using its sensor data; the latter could be achieved for some situations by using radar measurements of the platforms themselves or commonly held targets.

7.3.6.2 Data links: Because of the large volume of data produced by active and passive sensors, data links are required to support multiplatform data fusion. Even with data links, however, the amount of data that can be transferred over a period (bandwidth) will be limited.

Given such limitations, it is unlikely that the data received through the data link will be at the same rate as an equivalent onboard sensor. Also, only the most important parameters on each track will be transmitted in order to save data bits. As a result of these factors, the status of data link received tracks will not be quite the same as own-ship generated tracks and this may affect the organisation of the multiplatform data fusion.

7.3.6.3 Multiplatform data fusion organisation: There are two obvious ways of including the various sensors in a multiplatform data fusion system. One way would be to take each piece of data and attempt to correlate it with all the other sensor data of both the same type and of different types; this would give a flat or non-hierarchical organisation. Another way would be to combine the sets of sensor data in some particular order, such as similar sensors first and then dissimilar sensors; this produces a hierarchical organisation. It seems that there are several advantages to the hierarchical organisation, not least the practical advantage of separating out some stages of processing that might be processed in parallel.

Figure 7.3.17 illustrates a possible hierarchy. It must be borne in mind that, in different situations, different sets of sensors will be available so that the structure must not be dependent on any one type for its useful operation. Of course, lack of some type of sensor will degrade the performance of the data fusion system but it should not render it totally useless.

The principles adopted in the suggested organisation are:

1. Like sensors on own ship should be combined first; this particularly applies to radar and its closely associated IFF system.
2. Radar pictures from different platforms should be combined next to establish a force-wide radar picture and at the same time to estimate any sensor platform position errors if necessary.
3. Like passive sensors, such as ESM and passive sonars should be combined next in order to reduce position uncertainty of contacts as far as possible.
4. Finally, the combined pictures from each type of sensor are fused together in the type of system developed in the stage 1 demonstrator.

The principle of combining like sensors first seems the best choice because like sensors have very similar parameters to compare whereas dissimilar sensors, in

364 An AI approach to data fusion and situation assessment

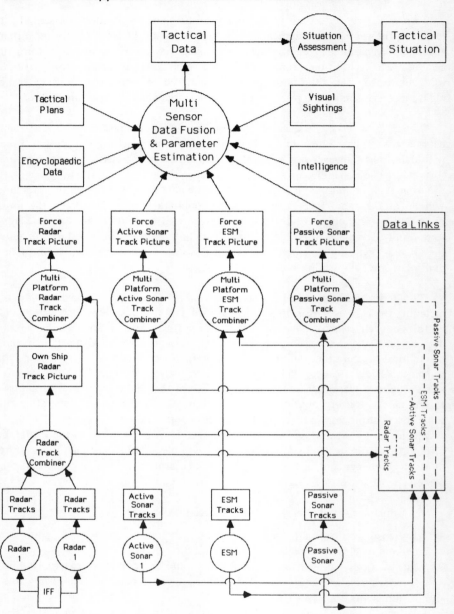

Fig. 7.3.17 *Multiplatform data fusion process*

general, only have a few parameters in common. This should mean that fewer ambiguities arise in the system as a whole.

7.3.7 Parameter estimation — or combination of evidence

This section examines the problem of deriving the most likely set of parameters for each individual vehicle or group of vehicles from the sensor and other data collected.

The result of the data fusion stage is a collection of evidence for each vehicle or group of vehicles. The set of evidence might include:

1. Position estimates and track history
2. Velocity estimates — course and speed, maximum and minimum observed speed, observed manoeuvring
3. IFF responses
4. ESM emitter characteristics
5. Intelligence reports
6. Plan details for friendly platforms
7. Visual sightings.

Given a set of evidence for a particular object, the first stage of interpretation is to determine the most likely values for each parameter of interest.

The parameters that are required include:

1. Current position
2. Current velocity
3. Platform type (ship, helicopter, aircraft, submarine)
4. Hostility (friend, neutral, hostile)
5. Identity — name or reference code
6. Mission — purpose and intended actions
7. Weapon load
8. Current mode of operation, for example surveillance, targeting, weapon delivery etc.

Some parameters may be known exactly, for example the name of a friendly ship. Others will require estimation, such as position, course and speed, or more complex inference, for example the likely mission of a hostile aircraft. The objective must be to build up a story for each object which explains where it came from, what and where it is and what it is likely to do in the future.

Some evidence will be grouped together either because the sensors cannot resolve the individuals or through the action of clustering rules designed to recognise groups with common properties, e.g. velocity, hostility. Evidence about a group of objects can be treated in a similar way to that about individuals, but many of the parameters will be collective and there will be the additional parameters of group size and perhaps shape.

This is the last stage in the data fusion process; further stages of inference come under the heading of situation assessment.

7.3.8 Situation assessment

The result of data fusion and parameter estimation is a likely set of conclusions for each perceived vehicle or, in some cases, group of vehicles. Situation assessment is the process of interpreting these conclusions to form a consistent view of the tactical position in terms of the effectiveness of resources deployed under the prevailing conditions. As well as an up-to-date statement of the deployment of resources on both sides, situation assessment will also include estimates of sensor and weapon coverage for friendly and hostile units.

Environmental factors such as weather, geographic and oceanographic conditions will affect these assessments. Also, the political situation and rules of engagement will colour the judgments made.

This section discusses ideas for features that are currently being included in the situation assessment knowledge base.

7.3.8.1 Possible tactical situation assessments: The following list gives examples of aspects of a tactical situation which could be assessed:

1 Threats
2 Engagements
3 Weapon system geometries
4 Weapon states
5 Rules of engagement (ROE) and/or exclusion zone (EZ) infringements
6 Sensor coverage
7 Weapon coverage
8 Adherence to plans
9 Each side's perception of the other
10 Defence screen.

7.3.8.2 Threat assessment: Threats can be classified as direct, indirect and potential. Direct threats are observed hostile units in the act of attacking. Indirect threats are estimates of the attacks which observed hostile units are likely to make based on intelligence about the weapons they carry. Potential threats are an assessment of threats which might be encountered based on intelligence only.

Assessment of both direct and indirect threats would be useful function for an advisory system to carry out because a large amount of knowledge is required and many feasibility calculations have to be performed in a short time scale. Potential threats can be assessed on a much longer time scale and there is less need for machine assistance.

Rules for threat assessment will merge the currently perceived deployment of enemy units with encyclopaedic intelligence data on enemy weapon systems and tactics to generate possible attack scenarios. Parameters to be estimated include: type of attack, how soon it could be mounted, number of units involved, likely targets etc.

7.3.8.3 Assessment of engagements: The outcome of any engagement is of great importance to a military commander because it may determine the next course of action. During a battle, such an assessment must be produced very rapidly to be of any use. For example, to conserve missiles an air defence ship needs to know as soon as all the attacking missiles are destroyed or otherwise rendered harmless. Unfortunately the information on which this type of assessment is based is difficult to obtain in the required time scale, although some assessment of engagement might be possible by considering 'loss of contact' reports.

7.3.8.4 Weapon system geometries: This assessment would consider possible conflicts arising from the positioning of various weapon systems in relation to one another. It would periodically examine the weapon arcs and produce alerts if conflicts or dangerous situations which might cause accidents were discovered. Again there are difficulties with obtaining the accurate information to do this assessment, but it may be possible to achieve some results by restricting the assessment to activities on own ship or checks between, say, friendly aircraft movements and shipboard anti-air missile systems.

7.3.8.5 Weapon states: The state of weapon systems including operating condition and stocks of weapons is useful information at ship or group command level. An assessment of this kind is derived from logistic support and signal traffic between units.

7.3.8.6 Rules of engagement and/or exclusion zone infringements: When to respond to a threat is a very important decision in which rules of engagement (ROE) and exclusion zones (EZ) play a crucial part. The ROE and EZs are precise but the perception of the tactical situation and in particular the present threat are imprecise.

Well informed judgments must be made, often rapidly, which could be assisted by explicit application of expert knowledge to alert and explain the situation in relation to ROE.

7.3.8.7 Sensor and weapon coverage: Sensor and weapon coverage estimation would be useful for modifying defence screen and emission control plans. For example, it could assist the problem of deciding which radars, if any, to use in a task group to give the required cover or where to place a ship with surface-to-air missiles to protect other units. For passive sensors the assessment could help to modify unit dispositions for better ranging of threats from given directions.

Calculation of sensor and weapon coverage is certainly not easy because many factors have to be taken into account, some of which can only be estimated. The results might also be difficult to represent for a reasonable set of cases. Since sensor coverage is also an important input to the data fusion process, the dangers of cyclic reasoning need to be avoided.

7.3.8.8 Adherence to plans: The idea of this assessment is to constantly monitor whether the plans made are being followed and, if not, to alert the command to the discrepancies.

For groups of friendly units there will generally be a plan against which the sensed picture can be assessed. One function will therefore be to judge how well the plan is being followed and to what extent the plan can be confirmed as being the actual situation. Any discrepancies between the observed situation and the planned one can form the basis of a set of alerts to the user.

Groups of neutrals can also be assessed for adherence to expected routes such as air lanes or shipping lanes. Uncharacteristic behaviour such as close spacing between apparently commercial aircraft or deviations from expected routes could also be used to alert the command or invoke a reassessment of identity and mission.

Hostile groups, it must be assumed, are probably acting in consort to achieve their goals and must therefore be assessed as a whole. Rules are required here to attempt to deduce what their most likely plan is and to judge when changes in behaviour occur, perhaps indicating a change of mode from just surveillance to intrusion, provocation or attack, for example.

7.3.8.9 Each side's perception of the other: This assessment is an attempt to estimate the extent of our knowledge of the enemy forces and the enemy's knowledge of our forces. The former will be based on the volume swept by our sensors against a background of intelligence information about hostiles known or likely to be operating in the area. The latter will be a judgment based on known enemy surveillance systems either purely from intelligence data or a combination of intelligence and recently observed enemy surveillance.

Unless enemy surveillance units have been or are being observed then assessment of what the enemy knows about own force must be based on intelligence information. Where shadowing aircraft, ships or submarines have been detected then some estimate of knowledge based on observed active sensors and assumed passive sensors could be made.

The outcome of this assessment might assist judgments of whether to use active sensors which might give away more information to the enemy than provide information to own forces.

7.3.8.10 Defence screen: One of the major objectives of any task group must be to set up an effective defence screen against a number of different types of threat. The objective of the defence screen assessment would be to monitor the effectiveness of all aspects of this screen under the constantly changing tactical environment. Its aim would be to alert the command as soon as possible to any weak areas such as might be caused by units changing station, equipment failures and losses during action.

Defence screen assessment is required to make a useful statement regarding the effectiveness of the defence screen against the possible attack scenarios identified in the threat assessment. It should attempt to define likely targets for each possible attack and point out any weak points in the screen. Calculations of weapon system

An AI approach to data fusion and situation assessment

coverage and effectiveness are required to assess the vulnerability of each unit to give an overall statement of survivability.

7.3.9 Artificial intelligence machines for multisensor data fusion

This section is a discussion of the specific requirements for a machine to support an intelligent command and control system for a warship in real time.

Figure 7.3.18 illustrates the concept of a command and control system, based on intelligent or expert systems, using the principles outlined in the previous sections of this paper. From all available sources of information the system provides not only a recognised tactical picture but also advice to higher levels of command and data to weapon systems. The advice and directions are generated by referring to a dynamic model of the world which is built up from the inputs and draws on tactical expertise and encyclopaedic data. Full access to the reasoning will be available to the users so that the advice given may be explained to whatever detail required in the manner of a consultation. The system is split into three layers of processing for convenience, although the boundaries are somewhat fuzzy.

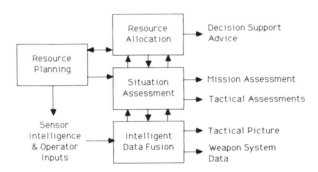

Fig. 7.3.18 *Intelligent command and control system*

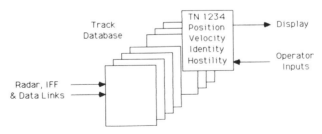

Fig. 7.3.19 *Current systems*

However the implementation of an AI-based command and control system is a good deal more complex than that of existing command and control systems and the processing is far more demanding. To understand the scale of the problem, first consider what current systems provide (Fig. 7.3.19). Current systems are

370   An AI approach to data fusion and situation assessment

centred around a simple database or track table which contains a track data record for each object detected by sensors, received over data links or inserted manually. Some of the parameters contained in each record are filled in automatically from sensor measurements, but many are completed manually by operators assigned to picture compilation tasks. Of course the software also provides a number of display, sensor, communications and weapon control functions.

The characteristics of this database are that it only represents the current set of tracks. There is little or no history, so that patterns of behaviour cannot be stored or recognised and out-of-date data such as intelligence cannot be matched with existing data. There is no allowance for uncertainty either in the correctness of the input data or in the way data from different sources is combined. It is left to the operators to ensure correctness. Owing to the small number of operators available, the uncertainty of the input data and the time scales, the database has been shown to contain many errors in practice. Current systems also rely heavily on having a good radar picture and provide little support for picture compilation using passive sensors.

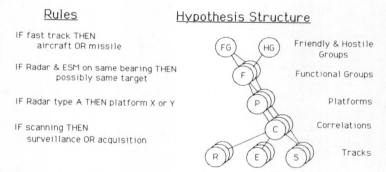

Fig. 7.3.20   *Expert system based command system*

The database represents only one level of information — that of tracks. It cannot therefore support the information required for situation assessment. To support a more intelligent command and control system a more complex multi-level data representation is required, such as the simplified illustrations shown in Fig. 7.3.20.

The objective is to assemble a body of knowledge in the form of rules which infer the information and advice that the system is required to produce. Along with the rules, simple examples of which are illustrated, goes a hierarchical data structure which represents possible conclusions from the application of the rules.

The system must create a more elaborate model of the world than a simple list of tracks in order to allow for the many possible ways of assembling and interpreting the data.

Several levels of hypotheses are used to represent intermediate stages in arriving at a model in terms of platforms and platform groupings which represent elements

An AI approach to data fusion and situation assessment

of the tactical world. Some of these hypotheses will contain track data similar to that in existing track tables, but there will also be:

(a) Hypotheses to represent correlations between tracks from different sensors
(b) Platforms with estimated parameters from the accumulated evidence
(c) Groups of functionally related platforms.

To be able to interpret the current state and predict future possible states, a story or script is required for each platform. To produce a likely story for each platform it is necessary to represent its entire history up to the present. Any new information can then be matched to this model, whether the data is timely or somewhat late arriving.

Given a representation of the world over the last few hours or days, it is possible to apply pattern matching techniques to discover what is happening even though the input data is fragmented. In the same way a human operator would use his memory of events to interpret the current situation. Such a model has considerable storage and processing implications.

To estimate accurately the size and power of a machine to support such a system is difficult at this stage because we have not yet implemented a sufficient set of rules. However, it is fairly obvious that if the system must retain data over a long period, represent several levels of inferences and support many alternative interpretations of the data, then the scale must be much greater than existing machines.

Given force-wide sensing systems plus intelligence there is potentially a huge volume of data to be handled. This can probably be cut down to about the level given below by reducing the amount of simple position update information of individual tracks. The following are estimates for the input rate and storage required:

Processing   Input data records per second: 100
Storage   Number of hypothesis records: 50 000. Each: 100 byte. Total: 5 Mbyte

The majority of input records will still be updates to existing tracks but a few per second will be new contacts — not necessarily new players in the volume of interest, but new fragments of tracks created by sensor coverage limitations. Each new input requires matching with the existing world model and may generate a new set of possible conclusions or at least require existing conclusions to be checked for validity.

The above storage estimate is very approximate since the number of hypotheses depends so much on the scenario, the sensors employed and to some extent the strategy used to create the model. If the tactical world of interest actually contained say 100 platforms, the number of sensor reports over one hour might run into several thousand. The possible correlations between the data, given typical uncertainties, could increase the number of hypotheses by a factor of 2 or more.

The above estimate of 5 megabytes of storage does not, at first sight, seem unmanageable since most computers these days can easily address memories greater than this and cheap high-density memory chips are available. The main problem lies in the processing required to update the world model from the constant stream of input data and carry out the pattern matching required to produce an up-to-date tactical interpretation of the model.

Conferring on a machine some of the information retention and pattern matching abilities of a human seems to require a 100-fold increase in storage and processing power over existing machines. So is there a suitable machine available or in research?

Figure 7.3.21 shows computing systems against two types of power — processing power or throughput, and representational power or ability to represent the problem conveniently and efficiently. Software can of course transform the representational power of any general purpose machine, but usually at the expense of processing power.

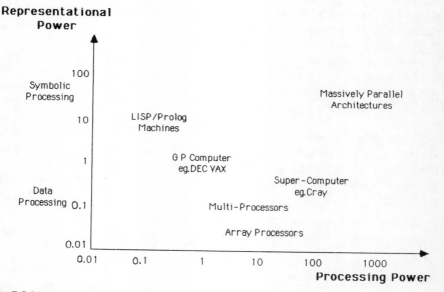

Fig. 7.3.21 *Computer architectures for data fusion*

The scales are set relative to an ordinary machine such as a DEC VAX© which is similar to the computers on board ships now. For AI purposes, representational power is usually most important and an AI machine to most people means a LISP machine. Because most AI work uses the LISP language and conventional machines do not run it efficiently, special machines have been developed to recover some of the lost performance. Machines for Prolog are also being developed for similar reasons. These machines improve the processing power while retaining the excellent programming environment of LISP, but are still sluggish compared with conventional machines and unsuitable for real-time systems.

An AI approach to data fusion and situation assessment 373

To increase processing power the obvious direction is some form of parallelism. Super-computers like Cray© attempt to hide this parallelism from the user as much as possible but rely to some extent on being able to process vectors in parallel. This works well for problems with simply structured data but may not work well with complex hypothesis structures. Also they largely support algorithmic processing rather than symbolic processing, and are very expensive.

Multiprocessors and array processors are much cheaper routes to increased processing power but put more of a burden on the programmer to organise them effectively. Array processors can be used effectively where there are few dimensions or types of variables and a simple algorithm that needs to be performed very rapidly, such as in signal processing applications. It is difficult to see how an AI application could be mapped on to an array processor because there are many variables to consider and the knowledge forms a tangled hierarchy rather than a fixed two- or three-dimensional matrix.

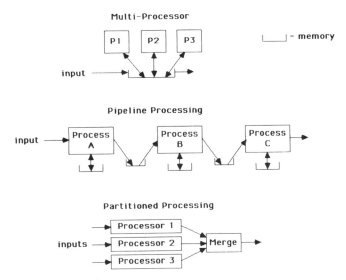

Fig. 7.3.22 *Parallel processing*

The obvious way to increase processing power beyond what is economically feasible in one processor is to employ several in parallel. However, this can only be an advantage in particular circumstances. Figure 7.3.22 shows three forms of parallelism. At the top is a simple multiprocessor with shared memory. The performance of this configuration is ultimately limited by the memory access time. The cheap memory used in most computers is limited by the available technology, which has increased in density but not much in speed. If you want very fast memory you have to use special technology, which is much more costly. The other problem with a multiprocessor is that it assumes the task can be divided into parallel processes, which is not easy with rule-based systems.

Pipelined processing is another alternative. It consists of a processing chain passing data between stages. This works well if the task can be divided into roughly equal serial stages and not too much data has to pass between the stages; again this may not suit real-time knowledge based systems.

Partitioned processing attempts to divide both processing and associated data into parallel streams which presumably have to be merged eventually. This might be feasible to some extent for the naval command system problem by partitioning on environment, e.g. air, surface and subsurface. Such divisions only work if there is little access required across the boundaries — which may not be the case.

In summary, conventional parallel processors only give advantages under particular conditions. It is not easy to see how they might support an intelligent command system in general, but some configuration might cope with a particular case.

A more likely future solution may be a form of what are generally termed 'massively parallel architectures', one form of which is the NETL machine (Fahlman, 1979). NETL stands for the network language which goes with a form of machine developed in the USA first at MIT and later at CMU. It takes a conventional AI knowledge representation scheme, i.e. a semantic network, and maps it on to a network of processors in which each concept and relation has its own processor. This gives it some human-like knowledge handling capabilities which could be used by other AI programs.

One particular ability is property inheritance, that is, the ability to connect all the relevant information to a given concept without incurring exponentially increasing search times. Another feature is its ability to intersect complex sets to find remote connections between pieces of information or best fit matches of complex patterns against a very large body of knowledge. In principle these are just the sort of operations required for data fusion and situation assessment.

The NETL machine uses a semantic network to represent real-world knowledge. Figure 7.3.23 illustrates a tiny example of a semantic network. Each concept is represented by a labelled node. Relations are shown as links connecting nodes. The unlabelled near-vertical links represent the 'is_a' relation, for example, 'a type 42 is_a warship', 'warship is_a ship' etc. Most of the nodes shown represent types of things but at the bottom of the hierarchy are nodes representing individuals such as HMS *Manchester*. Other relationships are shown as labelled links.

This representation is very general. In principle it would be possible to represent all the knowledge about the military world, both encyclopaedic and real-time, using such a scheme, although it would require a very large number of nodes and links. The problem with doing this on a conventional computer is that searching through the network is very time consuming, for instance when trying to find the connection between a set of remotely connected clues. AI programs that implement semantic networks have to incorporate special domain-specific rules to limit the search, and even then have slow and unpredictable response times. However, the NETL machine is able to do such things very quickly by using a simple processor for each node and link, making it possible to do the search with total parallelism. The search time is only dependent on the depth of the knowledge hierarchy and the number of clues, not the size of the knowledge bases.

The status of the NETL machine is that simulations have been built and tested with static examples, although it is some way from being applied to a real-time problem. It should be possible to build it using VLSI; hopefully many processors could be fitted on one chip so that a machine with, say, a million processors could be built. There are, however, serious problems with providing the links.

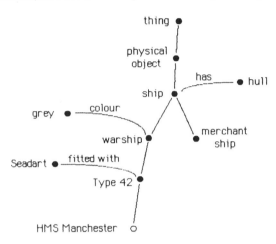

Fig. 7.3.23 *NETL machine — semantic network*

A similar though more general architecture called the 'connection machine' (Hillis, 1985) is under construction in the USA and some groups in the UK are looking at related schemes. NETL is not the only type of massively parallel architecture but it looks the most promising type for this application at the moment.

7.3.10 Conclusions and future work

This paper has described research aimed at applying expert system techniques to the problem of generating a tactical picture on board a warship and has made specific reference to our completed stage 1 data fusion demonstrator.

But data fusion is only the first stage in a sequence of inferencing to generate the higher levels of knowledge on which decisions can be taken. This process — of taking the detailed vehicle information and generating high-level descriptions appropriate for command and control — we call 'situation assessment'. Adding this to our data fusion model is the next major phase of our research, now under way.

In principle it would seem fairly straightforward to add the necessary rules to generate the required knowledge levels, but we can anticipate some implementation difficulties, mainly because of problem scale. For instance, in order to support the high-level rules for situation assessment, we will need to incorporate a great deal of encyclopaedic information about geography, military equipment details and tactics, which could be the subject of a very large database, and yet must be accessible to the rules efficiently in order to operate in the time scales required.

There are also a number of reasons why it may not be possible to separate the highly dynamic multisensor data fusion process from the situation assessment process. First, it is just the hypotheses that are generated by the multisensor data fusion which form the basis of situation assessment. Secondly, there may well be feedback from the situation assessment process in the form of context setting which will affect the data fusion rules. Thirdly, the command may be most interested in the high-level situation description, but he may also require explanations which will call up the details from the knowledge base. So it would seem that an intelligent knowledge-based system to support the higher levels of information management in command and control will require a very large knowledge base and a good deal of processing power.

Before a system could be procured for operational use, there are several important issues to be resolved. These are briefly mentioned below.

Identifying the user   Having demonstrated some capability in the laboratory, it will be necessary to identify the naval users at the level of command to which the system is most applicable. These users can help to specify the forms of output they require and should be able to add to the knowledge base from their expertise.

Real-time implementation   The hardware and software used so far to build the demonstrator are not nearly powerful or efficient enough to support an operational version. Whether available computers can be used with much more efficient software or special processing hardware is required needs to be determined. A careful study must be carried out to find the functions that need to be performed, the data sets involved and the required time scales.

Testing and acceptance   It is not clear how an expert system for military command and control can be tested thoroughly because of the infinite number of possible scenarios that could occur. Elaborate simulations may go some way to proving the potential performance, but to gain user acceptance the system must perform in a very explicit fashion rather than as a 'black box'. Explanation of reasoning is an essential feature for gaining user acceptance.

Scheduling   Given that there is a vast body of data available and only limited processing resources, scheduling is important to make sure that the system concentrates on the most immediate and important tasks. The problems are to represent the relative importance of all the possible tasks across all the different circumstances and to avoid creating overheads by the scheduling itself.

Inferencing and uncertainty   A great deal of research effort into expert systems is being devoted to inferencing systems, but no system has yet been devised that is generally applicable and always gives consistent results. The data fusion model uses the system of likelihoods provided by MXA to represent uncertainty, but only in a crude way. This subject needs to be properly addressed so that a particular method can be recommended.

Explanations   Explanations of reasoning are a feature of expert systems but, because few expert systems have been attempted for real-time problems, there are no well known techniques for implementation. The provision of explanations in time-critical domains requires careful consideration, since the user may not have time to read volumes of text. In some cases it may be more appropriate for the system to explain what it is likely to do before the event rather than during some critical action.

User overrides and modifications   In any system that attempts to make decisions on behalf of its users, the question of whether the user can override such decisions arises. If overrides are not allowed the user may ignore the system altogether if it makes a single mistake. However, if the user can override any decision without question then he may inadvertently upset the entire operation of the system. A possible compromise is the concept of an argumentative system which checks the consistency of a user override and presents the user with any evidence that conflicts with his decision. In the end, however, the system should comply with the user's view if he insists.

Adequacy of the technology   A fundamental question which this research project is attempting to answer is whether the expert system approach is sufficient for supporting one of the key elements of the command and control problem, in this case tactical picture compilation and situation assessment. The problems of incorporating sufficient knowledge to cover all situations and of proving the performance of the system may be too difficult. Of course the system can never be perfect, but as with other uses of computers it may be provably better than existing manual methods.

7.3.11 References

DUDA, R. O. (1980) 'The Prospector system for mineral exploration', SRI International, USA
FAHLMAN, S. E. (1979) *NETL: A System for Representing and Using Real-World Knowledge*, MIT Press, Cambridge, Mass.
HILLIS, W. D. (1985) *The Connection Machine*, MIT Press, Cambridge, Mass.
NII, H. P., FEIGENBAUM, E. A., ANTON, J. J. and ROCKMORE, A. J. (1982) 'Signal-to-symbol transformation: HASP/SIAP case study', *AI Magazine*, no. 3, 23–35
RICE, J. P. (1984) 'MXA – a framework for the development of blackboard systems', Proceedings of the third seminar on applications of MI to defence systems, RSRE, UK
SHEPHERD, A. M., WHITE, I., MILES, J. A. H. (1982) 'RATES: radar automatic track extraction system – a functional description', internal memorandum XCC82003, Admiralty Research Establishment
STAMMERS, R. A. (1983) *MXA Language Manual*, SPL International, UK

Chapter 7.4

Air defence threat assessment

S. Middleton
(Royal Signals and Radar Establishment)

7.4.1 Introduction
As C^3 systems increase in complexity the need for advanced computer processing methods, capable of tackling problems involving judgment and reasoning as well as data processing, becomes ever more apparent. A recent development is the expert or knowledge-based system (KBS). Expert systems are based on techniques developed by researchers in the subfield of computer science known as artificial intelligence (AI) and are now being applied to non-trivial, real-world problems. Such systems seem to offer much promise for aiding the higher-level processes of C^3, and the RSRE has been exploring their potential role in air defence command and control.

This paper describes investigations which have been carried out as part of a research programme into applying knowledge-based system techniques to part of the air defence command and control task. The area focused on is that of air defence threat assessment (TA). This problem is signficantly different from those for which most of the classic expert systems have been built. This has meant that building our experimental, prototype, threat assessment expert system has involved overcoming many design issues and evolving our own KBS framework to support the various types of expert reasoning found in the air defence domain.

Two versions of the threat assessment expert system were developed, and these are described. The first is based on blackboard architecture; the main features of this system, and the reasons for abandoning this approach for the TA problem, are outlined. The second system is based on a new object-oriented language called BLOBS, which combines useful features from both object-oriented and blackboard systems. BLOBS allows a combination of object-oriented, rule-based and procedural programming and provides the diversity of knowledge representation and reasoning mechanisms necessary to tackle the air defence threat assessment problem.

7.4.2 The problem

As its name implies, air defence threat assessment is the task of assessing the threat posed to the area being defended by any objects in the surrounding airspace. Its purpose is to ensure protection of the defended area by providing sufficient accurate and timely information about any potentital threats to allow suitable air defence assets to be allocated against them. Depending upon the situation and the level of threat posed by a target, this allocation response may range from increasing the readiness of aircraft on the ground in order to reduce the response time should the target later need to be intercepted, to ordering that the target be engaged.

It is obviously very important that assessments are as accurate as possible. The accuracy which can be achieved is limited by the quality of information available from which to make the assessments. The principal sources of information are usually various types of sensors such as radar and electronic surveillance measures (ESM). The reports from these are combined to form a tactical air picture which can then be used as the basis for threat assessment. Other sources of information, such as the flight plans of certain friendly and neutral aircraft, IFF transponder

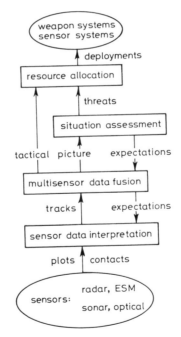

Fig. 7.4.1 *Levels of C^2 tactical decision making*

responses, intelligence reports etc., may also be available and can be used to aid tactical picture compilation and in helping to assess the threat. Figure 7.4.1 shows the levels of tactical decision making for command and control in general and the position of threat assessment, which is a subset of situation assessment.

7.4.2.1 What does an assessment consist of?

Ideally the air defence commander would like to know as much as possible about every object in the airspace. Some of the main features he will be trying to determine are:

Position   This is the most basic information, and is vital. Once an object has been detected it is highly desirable to keep track of its position. If the object stays within radar cover this is straightforward, but when an object appears and disappears it can be very difficult. Even a crude indication of position may be sufficient to allow an air defence aircraft to find the object.

Track identity   Once it has been established that some object is there and some form of track information is available, it is necessary to decide which basic category it falls into. Can the track be classified as friend, foe or neutral?

Track strength   How many objects are represented by a single track? Strength information is important in resource allocation when deciding upon the scale of response.

Object type   It may be possible to gather some information on likely object type; for example, whether it is likely to be a fighter, bomber or missile.

Aircraft characteristics   For example, aircraft performance and possible weapon fits. This will be based on collateral information and background knowledge.

Intentions   In order to assess the threat a track poses it is necessary to predict its likely intentions. This prediction may be based on past experience and/or expectations of certain types of behaviour.

7.4.2.2 Motivation and context for work

It is worth mentioning some of the reasons why we have focused on the problem of threat assessment rather than the lower-level problems of correlation, association and fusion of different types of sensor data:

(a) There is already research in progress into the problems of data fusion, using both knowledge-based and more conventional techniques, e.g. Lakin and Miles (Section 7.3) and Wilson (Section 7.2).
(b) In order for knowledge-based techniques to provide any significant power it is necessary for there to be either an 'expert' from whom to derive rules or a set of training data to allow rules to be derived. Neither of these was available for the data fusion aspects in our case.
(c) It was felt that knowledge-based systems, at the current state of the art, had a greater chance of making a successful contribution to the higher-level functions of command and control where data rates are lower and the comparative inefficiency of knowledge-based programming techniques would not prove prohibitive.

The threat assessment expert system thus works at the tactical air picture level and assumes that tracks are being maintained by some lower-level part of the system.

7.4.3 Analysis of the problem

In this section the threat assessment problem is looked at in terms of an expert systems solution. Features of the TA problem which make many of the available expert systems tools inappropriate are highlighted.

7.4.3.1 What is an expert system?: An expert system is a computer program which embodies knowledge derived from one or more experts and employs techniques developed in artificial intelligence research, particularly symbolic inference, in order to perform a task requiring expertise and judgment. The current scale of interest in expert systems is partly due to the fact that they represent a new method of programming, often called knowledge-based or rule-based programming. This is based on the principle that an expert's knowledge about how to perform certain tasks can be captured and represented in a symbolic form, usually as a set of rules. The expert system provides a mechanism to 'reason' using the symbolic representation of the expert's knowledge and thereby emulates his problem-solving abilities. It is also capable of explaining its line of reasoning in terms of the symbolic rules used to reach a conclusion. Such explanations give the user a greater insight into the system's operation and help ensure confidence in the results being produced.

The power of this approach is that it is well suited to some types of problems which are not amenable to conventional programming techniques. Some problems for which precise algorithms are difficult to specify may be suited to a description in terms of rules of thumb and trial and error methods.

7.4.3.2 Problems in building a threat assessment expert system: Building an expert system for air defence threat assessment is far from straightforward. Though there has been a great deal of work on expert systems, few truly successful systems have been built. Stefik's excellent tutorial (Stefik *et al.*, 1982) categories expert tasks as: interpretation, diagnosis, monitoring, prediction, planning and design. Most expert systems that have been built fall into the first two categories. Threat assessment is a monitoring task which also involves some prediction; these types of problem are more difficult to tackle with expert system techniques and have been the subject of less research.

One of the most useful features of expert systems is that knowledge and heuristic rules elicited from an expert can be represented and manipulated in a high-level form, easily understood by the system's user. A great deal of time was spent early in the project interviewing officers experienced in TA. A rule representative of the level at which these officers wished to see the expert system work is:

Track is probably low threat if
    not approaching defended area and
    travelling at low speed

This presents certain problems when considering building an expert system:

(a) There is a gap between the concepts used in the rules and the data received by the system.

382   Air defence threat assessment

(b) There are many imprecise or fuzzy concepts, for example what does 'low speed' mean? What does 'approaching' mean?
(c) The rules assume certain background knowledge, e.g. where are the 'defended areas'?
(d) How is uncertain reasoning to be handled?
(e) Rules are particular to a single track. How can multiple versions of the same rule be executed in parallel?
(f) Tracks are moving continually; so, though a track is not approaching a defended area now, it may manoeuvre in the next minute. How often is it necessary to recalculate? How are perturbations eliminated to reveal long-term intentions?

In addition an expert system for TA must deal with real-time, continuous data from several different sources. These features make the underlying processing problem quite different from that encountered in, for example, a medical diagnosis system. Diagnostic systems usually obtain their data by asking their user a question or by consulting a database. A system for TA cannot, in most cases, solicit its data in a predetermined order but must process data in the order in which it arrives and must do so in real time. This means that the many expert system shells that have been developed for diagnostic types of application are totally unsuited to this problem; a very different expert system framework is needed.

7.4.3.3 Related work: When our work began in 1982 there had been little work in the UK on the application of expert system techniques to problems with the features of our threat assessment task. A survey of work in the USA revealed several pieces of work which seemed relevant, though none tackled exactly the same problem.

Hearsay II (Lesser *et al.*, 1977) was a speech recognition system developed at Carnegie Mellon University for DARPA. Though it was outperformed by other speech recognition systems, it introduced an important system architecture known as the blackboard architecture. This has been adopted as the basis for many other artificial intelligence programs, particularly those involving reasoning over time and co-operative problem solving.

One such system is HASP/SIAP (Nii *et al.*, 1982) whose task is to correlate passive sonar reports from multiple fixed sonar receivers. It tries to build up a current best hypothesis to account for the reports received so far, and therefore tackles a very similar problem to ours but with longer time intervals between reports.

7.4.4 Approach taken
With the features of our problem in mind it became apparent that, before a threat assessment expert system could be built, an expert systems framework suited to representing and manipulating the types of rules provided by our experts was needed. The framework had to run on a VAX 11-780 under the VMS operating

system, which meant that very few AI tools were available. The design and engineering of this framework became a major step in our work and was carried out, under a research contract, by Cambridge Consultants Ltd.

Two versions of the threat assessment expert system were built in the course of this work, and these are described below.

7.4.4.1 The first version of the ADX: From an initial survey of AI tools and architectures it seemed that the blackboard architecture could provide a suitable basis for the threat assessment expert system. A blackboard system framework and an experimental prototype threat assessment expert system, called the ADX (Bell, 1984), were built in the POPLOG environment (Hardy, 1984).

In order to allow the system to be run against a variety of scenarios a simulation facility was needed. It was decided to build this simulator using an object-oriented programming language, as had been done at Rand with ROSS (rule-oriented simulation system) (McArthur and Klahr, 1982) and SWIRL (simulating warfare in the ROSS language) (Klahr *et al.*, (1982). A simple object-oriented programming language was developed in POPLOG to support this simulation,

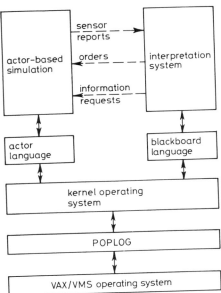

Fig. 7.4.2   *System overview*

and a facility that allowed scenarios to be set up and sensor and other reports to be generated was built using this language (Middleton, 1984).

The first version of the system therefore consisted of an air defence simulator based on a simple object-oriented programming language and the threat assessment expert system based on a blackboard architecture. Figure 7.4.2 shows the main components of this version of the ADX.

The main features of the blackboard architecture and object-oriented programming are described in the next two sections.

7.4.4.1.1 The blackboard architecture

The blackboard architecture supports a form of co-operative incremental problem solving. It has most commonly been used for problems where it is necessary to interpret the meaning of some low-level data, e.g. speech signals, sonogram data etc. The central feature of a blackboard system is a global data area (or blackboard) containing the input data and the current (or many alternative) interpretation(s) of the data. The blackboard is usually split into levels representing levels of abstraction meaningful in the particular domain. For example, the lowest level may contain radar plots, the next hypothesised radar tracks, the next the type of object represented by the track.

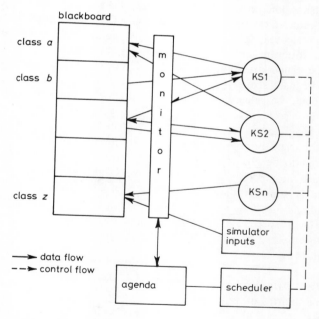

Fig. 7.4.3 *ADX blackboard structure*

The links between levels are created by knowledge sources; these are like situation→action rules. They monitor the blackboard for certain changes or pieces of information, and when such information is available they indicate that they can make a change to the blackboard and are placed on a queue or agenda. Knowledge sources may make forward or data-driven inferences, or may make backward or expectation-driven inferences. This allows a combination of backward and forward problem solving to take place.

There is usually some form of scheduling system that determines which changes to make and in what order, and this selects a knowledge source and allows it to change the blackboard. This process is called 'focusing attention' and allows the system to choose the most promising line of reasoning when many alternatives exist. The scheduling system itself often consists of a set of domain-specific rules. When an eligible knowledge source is run it changes the blackboard, for example, linking a new plot to a track. This may trigger other knowledge sources, so the process continues in a series of cycles. This type of problem solving is called 'opportunistic', as knowledge sources contribute to the overall problem solution as the opportunity arises.

Figure 7.4.3 shows the blackboard system structure used in the ADX.

7.4.4.1.2 Object-oriented programming

A theme that has emerged from both programming language research and artificial intelligence knowledge representation research is the idea of object-oriented programming languages. Objects are essentially a form of abstract data type in which a data structure and procedural code can be combined in a single program unit. The first object-oriented language was Simula67. This introduced the 'class' concept, where data structures could consist of both data and access procedures and classes could be related in hierarchies. Applications-driven work in artificial intelligence has also led to a diverse set of object-centred knowledge representation languages, which include structures variously called frames, scripts, flavours and units (see Stefik and Bobrow, 1985 for an overview). Objects allow related pieces of information to be grouped within a single structure, which often seems more natural, and is more compact, than representing all items as individual facts or relationships. In some of these languages objects are viewed as passive structures to reason from/about. Other languages take the view of objects as active processes or 'actors' (Hewitt and de Jong, 1982).

The object-oriented language implemented to support the air defence simulator was based on the actor style of object-oriented programming.

7.4.4.1.3 Basic features of actor languages

In the simplest forms of actor languages an object is defined in terms of a number of properties, or variables, describing its state, and a number of behaviours, or blocks of code, describing how it reacts to certain conditions. For example, an aircraft object might have properties such as 'max_speed' and 'climb_rate' and behaviours such as 'take_off' and 'change_heading'.

Object definitions can be organised as a hierarchy, the top-level definitions being the most general and the lower-level ones adding specialised features or modifying those of the general class. For example, an object definition for a 'fighter_aircraft' might inherit all the properties and behaviours of an 'aircraft' object but it would modify the 'max_speed' property and have extra behaviours such as 're_arm', 'engage_enemy' etc. This method of relating objects in taxonomies results in a natural mapping between program objects and real-world objects.

Communication between objects is via message passing. When an object receives a message the appropriate behaviour for that message type is executed. For example, an airbase object may receive a message to scramble two fighters; this would cause its 'scramble' behaviour to be executed which would send a 'take_off' message to two of the airbase's 'fighter_aircraft' objects, which in turn would run their 'execute_flight_plan' behaviours. In this way complex systems and their behaviour can be defined.

These features make object-oriented languages ideally suited to simulation applications (Middleton and Zanconato, 1985).

7.4.4.1.4 Problems with the first approach

Certain problems with the blackboard architecture for the air defence problem became apparent (Bell, 1984).

Continuous reasoning Because data is arriving continuously and the space available for the blackboard is finite, some method is needed for removing old information and radar reports from the blackboard. Some knowledge about the value of various pieces of information is needed when deciding what to remove. Various methods were investigated, including deletion rules, least recently used algorithms, and a persistence mechanism whereby knowledge sources could tag a piece of information with a lifetime, making sure it was not deleted until this time had expired. The last method was the most successful but even so had problems and was cumbersome to use.

Consistency Problems of data consistency can arise when many separate agents are reading and updating the same global data area. The information which caused one knowledge source to become eligible to run can be changed by some other knowledge source before the first knowledge source is run.

Debugging As many knowledge sources are changing the blackboard it can be very difficult to locate the knowledge source responsible for incorrect updates.

No private data All state data is held on the global blackboard. This means that data which should be kept private can be read and written by other knowledge sources. This is highly undesirable so blackboards are often divided into panels or areas with special access permissions.

Efficiency 1 Scanning the entire blackboard for the piece of information required is normally inefficient. Most blackboard systems index entries according to criteria relevant for that particular problem, thus making searches of the blackboard more efficient. The air defence domain is characterised by complex predicates – often relating to geometrical or spatial relationships. Maintaining indexes of this complexity would cause severe efficiency problems in itself.

Efficiency 2 Typically, knowledge sources specify as simple predicates those blackboard changes which should cause them to run. When these predicates become true, the associated knowledge source can be run. In the air defence environment these predicates tend to be complex spatial relationships; determining the values of these predicates is often significantly more computationally expensive than the actual processing inside the knowledge source.

Some of these problems result from the blackboard architecture itself; others are due to particular features of the TA problem.

7.4.4.2 The second version of the ADX: Problems encountered with the blackboard architecture resulted in the need for a new approach. The blackboard and object-oriented systems both had useful features, and a new framework was created which aimed to provide features of both these architectures within a single language while allowing some of these problems to be overcome. This new framework was called BLOBS and was used to build the second version of the threat assessment expert system and simulator.

7.4.4.2.1 Features of BLOBS
BLOBS (Dickinson, 1985) is an object-oriented language designed to support both continuous reasoning expert systems and simulation applications. Its design is influenced by the actor view of object-oriented programming as found in ROSS. It differs from ROSS in several important respects, such as: the addition of public (i.e. globally accessible) data areas within objects; the typing of messages and objects; the use of a pseudo-real-time clock; the addition of facilities to assess the current computer processing load; and the addition of processing mechanisms known in AI as 'demons'.

The BLOBS language has been built in POP11 within the POPLOG environment. It extends the simple view of object-oriented languages with some additional features. Objects in BLOBS can include:

Inheritance information   This defines the superclass(es) to which this object (blob) belongs.
A set of variables   Variables may be public and readable (but not writeable) by other objects, or private and hidden from other objects.
A set of pseudo-variable definitions   These associate a function or expression with a variable name. The variable may be public or private and is accessed like any other, but results in the function or expression being evaluated in the context of the owner blob to return the value. This is useful for representing continuously changing variables; any access to the variable can cause an update procedure to run to supply the current value. This ensures that other objects are using up-to-date values and that calls to the update procedure are only carried out when required.
A set of procedures
A set of behaviours

The following is an example of part of a simple object definition:

    dynamic blob flying_object;
      inherit physical_object;

/*Following are public variables — they can be read by other objects: */

    public
        speed,
        heading,
        .
        time—airborne;

/*Public definitions associate a procedure call with access to a public variable.*/
/*Other objects will read the variable like any other but the read will cause the*/
/*associated procedure to run to supply the value returned.   */

    public definition current_position = *my* update_value_of("position");

    public definition current_heading = *my* update_value_of("heading");

/*Private variables cannot be seen or read by other objects: */

    private
        flight_plan,
        .
        sectortime;

/*Procedures, as found in standard languages, can be defined: */

    private procedure calculate_position(simtime, status, lasttime)
      → new_position;
      .
    endprocedure; /* calculate_position */

/*A number of behaviours can follow the procedure definitions. These define the*/
/*way the object will react to certain events, e.g. receipt of a message.   */

    on initialisation do;

/*The initialisation behaviour is run when the object is created: */

      .
    enddo;

/*Other behaviours follow, e.g. message behaviours: */

    on message new_flight_plan *with* flight_plan → *my* flight_plan *do enddo*;

endblob;

Communications in BLOBS may be:

By message passing   A blob can 'send a message' to another object which causes that object's message behaviour for that message type to execute. Messages can be sent at a particular time or after a specified delay. For example:

send execute_next_leg to (my id) after (my leg_time);

This sends the message 'execute_leg' (of flight plan) to the object itself after the delay specified by 'my leg_time'.

Unlike most AI languages BLOBS tries to carry out as much compile-time program checking as possible and message formats must be declared before use. They also contain a number of named fields allowing parameters to be passed by field name.

By reading another object's public variables   For example:

the current_position of fighter17 → his_position;

In reaction to changes to some object's public variables   A blob can contain a behaviour which monitors changes in another object's (or its own) public variables. For example:

on_change_in the height of any flying_object do

enddo;

In reaction to the birth (creation) or death (deletion) of an instance of another object   A blob can specify a behaviour which will run when objects of a certain class are created/deleted.

The last two forms of communication and pseudo-variables are known as demons. A demon is a block of code which is executed when a specific event occurs. In the cases above, the events causing demon code to execute are changes to specified public variables, the creation/deletion of objects or an attempt to access a psuedo-variable. The object/event causing a 'demon to fire' has no knowledge of its effect. Therefore demons provide a very useful mechanism for monitoring the state of other objects.

There are three main types of objects in BLOBS, called static, dynamic and mixer. There can only be one instance of a static blob and this will exist throughout the life of the program (it cannot be created or deleted). Static blobs are used to represent global and persistent types of objects. There can be many instances of a dynamic blob. Dynamic blobs can be created and deleted and are used to represent objects such as tracks or aircraft, which might only exist for a limited time. Mixer blobs are used to define some general features of a class of objects. They do not completely define an object and cannot exist in their own right; they merely provide inheritance information. For example, a 'flying object' mixer might contain all the primitives for calculating flight parameters and could be inherited by all types of aircraft, birds etc.

7.4.4.2.2 The rule language based on inferno

The BLOBS object-oriented framework does not provide a classical rule-based language system. Though 'situation → action' rules often have a natural mapping on to demons, there is no mechanism for representing 'if conditions then conclusion with certainty' rules directly in BLOBS. It was clear that many of the rules from our experts contained uncertainty and therefore a mechanism capable of representing and propagating this uncertainty was needed.

The area of uncertain reasoning within experts systems is still under active research. If statistical data is available, well understood techniques such as Bayesian inference may be used. For most problems being tackled by an expert systems approach such data is not available and a number of *ad hoc* techniques have arisen such as subjective Bayesian inference, fuzzy logic etc. Many of the available methods were investigated (see Appendix C of Bell *et al.*, 1985) and the Inferno approach (Quinlan, 1983) was chosen.

Inferno is a cautious approach to uncertain inference. Unlike many of the available methods, it makes no assumptions about the probability distributions of propositions and uses provable constraints when deriving inferences. The lack of assumptions leads to weaker consequential beliefs than could be obtained with some of the other methods but ensures that all inferences are correct and that any errors that arise come from the data. In cases where relationships between propositions, such as independence, are known, Inferno allows these relationships to be asserted and the resulting inferences to be strengthened.

Inferno uses a probability interval, rather than a single point probability, to represent belief in a proposition. This means that evidence for and against a proposition A are used to derive bounds $t(A)$ and $f(A)$ on the probability of A, where $P(A) >= t(A)$ and $P(\tilde{A}) >= f(A)$. This allows the degree of uncertainty about a proposition to be measured. For example, an interval $\{t(A),f(A)\}$ with values $\{0.4, 0.2\}$ means that the probability lies between 0.4 and $(1-0.2) = 0.8$. Inferno uses such intervals to ensure that belief values are consistent. If a propagation would result in a total belief value $t(A) + f(A)$ greater than 1, then the inconsistency is flagged.

A version of Inferno was implemented in the BLOBS language (Zanconato, 1985). The primitive relations used in Inferno were considered to be rather low level and not easily understood by the domain experts. Therefore a higher-level syntax was developed and rules in this format mapped on to the basic Inferno primitives. Numeric values for the necessity and sufficiency of evidence for a proposition could be replaced by symbols. For example, the rule given in Section 7.4.3.2 was represented as:

independently
    not approaching_defended_area and
    speed_low
are weakly required to strongly infer low_threat

Facilities were also provided to specify mappings rather like fuzzy sets. Values such as a track speed in knots needed to be mapped on to belief values for certain propositions in the Inferno rules. This facility allowed a value, such as a speed, to be used to update the belief value for a set of nodes

Proposition name	Input value	Belief value
speed_very_low	{ 0 100}	{1 0}
speed_low	{ 0 250}	true
speed_medium	{ 200 510}	true
speed_high	{ 500 9999}	true
speed_very_high	{1000 9999}	true
speed_unknown	undef	true

In this example an input value which lies within the range specified for a named proposition will cause the belief value for that proposition to be set to the specified belief value. In this example all the belief values are the same, i.e. {1 0} (or equivalently 'true'). Any legal belief interval can be specified. Thus a speed of 90 knots would cause belief in propositions 'speed_very_low' and 'speed_low' to be updated. The meaning of 'speed_low' will vary from context to context, so different fuzzy set mappings for the same names can be defined for different groups of rules.

Facilities were also provided for producing explanations for any updates in terms of the rules used and their belief values.

The Inferno system can be used by any object in a BLOBS program where this type of rule-based reasoning is required. There is a natural tendency to group rules into sets of related function called rule sets. Each rule set defines an inference network. The blobs implementation allows many run-time versions of the same inference network to be used, i.e. a single rule set can be used to reason about several different tracks at the same time. Only one version of the compiled inference network is needed and this is held in a static blob, whereas the particular belief values for different users of the rule set are held in separate dynamic blobs.

7.4.4.2.3 The threat assessment expert system in BLOBS

The threat assessment expert system consists of a network of objects or blobs. The numbers of objects present in the network varies over time as new objects are created to reason about new tracks, and objects for reasoning about old tracks are removed. A snapshot of a part of the network is shown in Fig. 7.4.4.

392 Air defence threat assessment

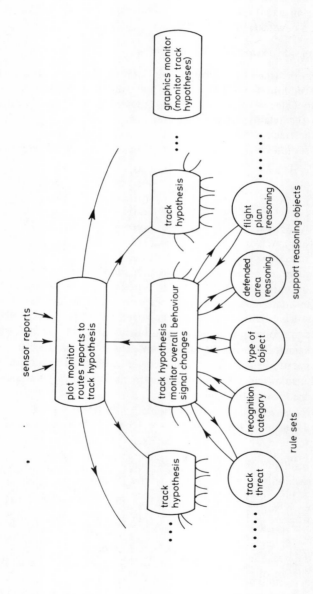

Fig. 7.4.4 *Part of ADX reasoning network*

Plot monitor

A static blob called the plot monitor is responsible for inspecting all reports from sensors. When a new report is received this object will decide whether it is an update to an existing track or whether a new track hypothesis should be created. If it is an update to an existing track the report is routed to the appropriate track hypothesis; if not, a new track hypothesis object is created.

Track hypothesis blobs

The track hypothesis object is responsible for co-ordinating the reasoning about one particular track over time. It maintains current information about the track's position, height, heading, strength, any associated information such as IFF responses etc. It also holds the results of reasoning about the track, for example, current belief values for the track threat, its recognition category etc. The actual reasoning is carried out by several other support blobs or objects, which are created by the track hypothesis object.

Support blobs

The support objects are of various types. The main types are: objects containing Inferno rule sets; utility reasoning objects, which provide the belief values necessary to update the Inferno networks; and object-oriented reasoning objects, which are used to represent rules that do not fit into the Inferno framework, e.g. rules involving reasoning with time, or creating expectations.

Rule sets and their utility reasoning objects

In the ADX, rule sets are created for: deciding a track's recognition category; trying to deduce the type of object, e.g. fighter, bomber, missile etc.; and deciding the threat level of the track. Rule set objects contain a set of Inferno rules, supplied by the air defence experts, held in an Inferno inference network. The gaps between propositions in the rules and the data available to the system must be bridged by a number of mechanisms so that belief values for nodes in the inference network can be supplied. This is done in a number of ways:

1. Directly from input reports, e.g. the current alert state.
2. Through fuzzy set operations, e.g. 100 knots is mapped on to a fuzzy set for speeds and the results used to update the inference network.
3. Using utility reasoning objects and background knowledge, e.g. 'approaching a defended area'. The rules for air defence threat assessment contain a number of imprecise concepts which require a spatial reasoning. In this example it is necessary to observe the behaviour of a track over time and decide whether it is approaching any defended area. In order to do this the expert system must 'know' where the defended areas are. A simple method of representing this type of geographical background knowledge is in an object:

>   dynamic blob defended_area
>   public name;

```
       public perimeter;
       .
       on_message remove_this_defended_area do
           suicide;
       enddo;
    endblob;
```

The public variables hold information about the area and the object can contain behaviours to carry out various functions. In this case there is a behaviour to remove the defended area. Therefore background knowledge can be added and deleted dynamically and can be accessed by the rest of the system in a uniform manner.

The utility reasoning blob, which has to decide whether a track is approaching a defended area, can therefore access all defended area objects, carry out calculations on relative positions, headings etc., and decide if the track is approaching any area and if so which one it is nearest to. The following is a demon which is triggered by significant changes in the hypothesised position or heading of the track being monitored. The 'for statement' creates a list of all the objects of type 'defended area' that currently exist.

```
   on_change_in the rough_position, the rough_heading of (my creator) do
       for all defended_area do
       ;;; carry out calculations
       .
       endfor;
       .
   enddo;
```

This may sound simple, but the meaning of 'approaching' is rather vague. For example, does the track's current heading have to intersect with the defended area for it to be 'approaching', or is a heading within a wider tolerance still approaching? What about tracks that are heading straight for the area but are a great distance away? It takes a good deal of tuning to create an algorithm which, given a set of tracks and a set of defended areas, always agrees with the expert's idea of 'approaching'.

Object-oriented reasoning

Certain types of rule in the TA system cannot be represented in Inferno, for example rules involving time, such as:

'If a new track has not been assigned a recognition category after 5 minutes then assign a default category.'

'Increase expectation of attack within 1 hour of sunrise.'

These can often be mapped naturally on to message sending code, which schedules a message for the required event time. When the message is received, the message behaviour carries out the necessary actions, e.g. checking the recognition category of a track hypothesis.

There are also forms of reasoning such as those carried out in track repair. If a track disappears from radar, the track hypothesis object can set its state to 'lost track' and can contain code to predict its likely position based on last known information. A track repair object can then look at all lost tracks within the vicinity of a newly appeared track and try to decide if the new track is any of the lost tracks. A lost track will have a 'lifetime', and if it is not claimed within this lifetime it will disappear.

Another form of reasoning, often necessary in time-limited problem solving, is the ability to take different actions based on the amount of available time. BLOBS provides an operator called IDLENESS which measures the current load on the system; it also provides the ability for an object to boost its priority to cause itself to be run in preference to other objects. Using these features, code can be written to check the current processing load, estimate how much time is available and choose the most suitable option, e.g. slow-but-sure, rought-and-ready or emergency.

7.4.4.2.4 Relations to the blackboard-based system

There are many similarities with the old blackboard system, but some of the problems have been overcome.

Continuous reasoning The problem of deciding when to delete information has not been overcome entirely, i.e. it is not always possible to predict that a piece of information is going to be useful in the future. However, the responsibility for deleting information is now clearly that of the owning object, e.g. the track hypothesis object can decide how much track history to maintain.

Consistency and debugging For similar reasons, consistency is no longer a problem. Only the owner of a piece of information can change it. There is clear responsibility for every piece of information. This also helps in debugging.

Private data BLOBS private variables allow data to be hidden from other objects.

Efficiency 1 A great deal of the underlying processing in the air defence problem involves detailed and expensive spatial/geometric calculations. The track hypothesis object makes sure that these calculations are only carried out when significant events have occurred, for example a significant change in track heading. Minor perturbations are ignored. The parts of the inference net that are interested in the results of these calculations also declare an interest by defining demons. This means that the rules are only rerun on significant changes and no search of rules that do not match the current situation is made.

Efficiency 2 One of the most useful features of the blackboard architecture is its flexible search control. This allows only the most promising parts of a large search space to be expanded. For this problem there is no need to choose only the most promising updates; the search space is small enough to allow all updates to be made. The ordering of doing updates can be changed by objects resetting their priorities.

7.4.4.2.5 The ADX MMI

The current version of the system displays a track picture giving the system's interpretation of the situation in terms of track position, strength, recognition category etc. A VDU displays text output by the system when a significant change has occurred, for example if the recognition category of a track has changed. Simple explanations, in terms of the rules used and their degree of support, are also displayed. A joystick device allows information about any particular track and the reasons for deduced information to be displayed on the VDU at the user's request.

7.4.5 Current status

In the prototype system there are: three rule set blobs, containing a total of approximately 60 rules; three object-oriented reasoning blobs; and a master monitor blob, created for each track. We expect to require significantly more rules in the full threat assessment system, as we have concentrated more on the development of the framework than on producing a 'good' set of rules.

The current system has been tested against a variety of scenarios. It performs reasonably well in assessing situations, though as yet lacks the variety of rules which would be required to emulate expert performance. The limiting factor in computing performance has been space rather than time. Over 50% of computing capacity is used by the simulator, so performance could be improved by separating the two parts of the system.

The current user interface is rather basic and more work is needed on producing explanations and presenting results to the user. The parts of the system which use object-oriented rather than rule-based reasoning do not have a uniform mechanism for producing explanations. Another problem is the volume of information produced, which can quickly fill a VDU display.

7.4.6 Summary and conclusions

This paper has described the approach taken in building an expert system for air defence threat assessment. The nature of this problem has required the development of a new programming language called BLOBS. A prototype version of the threat assessment expert system has been built using this framework with some success but contains only a subset of the rules that will be required in the full system. The nature of the threat assessment task is such that experts have difficulty in verbalising certain parts of their knowledge. This is likely to mean that gaps will exist even in future versions of the system. This contributes to our basic view that initial use of the system is more likely to be as an adviser rather than as an autonomous system.

Techniques from artificial intelligence have proved useful:

(a) In allowing a high-level description of the problem
(b) By providing a mechanism for uncertain inference
(c) By providing mechanisms to explain reasons for decisions to the user
(d) In allowing the incremental development of the system
(e) In developing a simulator to exercise the system

(f) By allowing a 'divide and conquer' problem-solving strategy.

AI has not provided any new methods for dealing with some of the basic calculations which must be carried out, and this has led to the requirement for a mixed algorithmic and rule-based system. One of the main problems in extending the system is in adding good algorithms to implement some of basic concepts expressed in rules, especially those involving imprecise geometrical concepts.

BLOBS has proved to be a general purpose tool, useful for a wide range of applications. It has been used by other groups both for building simulations and for expert systems applications.

7.4.7 References

BELL, M. (1984) 'The ADX – an expert system framework for experimentation in air defence', Proceedings of the 3rd seminar on the applications of machine intelligence to defence systems, RSRE, Malvern

BELL, M., ZANCONATO, R. and BENNETT, M. E. (1985) 'Technical Report on phase I of expert systems techniques for real-time applications in air defence', Cambridge Consultants Ltd, CCL-C1924-TR-001/1a

DICKINSON, I. J. (1985) 'BLOBS user manual', Cambridge Consultants Ltd, CCL-C1924-UM-001/1c

HARDY, S. (1984) 'A new software environment for list processing and logic programming', in O'Shea, T. and Eisenstadt, M. (eds) *Artificial Intelligence: Tools, Techniques and Applications*

HEWITT, C. and DE JONG, P. (1982) 'Open systems', AIM 691, AI Laboratories, MIT

KLAHR, P. *et al.* (1982) 'SWIRL: simulating warfare in the ROSS language', Rand Corporation note N-1885-AF

LESSER, V. R. *et al.* (1977) 'A retrospective view of the Hearsay-II architecture', Proceedings 5th international joint conference on AI, pp. 790–800

MCARTHUR, D. and KLAHR, P. (1982) 'The ROSS language manual', Rand Corporation note N-1854

MIDDLETON, S. (1984) 'Actor based simulation of air defence scenarios in POPLOG', Proceedings of the 3rd seminar on the applications of machine intelligence to defence systems, RSRE, Malvern

MIDDLETON, S. and ZANCONATO, R. (1985) 'BLOBS: an object-oriented language for simulation and reasoning', Proceedings of the working conference on AI in simulation, Ghent, Belgium

NII, H. P. *et al.* (1982) 'Signal-to-symbol transformation: HASP/SIAP case study', *AI Magazine*, 3(2), pp. 23–35

QUINLAN, J. R. (1983) 'INFERNO: a cautious approach to uncertain inference', *Computer Journal*, **26**(3)

STEFIK, M. *et al.* (1982) 'The organisation of expert systems: a prescriptive tutorial', Xerox, Palo Alto Research Center

STEFIK, M. and BOBROW, D. G. (1985) 'Object-oriented programming themes and variations', *AI Magazine*, **4**(6), pp. 40–62

ZANCONATO, R. (1985) 'INFERNO reasoning subsystem', Cambridge Consultants Ltd, CCL-C1924-UM-002

Copyright © Controller HMSO, London, 1987.

Index

Abbreviations, 283, 294
Abstract Syntax Notation One (ASN1), 229, 231
Active and passive sonar, 345
Actor languages, 385
ADA, 8
Adaptability, 88
ADX, 311, 383
    MMI, 396
Air Command and Control System (ACCS), 50, 57
Air defence threat assessment, 378
Air power deployment, 60
Analysis and specification methods, 3
Application layer standards, 229, 231, 258
Area Distributed Data Management System (ADDAM) 150
ARPANET, 170, 181
Artificial Intelligence (AI), 8, 308, 339, 378
    for data fusion, 369
Availability, 95
AWACS, 64

Backchaining paradigms, 312
Bartlett, J R, 183
Bayes transform, 331
Behaviour, 270, 335
Blackboard expert systems, 12, 310, 347, 384, 395
BLOBS, 311, 387

Bird, D F, 211

Catt, R H L, 129
CCITT network protocol, 186, 221
    X 75 protocol, 190, 217
    X 25 protocol, 191, 215
    X 400 message handling protocol, 236
Channel access, 165
Chessman, A S, 129
Combat net radio, 159
    packet radio, 162
Combat system design, 139
Command and Control theory, 1
    model, 19
Command and complexity, 30
    decisions, 34
    distributed, 37
    hierarchy, 26
    message set, 40
    process (C process), 23, 28
Commander, 266
Commitment, Concurrency and Recovery (CCR) standards, 231
Common Application Service Elements (CASE) protocols, 228
Communications, 71, 159
    architecture, 140
    seven layer reference model, 140, 214, 227
    TDMA, 72
Complexity, 308

Computer based decision making, 271
Co-ordination, 40
Confidentiality, 339
Connectivity, 124, 126, 132, 171
Consistency, 386
Constraints, 43
Continuous scheduler, 166
    reasoning, 386, 395
Control system structure and design, 315
CORE action diagrams, 281
    design, 280
Correlation of data, 323

DARPA internet protocol, 218
    packet radio, 162
Database C^3 systems, 107, 235
    performance, 149
    protocols, 235
Data communications, 160
    Exchange Specification (DES), 143
Datagrams, 163
Data fusion, 321, 339, 380
    links, 343, 345
    management, 129
    origin authentication, 240
    representation, 133
    transfer protocols/security, 226
Davies, B H, 159
Davies, T R, 159
Decision aiding, 268, 308
Debugging, 386
Dialogues, 296
Digital Elevation Model (DEM), 119
Descriptive methods, 3
Direct memory access, 141
Distributed channel access control, 165
    command, 37
    database integrity, 148
    database management, 145
    data exchange, 134
    systems architecture, 134, 161
    systems security, 243
Effectiveness, 84

criteria, 25
Electronic Counter Measures (ECM), 344
Electronic Support Measures (ESM), 344
Encipherment Algorithm, 239
Enhancement Algorithm, 160
Entity — hierarchy, 94
    relationship, 112
Environment data, 343, 345
    tests, 197
Equipment database, 343, 346
Expert systems, 11, 307, 346, 381

Fighting mirror, 81
File Access Data Units (FADU), 233
    Transfer, Access and Management (FTAM), 233
Flexibility, 88
Forward chaining, 36
Functional architecture, 135, 139
    decomposition, 46
    testing and validation, 195, 197

Gardner, A, 279
Gateways, 217
Generic entities, 93
Goal, 44
    architecture, 2
    specification, 282
Graphical ikons, 124
    Kernal System (GKS), 235
Graphics standards, 11, 235
Grammar, 294

Harris, C J, 307
Help facilities, 300
Hierarchic command, 26
Hill, J S, 247
Histogram of merit, 90
Hitchins, D K, 50
Human behaviour, 269
    decision making, 266, 307
    factors, 279
    performance/psychology, 272

Index

skill theory, 275
Hybrid tasking, 26

Identification, Friend or Foe (IFF), 332, 344
Identity information, 326
Imagery, 290
Implementation strategy, 2, 6
Inferencing, 376
Influence diagrams, 85
Information flows, 90
   modelling, 112
   shocks, 97
   Technology Standards Unit (ITSU), 218
Integrity, 240
Integration testing, 197
Intelligence, 343
Intelligent dialogue, 124
   Knowledge Based Systems (IKBS), 11, 308, 337
Interactive Spatial Information Systems (ISIS), 109
Interface Cognitive models, 289
   languages, 293
International Standards Organisation (ISO), 5, 121
   CASE standards, 231
ISO reference model, 213, 221
ISO/CCITT communications standards, 213
ISO/OSI seven layer model, 289
Internet protocol, 218
Internetworking, 164
Interoperability, 69, 88, 212
Interpret objectives, 27
Inter-process data areas, 117
Intervisibility analysis, 119

Job Transfer and Manipulation (JTM), 234
JTIDS, 64

Knowledge Based Systems (KBS), 11, 308, 378, 380
   base, 312
   control system design, 314
   representation, 126
Knowles, T, 238

Lakin, W L, 339
Lammers, G H, 84
Latent Command, 41
Layered protocol dialogue design, 121, 125
Layers, 121
Linguistic and spatial protocols, 122
LISP, 372
Local Area Network (LAN) 129, 132, 212, 248
   defence standard 00-19, 141

Man-machine interface (MMI), 9, 265, 279
   design, 270
Massively parallel architecture, 374
Maintenance, 13
MASCOT, 30, 115, 280
McCann, C A, 107
Message handling systems, 236, 258
   orientated text-interchange systems (MD TIS), 236
Middleton, S, 378
Miles, J A H, 339
Mission planning, 22
Multilevel interfaces, 293
Multiplatform data fusion, 362
Multiple Expert Archicture (MXA), 311, 348
Multisensor data fusion, 311, 340, 342

Nato Air Defence Ground Environment (NADGE), 50
   Standardisations Agreement (STANAG) 4222, 142, 225, 251
   (STANAG) 4250, 220
   digital data links, 224
Naval Engineering Standards (NES),

1026, 142, 252
1024, 142, 252
1028, 143, 252
Network control, 169, 172, 253
  management, 192
  measurement algorithm, 170
  model, 201
  protocols, 212, 189
Notaries, 244

Objective, 36
Object-orientated programming, 385
  reasoning, 394
Open Systems Interconnection (OSI)
  reference model, 213, 223, 251
  security addendum, 242
Overall Command System, 20

Packet switched radio, 159, 162, 184
  switching overlays, 188
Passwords, 241
Peer entity authentication, 240
Pearson, H J, 233
Penalty, 44
Performance testing and validation,
  195, 197
Perception process, 22
Physical layer encipherments, 244
  security measures, 241
Picture builders, 267
Plan and command information, 343,
  345
Planning, 36, 43
Poplog, 383
Positional information, 322, 341
Posteriori probabilities, 329
Presentation layer, 229
Primary packet switches (PPS), 189
Probability matrix, 328
Programmers hierarchical interactive
  graphics system (PHIGS), 236
PROLOG, 13
Protection mechanisms, 240

Protocols, 121, 190, 226, 258
  enhancements, 244
  testers, 199
Prototyping, 4, 283
Pseudo-variable definition, 387
Ptarmigan packet switched network,
  183, 194, 201

Quality of communications service,
  221

Radar, 343
Real time sensor data, 341
Relational database management
  system, 112
  data model 112,
Reliability protocol, 152
Requirements proving, 194
Resource allocation, 45
Richards, F A, 247
Route calculation and dissemination
  algorithm, 170
Routing, 169

Scheduling, 376
SDI-fighting mirror, 81
Secondary Packet switches (SPS), 189
Security, 9, 88, 226, 238, 243
  addendum, 239
  architecture, 243
  alarms, 241
  audits, 241
  OSI networks, 238
  protection, 239
Self-configurability, 172
Sensitivity analysis, 97
Sensors, 343
Session layer, 229
Seven layer reference model, 227, 221
Shafer-Dempster approach, 332
SHORE paradigm, 12, 309
Signal processing, 176
Simulation, 98
Singleton, W T, 265

Situation assessment, 339, 362
    database, 313
Skeletal plans, 316
Software, 7, 253
    reusability, 9
    tools, 119
Spatial databases, 107
    problem solving tools, 126
Specification and design methods, 8
Standards, 211, 247
    bodies, 220
    compatibility, 224
    databases, 150
    interoperability, 5
    OSI, 5
STANAG, 220, 225, 257
Stochastic scripts, 317
Survivability, 88
Symons, B J, 183
System design, 129, 280, 302
    dynamics modelling, 85
    proving for Ptarmigan, 193

Tactical area communications, 183, 216
    data interchange language, 70
    environmental, 187
    planning, 110
Tasks, 43, 283
    allocation, 284, 286
    analysis, 272
    methodology, 285
Taylor, M M, 107

Testable system goals, 279
Test tools, 199
Threat assessment, 366
    air defence, 378
    expert systems, 381, 391
Time dependent data, 115
Topographical information, 120
Traffic forwarding algorithm, 170
Transparency, 290
Trusted functionality, 244
Turoi, M I, 107

Uncertainty, 331, 376
Upper layers of reference model, 227
User centred design, 280
    computer interfaces, 279, 286
    design/protocols, 289
    interface, 113
    services, 163

Validation, 95
Virtual machines, 292
    metaphors, 292
    terminal protocols, 234
Vocabulary, 293

Warren, C S, 183
Wavel, 208
Wells, S G, 183
White, I, 1
Wilson, G B, 321
World model, 27, 46
    state space, 34